The Application of Spectral Techniques in Agriculture and Forestry

The Application of Spectral Techniques in Agriculture and Forestry

Guest Editor

Youzhen Xiang

MDPI

Basel • Beijing • Wuhan • Barcelona • Belgrade • Novi Sad • Cluj • Manchester

Guest Editor
Youzhen Xiang
Key Laboratory of
Agricultural Soil and Water
Engineering in Arid and
Semiarid Areas of Ministry of
Education
Northwest A&F University
Yangling
China

Editorial Office
MDPI AG
Grosspeteranlage 5
4052 Basel, Switzerland

This is a reprint of the Special Issue, published open access by the journal *Plants* (ISSN 2223-7747), freely accessible at: https://www.mdpi.com/journal/plants/special_issues/9LA0230A2H.

For citation purposes, cite each article independently as indicated on the article page online and as indicated below:

Lastname, A.A.; Lastname, B.B. Article Title. *Journal Name* **Year**, *Volume Number*, Page Range.

ISBN 978-3-7258-3669-7 (Hbk)
ISBN 978-3-7258-3670-3 (PDF)
https://doi.org/10.3390/books978-3-7258-3670-3

Contents

Youzhen Xiang
Introduction to the Special Issue of *Plants* on "The Application of Spectral Techniques in Agriculture and Forestry"
Reprinted from: *Plants* **2024**, *13*, 2632, https://doi.org/10.3390/plants13182632 . . . . **1**

Tao Sun, Zhijun Li, Zhangkai Wang, Yuchen Liu, Zhiheng Zhu, Yizheng Zhao, et al.
Monitoring of Nitrogen Concentration in Soybean Leaves at Multiple Spatial Vertical Scales Based on Spectral Parameters
Reprinted from: *Plants* **2024**, *13*, 140, https://doi.org/10.3390/plants13010140 . . . . **5**

Wanli Yang, Zhijun Li, Guofu Chen, Shihao Cui, Yue Wu, Xiaochi Liu, et al.
Soybean (*Glycine max* L.) Leaf Moisture Estimation Based on Multisource Unmanned Aerial Vehicle Image Feature Fusion
Reprinted from: *Plants* **2024**, *13*, 1498, https://doi.org/10.3390/plants13111498 . . . . **21**

Hongzhao Shi, Xingxing Lu, Tao Sun, Xiaochi Liu, Xiangyang Huang, Zijun Tang, et al.
Monitoring of Chlorophyll Content of Potato in Northern Shaanxi Based on Different Spectral Parameters
Reprinted from: *Plants* **2024**, *13*, 1314, https://doi.org/10.3390/plants13101314 . . . . **36**

Zilin Ye, Jiangping Long, Tingchen Zhang, Bingbing Lin and Hui Lin
L-Band Synthetic Aperture Radar and Its Application for Forest Parameter Estimation, 1972 to 2024: A Review
Reprinted from: *Plants* **2024**, *13*, 2511, https://doi.org/10.3390/plants13172511 . . . . **53**

Yujin Wang, Yule Lu, Ning Yang, Jiankun Wang, Zugui Huang, Junying Chen and Zhitao Zhang
Influence of Time-Lag Effects between Winter-Wheat Canopy Temperature and Atmospheric Temperature on the Accuracy of CWSI Inversion of Photosynthetic Parameters
Reprinted from: *Plants* **2024**, *13*, 1702, https://doi.org/10.3390/plants13121702 . . . . **68**

Hao Liu, Youzhen Xiang, Junying Chen, Yuxiao Wu, Ruiqi Du, Zijun Tang, et al.
A New Spectral Index for Monitoring Leaf Area Index of Winter Oilseed Rape (*Brassica napus* L.) under Different Coverage Methods and Nitrogen Treatments
Reprinted from: *Plants* **2024**, *13*, 1901, https://doi.org/10.3390/plants13141901 . . . . **95**

Roman Danilov, Oksana Kremneva, Igor Sereda, Ksenia Gasiyan, Mikhail Zimin, Dmitry Istomin and Alexey Pachkin
Study of the Spectral Characteristics of Crops of Winter Wheat Varieties Infected with Pathogens of Leaf Diseases
Reprinted from: *Plants* **2024**, *13*, 1892, https://doi.org/10.3390/plants13141892 . . . . **111**

Hanmi Zhou, Yumin Su, Jiageng Chen, Jichen Li, Linshuang Ma, Xingyi Liu, et al.
Maize Leaf Disease Recognition Based on Improved Convolutional Neural Network ShuffleNetV2
Reprinted from: *Plants* **2024**, *13*, 1621, https://doi.org/10.3390/plants13121621 . . . . **134**

Juan C. Nowack, Luz K. Atencia-Payares, Ana M. Tarquis and M. Gomez-del-Campo
Application of Unmanned Aerial Vehicle (UAV) Sensing for Water Status Estimation in Vineyards under Different Pruning Strategies
Reprinted from: *Plants* **2024**, *13*, 1350, https://doi.org/10.3390/plants13101350 . . . . **152**

Hengjia Zhang, Tao Chen, Shouchao Yu, Chenli Zhou, Anguo Teng, Lian Lei and Fuqiang Li
Optimizing the Mulching Pattern and Nitrogen Application Rate to Improve Maize Photosynthetic Capacity, Yield, and Nitrogen Fertilizer Utilization Efficiency
Reprinted from: *Plants* **2024**, *13*, 1241, https://doi.org/10.3390/plants13091241 . . . . **177**

Giovanni Bitella, Rocco Bochicchio, Donato Castronuovo, Stella Lovelli, Giuseppe Mercurio, Anna Rita Rivelli, et al.
Monitoring Plant Height and Spatial Distribution of Biometrics with a Low-Cost Proximal Platform
Reprinted from: *Plants* **2024**, *13*, 1085, https://doi.org/10.3390/plants13081085 . . . . **199**

Liang Zhong, Shengjie Yang, Yicheng Rong, Jiawei Qian, Lei Zhou, Jianlong Li and Zhengguo Sun
Indirect Estimation of Heavy Metal Contamination in Rice Soil Using Spectral Techniques
Reprinted from: *Plants* **2024**, *13*, 831, https://doi.org/10.3390/plants13060831 . . . . **217**

plants

MDPI

Editorial

Introduction to the Special Issue of *Plants* on "The Application of Spectral Techniques in Agriculture and Forestry"

Youzhen Xiang [1,2]

[1] Key Laboratory of Agricultural Soil and Water Engineering in Arid and Semiarid Areas of Ministry of Education, Northwest A&F University, Yangling 712100, China; youzhenxiang@nwsuaf.edu.cn

[2] Institute of Water-Saving Agriculture in Arid Areas of China, Northwest A&F University, Yangling 712100, China

This Special Issue, titled "Applications of Spectral Technology in Agriculture and Forestry", presents a collection of cutting-edge research findings exploring various applications of spectral analysis in agricultural and forestry environments. The papers in this issue collectively examine the use of advanced spectral methods across key domains, including crop health monitoring, disease detection, forest parameter estimation, soil quality assessment, water stress analysis, and nutrient management. These studies not only highlight advances in their respective fields but also reveal the complex interplay between spectral technologies, machine learning, and sustainable resource management in agricultural ecosystems. Through the research presented, this Special Issue showcases an evolving paradigm where precision agriculture and forestry practices increasingly rely on sophisticated spectral data analysis for information acquisition and decision optimization. This Special Issue compiles research from around the world, covering diverse applications of spectral technologies in agriculture and forestry across different climates, ecosystems, and crop types. The twelve papers included demonstrate the broad applicability of these technologies in varying geographical regions and crops, emphasizing the efforts of scientists from multiple countries, including regions such as Europe and Asia, to promote precision agriculture and forestry practices. The following is an overview of each paper, providing insights into how they collectively advance the development of precision agriculture and forestry.

A common theme across these studies is the use of advanced spectral indices and remote sensing techniques to monitor various physiological parameters of plants. For example, Liu et al. (2024) [1] proposed a novel spectral index designed to overcome the angular effects on the estimation of the leaf area index (LAI) in winter rapeseed. Their method utilizes multi-angle hyperspectral data to test the stability of 16 traditional vegetation indices (VIs) in monitoring LAI from different observation angles. The study found that the OPIVI index exhibited the highest correlation in LAI estimation, providing valuable guidance for the selection of vegetation indices in future UAV and satellite applications. Shi et al. (2024) [2] focused on using hyperspectral data to monitor chlorophyll content in potato crops, demonstrating how differential transformations of spectral indices can effectively estimate chlorophyll levels. They constructed several machine learning models, including Support Vector Machine (SVM), Random Forest (RF), and Backpropagation Neural Network (BPNN) models, to predict potato chlorophyll content, highlighting the versatility of hyperspectral data in monitoring different physiological parameters. Both studies suggest that combining multi-angle and differential spectral indices with machine learning algorithms is an effective approach to capturing key physiological features of crop growth. Liu et al. (2024) and Shi et al. (2024) [1,2] provide complementary insights into the application of spectral data in precision agriculture, recommending the integration of various spectral indices with machine learning to construct a robust, non-destructive crop monitoring framework.

Citation: Xiang, Y. Introduction to the Special Issue of *Plants* on "The Application of Spectral Techniques in Agriculture and Forestry". *Plants* **2024**, *13*, 2632. https://doi.org/10.3390/plants13182632

Received: 16 September 2024
Accepted: 19 September 2024
Published: 20 September 2024

In the area of crop disease detection, spectral technology also plays a critical role. Danilov et al. (2024) [3] investigated the effects of disease development on the spectral characteristics of winter wheat varieties, revealing how disease severity alters wheat spectral reflectance, particularly in the near-infrared range. Their study demonstrated significant differences in the spectral characteristics of winter wheat varieties under disease influence, offering new possibilities for monitoring crop health and disease progression. In contrast, Zhou et al. (2024) [4] utilized an improved convolutional neural network (CNN) model, ShuffleNetV2, to identify maize leaf diseases. They introduced the SimAM attention mechanism to enhance the model's accuracy in complex backgrounds. The results showed that the model achieved an accuracy of 98.40% on the maize leaf disease dataset, with a more compact model structure. Both studies underscore the importance of spectral data in disease detection, with the former focusing on near-infrared spectral monitoring of disease severity and the latter demonstrating the efficiency of deep learning models in disease identification.

The issue also discusses the application of spectral technology in forest parameter estimation and soil quality assessment. Ye et al. (2024) [5] provided a comprehensive review of L-band synthetic aperture radar (SAR) technology for forest canopy penetration and vertical structure parameter estimation, summarizing the application of L-band SAR in estimating forest height, moisture, and biomass. The study explored the challenges and future research directions of L-band SAR in forest resource management. Zhong et al. (2024) [6] studied how rice leaf spectra could be used to indirectly estimate heavy metal contamination in soil, utilizing a genetic algorithm-optimized partial least squares regression (GA-PLSR) model for soil quality monitoring. Despite focusing on different application areas, with Ye et al. concentrating on SAR technology in forestry and Zhong et al. (2024) [6] on spectral technology for agricultural soil monitoring, both studies emphasize the importance of remote sensing as an environmental assessment tool, showcasing how spectral technology can provide critical data for resource management.

Water stress analysis and nutrient management represent another field where spectral technology is making significant contributions. Wang et al. (2024) [7] investigated the impact of the time-lag effect between canopy temperature and atmospheric temperature on the accuracy of the Crop Water Stress Index (CWSI). They quantified the time-lag parameter for winter wheat and improved the predictive accuracy of CWSI using a genetic algorithm–support vector machine (GA-SVM) model. The study's results showed that accounting for the time-lag effect effectively enhanced the correlation between CWSI and photosynthetic parameters, providing theoretical support for the application of thermal infrared remote sensing in crop water stress diagnostics. The study by Yang et al. (2024) [8] also focuses on crop water status diagnosis. They utilized UAV multispectral technology to estimate soybean leaf moisture through a comprehensive analysis of vegetation indices, canopy texture features, and randomly extracted texture indices. By employing Extreme Learning Machine (ELM), Extreme Gradient Boosting (XGBoost), and BPNN models, they achieved significant results, with the XGBoost model demonstrating the highest accuracy in leaf moisture monitoring. Similarly, Sun et al. (2024) [9] utilized spectral parameters to monitor nitrogen concentration in soybean leaves, finding the highest correlation between spectral parameters and nitrogen concentration in the upper leaves of the crop. They constructed several machine learning models, with the Random Forest (RF) model exhibiting the highest accuracy in estimating soybean leaf nitrogen concentration. Both studies highlight the integration of spectral data and machine learning to improve the accuracy of crop water and nutrient monitoring. Additionally, Nowack et al. (2024) [10] explored the use of UAV-mounted multispectral sensors to estimate vineyard water status under different pruning strategies, finding that red light and red-edge bands effectively predicted vine water status. This study further emphasizes the value of high-resolution multispectral imaging in crop water management. Zhang et al. (2024) [11] conducted field experiments to explore the effects of optimizing mulch type and nitrogen application rate on maize photosynthetic capacity, yield, and nitrogen use efficiency, discovering that using biodegradable plastic mulch combined with moderate nitrogen application significantly improved maize photosynthetic efficiency and yield. These studies highlight the potential of spectral technology in various nutrient conditions and farming practices.

Lastly, Bitella et al. (2024) [12] proposed a low-cost, near-ground platform for monitoring crop height and spatial distribution using ultrasonic sensors and spectral data, achieving precise monitoring of plant growth characteristics across different cropping systems. This research not only demonstrates the potential of low-cost remote sensing platforms in agriculture but also complements the studies by Liu et al. (2024) and Shi et al. (2024) [1,2], which utilize multi-angle and hyperspectral data to monitor crop growth, providing diverse technological pathways for precision agriculture.

In summary, the papers in this Special Issue provide a deeper understanding of the applications of spectral technology in precision agriculture and forestry management, expanding the research scope of this field. The topics covered, including crop health monitoring, disease detection, forest parameter estimation, soil quality assessment, water stress analysis, and nutrient management, reveal the diversity and practicality of spectral technology while emphasizing its crucial role in promoting sustainable development of agricultural ecosystems. These studies point to a rapidly evolving scientific frontier, where the deep integration of spectral data and machine learning techniques is set to become the core driving force for future precision agriculture and forestry development. The papers in this Special Issue draw on each other's findings, employing multi-source data fusion, machine learning modeling, remote sensing, and hyperspectral analysis to establish a comprehensive and flexible analytical framework capable of real-time, accurate monitoring of crop and forest ecosystem dynamics. This framework provides both the theoretical foundation and practical pathways for addressing the increasingly complex challenges in agriculture and forestry. More importantly, these research findings not only provide a solid theoretical basis for the application of spectral analysis in multidisciplinary fields but also offer valuable guidance for scholars and practitioners in precision agriculture and forestry. Through this compilation, we witness the immense potential of spectral technology in data-driven decision making, sustainable resource management, and ecosystem health assessment, laying a solid foundation for future in-depth research and practical applications in related fields.

Acknowledgments: We wish to express our heartfelt thanks to all the authors for their valuable contributions to this Special Issue. We are also immensely grateful to the *Plants* Editorial Office for their unwavering support, without which this Special Issue would not have come to fruition. I would also like to extend my sincere gratitude to all my graduate students for their hard work in researching the applications of remote sensing technology in agriculture. I hope you continue to achieve even greater success in the field and contribute significantly to its advancement.

Conflicts of Interest: The author declares no conflicts of interest.

References

1. Liu, H.; Xiang, Y.; Chen, J.; Wu, Y.; Du, R.; Tang, Z.; Yang, N.; Shi, H.; Li, Z.; Zhang, F. A new spectral index for monitoring winter oilseed rape leaf area index under different coverage methods and nitrogen treatments: Overcoming view zenith angle effect. *Plants* **2024**, *13*, 1901. [CrossRef] [PubMed]
2. Shi, H.; Lu, X.; Sun, T.; Liu, X.; Huang, X.; Tang, Z.; Li, Z.; Xiang, Y.; Zhang, F.; Zhen, J. Monitoring of chlorophyll content of potato in northern Shaanxi based on different spectral parameters. *Plants* **2024**, *13*, 1314. [CrossRef] [PubMed]
3. Danilov, R.; Kremneva, O.; Sereda, L.; Gasiyan, K.; Zimin, M.; Lstomin, D.; Pachkin, A. Study of the spectral characteristics of crops of winter wheat varieties infected with pathogens of leaf diseases. *Plants* **2024**, *13*, 1892. [CrossRef] [PubMed]
4. Zhou, H.; Su, Y.; Chen, J.; Li, J.; Ma, L.; Liu, X.; Lu, S.; Wu, Q. Maize Leaf Disease Recognition Based on Improved Convolutional Neural Network ShuffleNetV2. *Plants* **2024**, *13*, 1621. [CrossRef] [PubMed]
5. Ye, Z.; Long, J.; Zhang, T.; Lin, B.; Lin, H. L-band SAR and its application for forest parameter estimation 1972 to 2024: A review. *Plants* **2024**, *13*, 2511. [CrossRef] [PubMed]
6. Zhong, L.; Yang, S.; Rong, Y.; Qian, J.; Zhou, L.; Li, J.; Sun, Z. Indirect Estimation of Heavy Metal Contamination in Rice Soil Using Spectral Techniques. *Plants* **2024**, *13*, 831. [CrossRef] [PubMed]

7. Wang, Y.; Lu, Y.; Yang, N.; Wang, J.; Huang, Z.; Chen, J.; Zhang, Z. Influence of time lag effect between winter wheat canopy temperature and atmospheric temperature on the accuracy of CWSI inversion of photosynthetic parameters. *Plants* **2024**, *13*, 1702. [CrossRef] [PubMed]
8. Yang, W.; Li, Z.; Chen, G.; Cui, S.; Wu, Y.; Liu, X.; Meng, W.; Liu, Y.; He, J.; Liu, D.; et al. Soybean Leaf Moisture Estimation Based on Multi-Source Unmanned Aerial Vehicle Image Feature Fusion. *Plants* **2024**, *13*, 1498. [CrossRef]
9. Sun, T.; Li, Z.; Wang, Z.; Liu, Y.; Zhao, Y.; Xie, W.; Cui, S.; Chen, G.; Yang, W.; Zhang, Z.; et al. Monitoring of nitrogen concentration in soybean leaves at multiple spatial vertical scales based on spectral parameters. *Plants* **2024**, *13*, 140. [CrossRef] [PubMed]
10. Nowack, J.; Atencia, L.; Tarquis, A.; Campo, G. Application of Unmanned Aerial Vehicle (UAV) Sensing for Water Status Estimation in Vineyards under Different Pruning Strategies. *Plants* **2024**, *13*, 1350. [CrossRef] [PubMed]
11. Zhang, H.; Chen, T.; Yu, S.; Zhou, C.; Teng, A.; Lei, L.; Li, F. Optimizing the Mulching Pattern and Nitrogen Application Rate to Improve Maize Photosynthetic Capacity, Yield, and Nitrogen Fertilizer Utilization Efficiency. *Plants* **2024**, *13*, 1241. [CrossRef] [PubMed]
12. Bitella, G.; Bochicchio, R.; Castronuovo, D.; Lovelli, S.; Mercurio, G.; Rivelli, A.; Rosati, L.; Antonio, P.; Casiero, P.; Laghetti, G.; et al. Monitoring plant height and spatial distribution of biometrics with a low-cost proximal platform. *Plants* **2024**, *13*, 1085. [CrossRef] [PubMed]

plants

MDPI

Article

Monitoring of Nitrogen Concentration in Soybean Leaves at Multiple Spatial Vertical Scales Based on Spectral Parameters

Tao Sun [1,2], Zhijun Li [1,2,*], Zhangkai Wang [1,2], Yuchen Liu [1,2], Zhiheng Zhu [1,2], Yizheng Zhao [1,2], Weihao Xie [1,2], Shihao Cui [1,2], Guofu Chen [1,2], Wanli Yang [1,2], Zhitao Zhang [1,2] and Fucang Zhang [1,2]

1 Key Laboratory of Agricultural Soil and Water Engineering in Arid and Semiarid Areas of Ministry of Education, Northwest A&F University, Xianyang 712100, China; 2021050986@nwsuaf.edu.cn (T.S.); wangzhk2023@163.com (Z.W.); 18729390136@163.com (Y.L.); guci_owo@163.com (Z.Z.); zhao24989@nwafu.edu.cn (Y.Z.); 2021012220@nwsuaf.edu.cn (W.X.); 2022012353@nwafu.edu.cn (S.C.); 2022012367@nwafu.edu.cn (G.C.); 2022012387@nwafu.edu.cn (W.Y.); zhangzhitao@nwafu.edu.cn (Z.Z.); zhangfc@nwsuaf.edu.cn (F.Z.)

2 Institute of Water-Saving Agriculture in Arid Areas of China, Northwest A&F University, Xianyang 712100, China

* Correspondence: lizhij@nwsuaf.edu.cn

Citation: Sun, T.; Li, Z.; Wang, Z.; Liu, Y.; Zhu, Z.; Zhao, Y.; Xie, W.; Cui, S.; Chen, G.; Yang, W.; et al. Monitoring of Nitrogen Concentration in Soybean Leaves at Multiple Spatial Vertical Scales Based on Spectral Parameters. *Plants* **2024**, *13*, 140. https://doi.org/10.3390/plants13010140

Academic Editor: Anis Limami

Received: 12 December 2023
Revised: 28 December 2023
Accepted: 1 January 2024
Published: 4 January 2024

Abstract: Nitrogen is a fundamental component for building amino acids and proteins, playing a crucial role in the growth and development of plants. Leaf nitrogen concentration (LNC) serves as a key indicator for assessing plant growth and development. Monitoring LNC provides insights into the absorption and utilization of nitrogen from the soil, offering valuable information for rational nutrient management. This, in turn, contributes to optimizing nutrient supply, enhancing crop yields, and minimizing adverse environmental impacts. Efficient and non-destructive estimation of crop LNC is of paramount importance for on-field crop management. Spectral technology, with its advantages of repeatability and high-throughput observations, provides a feasible method for obtaining LNC data. This study explores the responsiveness of spectral parameters to soybean LNC at different vertical scales, aiming to refine nitrogen management in soybeans. This research collected hyperspectral reflectance data and LNC data from different leaf layers of soybeans. Three types of spectral parameters, nitrogen-sensitive empirical spectral indices, randomly combined dual-band spectral indices, and "three-edge" parameters, were calculated. Four optimal spectral index selection strategies were constructed based on the correlation coefficients between the spectral parameters and LNC for each leaf layer. These strategies included empirical spectral index combinations (Combination 1), randomly combined dual-band spectral index combinations (Combination 2), "three-edge" parameter combinations (Combination 3), and a mixed combination (Combination 4). Subsequently, these four combinations were used as input variables to build LNC estimation models for soybeans at different vertical scales using partial least squares regression (PLSR), random forest (RF), and a backpropagation neural network (BPNN). The results demonstrated that the correlation coefficients between the LNC and spectral parameters reached the highest values in the upper soybean leaves, with most parameters showing significant correlations with the LNC ($p < 0.05$). Notably, the reciprocal difference index (VI6) exhibited the highest correlation with the upper-layer LNC at 0.732, with a wavelength combination of 841 nm and 842 nm. In constructing the LNC estimation models for soybeans at different leaf layers, the accuracy of the models gradually improved with the increasing height of the soybean plants. The upper layer exhibited the best estimation performance, with a validation set coefficient of determination (R^2) that was higher by 9.9% to 16.0% compared to other layers. RF demonstrated the highest accuracy in estimating the upper-layer LNC, with a validation set R2 higher by 6.2% to 8.8% compared to other models. The RMSE was lower by 2.1% to 7.0%, and the MRE was lower by 4.7% to 5.6% compared to other models. Among different input combinations, Combination 4 achieved the highest accuracy, with a validation set R^2 higher by 2.3% to 13.7%. In conclusion, by employing Combination 4 as the input, the RF model achieved the optimal estimation results for the upper-layer LNC, with a validation set R^2 of 0.856, RMSE of 0.551, and MRE of 10.405%. The findings of this study provide technical support for remote sensing monitoring of soybean LNCs at different spatial scales.

Keywords: soybean; remote sensing; hyperspectral; leaf nitrogen content; spectral parameters

1. Introduction

Soybean, as one of the world's top five crops, serves as a crucial source of high-quality protein and edible oil for humanity and plays a vital role in global food security [1]. In China, the soybean planting area accounts for approximately 7.7% of the world's total, yet its yield is only 70.7% of the world average [2]. Ensuring high soybean yields is imperative. Nitrogen is a fundamental component of plant proteins, nucleic acids, amino acids, and other essential biomolecules, playing a key role in plant growth and development [3,4]. Monitoring leaf nitrogen concentration (LNC) allows an assessment of whether plants receive sufficient nitrogen supply [5,6], facilitating the adoption of appropriate measures to promote or adjust plant growth. Simultaneously, adequate nitrogen directly influences crop yield and quality. Proper fertilization and monitoring of LNCs assists farmers in effectively managing nitrogen supply, avoiding over- or under-fertilization, thereby enhancing crop productivity and quality [7,8]. Moreover, excessive nitrogen fertilization can lead to nitrogen leakage into soil and water, causing environmental issues such as nutrient enrichment of water bodies and soil acidification [9,10]. Monitoring LNCs enables precise determination of plant nitrogen requirements, helping to reduce over-fertilization and mitigate environmental impacts. Hence, the importance of monitoring crop LNCs is evident [11,12].

Traditional laboratory methods for collecting leaf samples and detecting LNC, such as the Kjeldahl method [13,14], are characterized by high sensitivity and accuracy. However, these methods involve destructive sampling, resulting in complex sample handling, low efficiency, and high costs. This approach may potentially damage plants, adversely affecting crop growth and development [15,16]. Efficient and non-destructive monitoring of crop growth is central to modern precision agriculture [17]. Current manual methods for measuring LNC suffer from limitations, including small measurement areas, high workload, and poor data representativeness, making them unsuitable for large-scale precision management of field crops [18]. With the widespread application of remote sensing technology in agriculture, timely and non-destructive monitoring of crop LNCs has become feasible [19,20].

Spectral remote sensing technology, with its ability to provide rich spectral information and conduct large-scale non-contact monitoring, has gained attention in the study of crop physiological growth indicators [21,22]. Numerous studies have been conducted on using spectral remote sensing technology to monitor crop LNC [23–25]. For example, Fan et al. (2019) [23] employed a continuous projection algorithm to select spectral parameters with optimal performance for LNC estimation, achieving a verification set determination coefficient (R^2) exceeding 0.75, indicating excellent estimation accuracy. Zhao et al. (2021) [24], through field experiments over four growing seasons, developed a novel method based on remote sensing data to calculate nitrogen parameters for winter wheat. The model exhibited good stability. Shu et al. (2023) [25], using high-resolution spectral imaging obtained from unmanned aerial vehicles, estimated the nitrogen status of corn leaves using spectral decomposition methods, significantly improving the accuracy of nitrogen status estimation in corn leaves based on unmanned aerial vehicle high-resolution spectral images. These studies suggest that spectral information has the capability to monitor crop LNC, and although constructing fixed-wavelength spectral indices can accurately monitor nitrogen conditions to some extent, the different physiological information of crops due to factors such as their growth environment and growth stage can lead to variations in their spectral characteristics [26,27]. In such cases, using the same wavelengths may result in inadequate utilization of spectral data, limiting the effectiveness of the calculated spectral index inversion model and reducing model accuracy [28]. Furthermore, for crops, there is a lack of comparison and discussion on the prediction effects of different vertical scales of LNCs. Models are often built and analyzed using LNCs from a specific site without comparison and optimization, which limits the spatial applicability of the established prediction models [29].

To address these issues, this study aims to construct three categories of spectral parameters for estimating LNCs at various soybean leaf layers: (1) empirical spectral indices with good correlations to crop parameters from previous studies; (2) optimal spectral indices, i.e., the best combination of indices within the wavelength range of 350–1830 nm with the highest correlation to the LNC at various soybean leaf layers; and (3) three-edge spectral indices, involving blue, yellow, and red edge areas. These parameters, often associated with red, blue, and green edge areas, provide valuable insights into the spectral characteristics of the studied vegetation. The four-node stage (V4) of soybeans, occurring when the fourth true leaf unfolds after planting, is a critical period for soybean growth and development [7]. During this stage, plants begin rapid growth and establish initial structures, such as leaves and stems. The health of plants at this stage directly affects subsequent growth and development. Additionally, soybeans require sufficient nutrients to support their growth and development at this stage. Therefore, monitoring LNC during this period is of paramount importance for field management to ensure soybeans receive the necessary nutrition for stable and high yields. In this study, different soybean leaf layers' LNCs under various treatments at the V4 stage were selected as the research objects. We constructed different types of spectral parameters and established soybean LNC estimation models based on the partial least squares regression (PLSR), random forest (RF), and backpropagation neural network (BPNN) algorithms. We compared and analyzed the estimation effects and stability of the models, aiming to establish LNC prediction models at different vertical scales for soybeans to achieve non-destructive and rapid LNC estimation.

2. Materials and Methods

2.1. Research Area and Test Design

The experiment was carried out at the water-saving irrigation experimental station (34°14′ N, 108°10′ E, altitude 521 m) of the Water-Saving Agriculture Research Institute of Northwest A&F University (Figure 1) in June–September 2021 and June–September 2022. In the experiment, 24 experimental plots were set up, each of which was 6 m long and 4 m wide. The experiment set four nitrogen application levels, 0 kg ha^{-1} (N0), 60 kg ha^{-1} (N1), 120 kg ha^{-1} (N2), and 180 kg ha^{-1} (N3), and two seed dressing treatments, rhizobium inoculation (R) and water seed dressing (unmarked). The experiment was conducted in a completely randomized design with three replicates. In order to reduce the influence between experimental treatments, a 2 m wide isolation belt was set between adjacent cells. The amount of phosphate and potassium fertilizer in each experimental plot was 30 kg ha^{-1}. The nitrogen fertilizer used in the experiment was urea (46% N), the phosphorus fertilizer was calcium superphosphate (16% P), and the potassium fertilizer was potassium chloride (62% K).

The seed dressing method was used to inoculate rhizobium (in line with the national industry standard GB20287-2006 [30]). The seed dressing method was used and 25 g of rhizobium powder was added into 500 ml water and stirred evenly. Before soaking, the rhizobium was fully shaken to ensure that the rhizobium was evenly attached to the seed surface. The seeds were dried after seed dressing, and the dried seeds were sown on 18 June 2021 and 10 June 2022, respectively. The planting density was 300,000 plants ha^{-1}, the row spacing was 50 cm, and the plant spacing was 10 cm. Soybeans were harvested on 30 September 2021 and 20 September 2022, respectively.

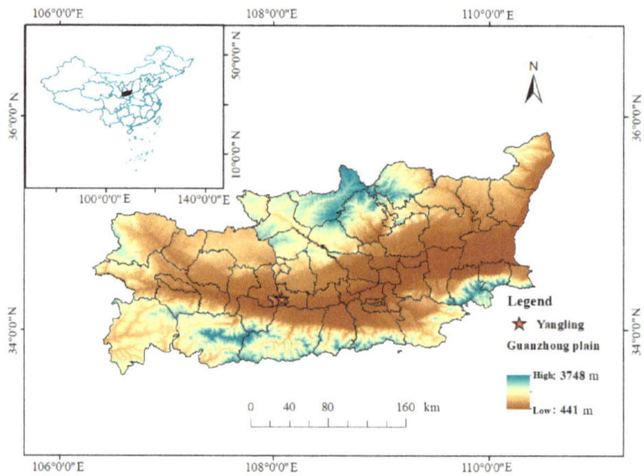

Figure 1. Study area.

2.2. Data Collection

The LNC and hyperspectral data were obtained in the experimental plots at V4 (17 July 2021 and 15 July 2022). In the two-year experiment, 48 groups of leaf nitrogen concentration and hyperspectral reflectance samples were obtained, respectively. After removing the outliers, there were 48 sets of data samples, and 2/3 of the samples were selected as the modeling set, with the remainder of the samples used as the validation set.

2.2.1. Measurement of Leaf Nitrogen Concentration

In our experiment, the root leaf nitrogen concentration (N_{RL}), lateral leaf nitrogen concentration (N_{LL}) and canopy leaf nitrogen concentration (N_{CL}) of the same soybean plant were collected at the same time, and the measurement sites are shown in Figure 2. In each plot, 10 soybean plants were randomly selected to determine their nitrogen concentration, and the average value was used to represent the nitrogen concentration value of different leaf layers in the plot. These leaves were dried in an oven at 105 °C for 30 min and then extra dried at 75 °C until a constant weight was achieved. Then, the dried samples were ground through a 1 mm sieve, digested with H_2SO_4–H_2O_2, and LNC was analyzed by the Kjeldahl method. The detailed determination process used Cheng et al. (2022) as a reference [13].

2.2.2. Acquisition of Hyperspectral Data

The spectral acquisition instrument used was a FieldSpec3 hyperspectrometer produced by the ASD company in the United States. The wavelength range of the instrument was 350~1830 nm. The spectral resolution of 350~1000 nm was 3 nm, and the sampling interval was 1.4 nm. The resolution of 1000~1830 nm was 10 nm, and the sampling interval was 2 nm. The instrument automatically interpolated the sampling data into a 1 nm interval output. The fiber length was 1.5 m and the field of view was 25°. The determination was carried out at 11: 00–13: 00 in sunny and windless weather. During the measurement, the spectrometer probe was about 15 cm away from the soybean canopy, always keeping 90° with the ground, and the field angle was 25°. Before the spectral determination, the instrument was corrected by a diffuse reflection reference plate with a reflectivity of 1. The instrument was optimized every 5 min, and the dark current was collected every 5 min to optimize the instrument. In each plot, the soybean around the sample point was measured by five-point plum blossom sampling, and the average value was taken as the final spectral value of the monitoring point. To reduce or eliminate the influence of useless information such as background noise, baseline drift, and stray light on the spectral reflectance curve, we used Savitzky–Golay convolution smoothing (9 points and 4 times) to preprocess the spectral data [2].

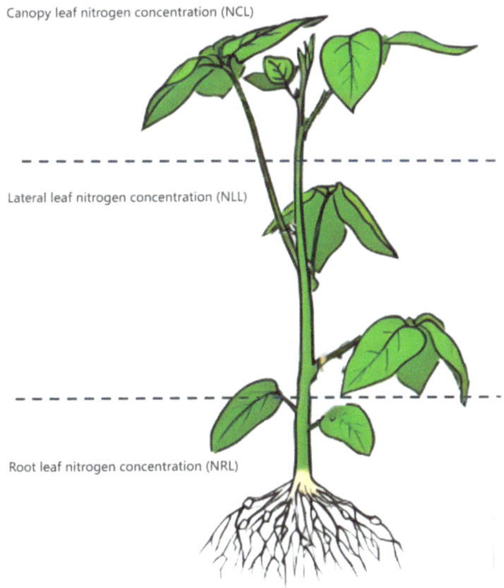

Figure 2. Leaf sampling details of each layer of the soybean plant.

2.3. *Techniques for Data Analysis*

The spectral index can reflect crop growth and nutritional status [26,27]. In this study, three types of spectral parameters were constructed to estimate the soybean's LNC: (1) the empirical spectral index with good correlation between previous studies and crop parameters; (2) the optimal spectral index with the highest correlation with the soybean's LNC was selected, that is, the best combination index in the range of 350–1830 nm; and (3) "trilateral" spectral parameters such as blue, yellow, and red edge areas. The selected spectral parameters were calculated as shown in Table 1. The calculation results of the spectral index were calculated by MATLAB R2022 (MathWorks, Inc., Natick, MA, USA). All the figures in this study were drawn by Origin Pro 2021 (OriginLab Corp., Northampton, MA, USA).

Table 1. Selection of spectral parameters, calculation formulas, and reference source.

Selected Spectral Parameter	Calculation Formula	Reference
Maximum first-order derivative value in the blue edge (490–530 nm) D_b	–	[31]
Maximum first-order derivative value in the yellow edge (462–642 nm) D_y	–	[31]
Maximum first-order derivative value in the red edge (670–760 nm) D_r	–	[32]
Maximum reflectivity of the green peak (510–560 nm) R_g	–	[33]
Minimum reflectivity of the red valley (650–690 nm) R_r	–	[33]
Blue edge (490–530 nm) area S_b	Sum of first-order derivatives within the blue edge wavelength range	[31]
Yellow edge (462–642 nm) area S_y	Sum of first-order derivatives within the yellow edge wavelength range	[31]
Red edge (670–760 nm) area S_r	Sum of first-order derivatives within the red edge wavelength range	[32]
Normalized red-blue amplitude difference (NDDr.b)	$(D_r - D_b)/(D_r + D_b)$	[34]
Normalized first-order red-blue amplitude difference (NDSDr.b)	$(SD_r - SD_b)/(SD_r + SD_b)$	[34]

Table 1. *Cont.*

Selected Spectral Parameter	Calculation Formula	Reference
Infrared percentage vegetation index (IPVI)	$R_{800} \times (R_{800} + R_{670})$	[35]
Optimized soil-adjusted vegetation index (OSAVI)	$(1+0.16)(R_{800}-R_{670})/(R_{800}+R_{670}+0.16)$	[36]
Normalized difference nitrogen index (NDNI)	$\frac{\left(\frac{1}{R_{1510}}\right)-\left(\frac{1}{R_{1680}}\right)}{\left(\frac{1}{R_{1510}}\right)+\left(\frac{1}{R_{1680}}\right)}$	[37]
Ashburn vegetation index (AVI)	$2 \times \frac{R_{800}}{R_{1100}} - \frac{R_{600}}{R_{700}}$	[38]
Difference 678/500 (D678/500)	R_{678}/R_{500}	[39]
Difference 800/550 (D800/550)	R_{800}/R_{550}	[40]
Difference 800/680 (D800/680)	R_{800}/R_{680}	[41]
Difference 833/658 (D833/658)	R_{833}/R_{658}	[42]
Differenced vegetation index MSS (DVIMSS)	$2.4 \times \frac{R_{800}}{R_{1100}} - \frac{R_{600}}{R_{700}}$	[43]
Double difference index (DD)	$(R_{749}-R_{720})-(R_{701}-R_{672})$	[44]
Ratio index (RI)	R_i/R_j	[27]
Difference index (DI)	$R_i - R_j$	[27]
Soil-adjusted vegetation index (SAVI)	$(1+0.16)\frac{R_i-R_j}{R_i+R_j+0.16}$	[28]
Normalized difference vegetation index (NDVI)	$R_i - R_j/R_i + R_j$	[28]
Triangular vegetation index (TVI)	$0.5 \times \left(120 \times (R_i - R_{550}) - 200 \times \left(R_j - R_{550}\right)\right)$	[28]
Modified simple ratio (mSR)	$R_i - R_{455}/R_j - R_{455}$	[28]
Modified normalized difference index (mNDI)	$R_i - R_j/R_i + R_j - 2R_{455}$	[28]
Product index (PI)	$R_i \times R_j$	[45]
Sum index (SI)	$R_i + R_j$	[45]
Reciprocal difference index (VI6)	$1/R_i - 1/R_j$	[45]

Note: R_i (i = 1,2,3) is the reflectivity at any band, and R445, R455, R500, R530, R531, R550, R570, R670, R680, R700, R705, R742, R750, and R800 represent the spectral reflectance of wavelengths 445 nm, 455 nm, 500 nm, 530 nm, 531 nm, 550 nm, 570 nm, 670 nm, 680 nm, 700 nm, 705 nm, 742 nm, 750 nm, and 800 nm.

Firstly, the correlation between the spectral parameters of the soybean leaves and LNCs in all 24 plots was analyzed, and four screening strategies were designed: (1) Combination 1, the empirical spectral indices with significant correlation coefficients ($p < 0.05$) with the LNC of each leaf layer were selected; (2) Combination 2, the spectral indices with significant correlation coefficients ($p < 0.05$) with the LNC of each leaf layer were selected from the random combination of dual-band spectral indices; (3) Combination 3, the spectral parameters that were significant ($p < 0.05$) to the LNC of each leaf layer were selected from the "trilateral" spectral parameters; and (4) Combination 4, all spectral parameters of Combination 1, Combination 2, and Combination 3 with significant correlation coefficients ($p < 0.05$) with the LNC of each leaf layer were selected as input variables of the model.

Subsequently, we used MATLAB R2022 to test three machine learning methods, namely PLSR, RF, and BPNN. In the construction of the PLSR model, a 10-fold cross-validation method was employed to determine the optimal number of latent variables (LVs) for estimating the LNC. In this study, the optimal number of latent variables for the LNC estimation model was determined to be 3. In the construction of the RF model, after parameter optimization and multiple training, the number of decision trees in the LNC model was set to 600. The hidden layer transfer function of the BPNN was set to "TANSIG", and the Levenberg–Marquardt (trainlm) algorithm based on numerical optimization theory was used as the network training function. After much training, the number of neurons in the middle layer was determined to be 15.

2.4. Techniques for Data Analysis

In order to verify the prediction accuracy and predictive ability of the models, three indicators were selected, the determination coefficient (R^2), root mean square error (RMSE), and mean relative error (MRE), to evaluate the model accuracy [6]. The R^2, RMSE, and MRE were calculated using the following equations:

$$R^2 = \frac{\sum_{i=1}^{n}(\hat{y}_i - \overline{y})^2}{\sum_{i=1}^{n}(y_i - \overline{y})^2} \quad (1)$$

$$RMSE = \sqrt{\frac{\sum_{i=1}^{n}(\hat{y}_i - y_i)^2}{n}} \quad (2)$$

$$MRE = \frac{1}{n}\sum_{i=1}^{n}\frac{|\hat{y}_i - y_i|}{y_i} \times 100\% \quad (3)$$

3. Results

3.1. Comparison of LNCs in Different Leaf Layers of Soybean and Division of Sample Set

The crops' LNCs exhibited spatial and vertical heterogeneity [27]. The statistical results of the LNC are depicted in Figure 3 and Table 2. It can be observed that the soybean leaf nitrogen concentration followed the size order of $N_{RL} < N_{LL} < N_{CL}$, indicating a gradual increase in LNC from the root system to the top of the soybean plant.

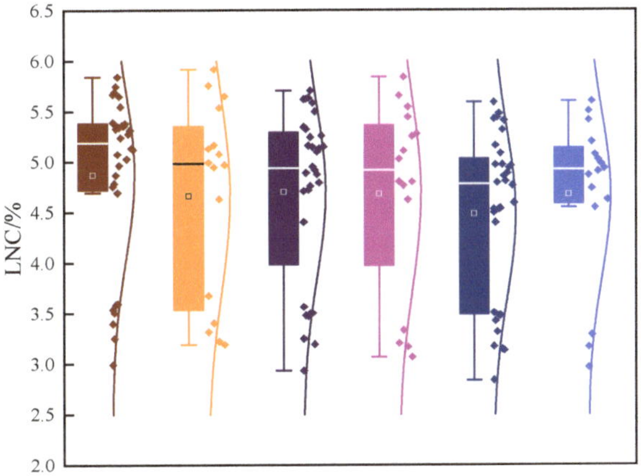

Figure 3. Statistics of the LNC in each leaf layer of soybean. The horizontal line in the box line diagram represents the median, and the white box represents the average value. The N_{RL} modeling set is dark brown and the validation set is light brown. The N_{LL} modeling set is dark purple and the validation set is light purple. The N_{CL} modeling set is dark blue and the validation set is light blue.

From Table 2, it is evident that different nitrogen application rates significantly impacted the LNCs of the soybeans ($p < 0.05$). In the majority of leaves, the inoculation of rhizobia also exerted a significant effect on the LNC ($p < 0.05$). Notably, as nitrogen application increased, leaf LNC increased accordingly. Furthermore, under the condition of the same nitrogen application rate, rhizobia inoculation promoted nitrogen absorption in the leaves, thereby augmenting the LNC.

3.2. Correlation Analysis between Spectral Parameters and LNCs of Soybean Leaf Layers

The correlation analysis results between the empirical spectral indices, "three-side parameters", and LNCs at different leaf layers are presented in Table 3. The process of wavelength combination selection for the calculated indices from pairwise spectral bands is detailed in Figure 4. The correlation coefficients and wavelength positions of the ten selected indices with different leaf layers are shown in Figure 4 and Table 4. The results indicate that the majority of "three-side parameters", arbitrary two-band spectral indices, and empirical spectral indices exhibited higher correlation coefficients with the N_{CL} compared to the N_{LL} and N_{RL}. Among the indices constructed with two bands in the N_{CL}, the highest correlation with the LNC was achieved, with all exceeding 0.65. Notably, the index VI6 demonstrated the highest correlation with N_{CL} at 0.732, with a wavelength combination of 841 nm and 842 nm. Simultaneously, most of the selected spectral parameters in this study showed a significant correlation level ($p < 0.05$) with the LNC at various leaf layers. For the empirical indices, the index D833/658 had the highest correlation with N_{CL} at 0.671, while the "three-side parameter" Sr had the highest correlation with N_{CL} at 0.669. In general, the correlation coefficient between the spectral parameters and LNC gradually increased from the root to the top of the soybean leaves.

Table 2. The values of the LNCs in each leaf layer under different treatments. R, the influence of different rhizobium inoculation methods on each index; N, the effect of different nitrogen applications on each index; N*R, the influence of the interaction between the rhizobium inoculation method and nitrogen application on each index. The different letters indicate the significance within the same year at the 5% level by the LSD test. ns, not significant, ($p > 0.05$); *, significant at $p < 0.05$; **, significant at $p < 0.01$.

	2021			2022		
	N_{CL}	N_{LL}	N_{RL}	N_{CL}	N_{LL}	N_{RL}
RN3	5.81 a	5.79 a	5.59 a	5.70 a	5.62 a	5.47 a
RN2	5.55 a	5.38 a	5.07 a	5.33 b	5.25 b	4.98 bc
RN1	5.16 ab	5.07 ab	5.01 a	4.97 d	4.82 c	4.59 d
RN0	4.09 bc	3.95 bc	3.82 b	3.48 f	3.41 e	3.37 e
N3	4.96 abc	4.88 abc	4.66 ab	5.44 b	5.34 b	5.19 b
N2	5.35 a	5.19 a	5.03 a	5.13 c	5.01 c	4.88 c
N1	4.98 abc	4.87 abc	4.73 ab	4.69 e	4.58 d	4.49 d
N0	3.84 c	3.72 c	3.64 b	3.13 g	3.06 f	2.98 f
			Significance level			
N	*	**	**	**	**	**
R	ns	*	ns	**	*	**
N*R	ns	*	ns	*	ns	*

Table 3. The correlation coefficients between the LNC and spectral parameters of soybean leaves. *, the correlation coefficient reached a significant level ($p < 0.05$, the same below). The bold represents the highest correlation coefficients.

Selected Spectral Parameter	Correlation Coefficient		
	N_{CL}	N_{LL}	N_{RL}
Maximum first-order derivative value in the blue edge (490–530 nm) D_b	**0.601** *	0.582 *	0.514 *
Maximum first-order derivative value in the yellow edge (462–642 nm) D_y	**0.601** *	0.582 *	0.514 *
Maximum first-order derivative value in the red edge (670–760 nm) D_r	**0.660** *	0.586 *	0.526 *
Maximum reflectivity of the green peak (510–560 nm) R_g	**0.495** *	0.501 *	0.460 *
Minimum reflectivity of the red valley (650–690 nm) R_r	**0.208**	0.259	0.258
Blue edge (490–530 nm) area S_b	**0.483** *	0.511 *	0.476 *
Yellow edge (462–642 nm) area S_y	**0.176**	0.255	0.260
Red edge (670–760 nm) area S_r	**0.669** *	0.604 *	0.543 *
NDDr.b	**0.069**	0.013	0.007
NDSDr.b	**0.114**	0.015	0.0004
IPVI	**0.667** *	0.630 *	0.582 *
Optimized soil-adjusted vegetation index (OSAVI)	**0.512** *	0.419 *	0.369 *
Normalized difference nitrogen index (NDNI)	**0.550** *	0.480 *	0.441 *
Ashburn vegetation index (AVI)	**0.670** *	0.631 *	0.574 *
Difference 678/500 (D678/500)	**0.580** *	0.510 *	0.439 *
Difference 800/550 (D800/550)	**0.664** *	0.603 *	0.547 *
Difference 800/680 (D800/680)	**0.670** *	0.605 *	0.544 *
Difference 833/658 (D833/658)	**0.671** *	0.609 *	0.547 *
Differenced vegetation index MSS (DVIMSS)	**0.669** *	0.631 *	0.574 *
Double difference index (DDI)	**0.449** *	0.363 *	0.362 *

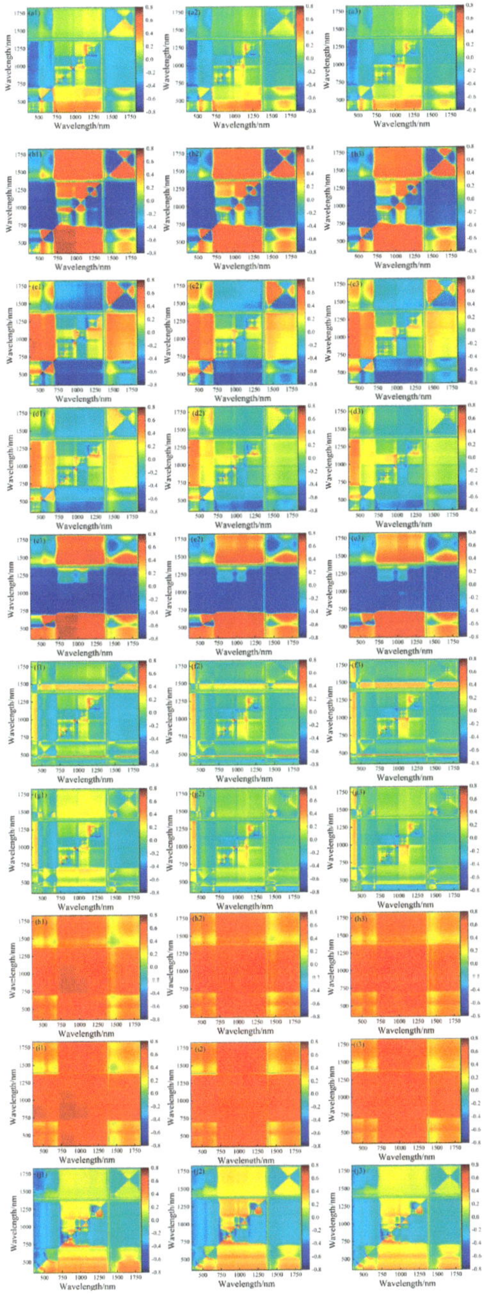

Figure 4. The correlation matrix diagrams of the spectral indices and soybean LNCs. (**a1**) RI and N_{CL}; (**a2**) RI and N_{LL}; (**a3**) RI and N_{RL}; (**b1**) DI and N_{CL}; (**b2**) DI and N_{LL}; (**b3**) DI and N_{RL}; (**c1**) SAVI and N_{CL}; (**c2**) SAVI and N_{LL}; (**c3**) SAVI and N_{RL}; (**d1**) NDVI and N_{CL}; (**d2**) NDVI and N_{LL}; (**d3**) NDVI and N_{RL}; (**e1**) TVI and N_{CL}; (**e2**) TVI and N_{LL}; (**e3**) TVI and N_{RL}; (**f1**) mSR and N_{CL}; (**f2**) mSR and N_{LL}; (**f3**) mSR and N_{RL}; (**g1**) mNDI and N_{CL}; (**g2**) mNDI and N_{LL}; (**g3**) mNDI and N_{RL}; (**h1**) PI and LNC_{CL}; (**h2**) PI and LNC_{LL}; (**h3**) PI and LNC_{RL}; (**i1**) SI and LNC_{CL}; (**i2**) SI and LNC_{LL}; (**i3**) SI and LNC_{RL}; (**j1**) VI6 and LNC_{CL}; (**j2**) VI6 and LNC_{LL}; and (**j3**) VI6 and LNC_{RL}. The colors from blue to red represent the negative correlation to positive correlation.

Table 4. The maximum correlation coefficients and wavelength positions between the spectral indices screened by any two bands and the LNCs of different leaf layers. *, the correlation coefficient reached a significant level ($p < 0.05$, the same below). The bold represents the highest correlation coefficients.

Selected Spectral Parameter	N_{CL}		N_{LL}		N_{RL}	
	Correlation Coefficient	Wavelength Position (*i,j*)	Correlation Coefficient	Wavelength Position (*i,j*)	Correlation Coefficient	Wavelength Position (*i,j*)
RI	**0.664** *	(840,843)	0.567 *	(1043,1046)	0.574 *	(839,844)
DI	**0.701** *	(1625,1637)	0.684 *	(1221,1267)	0.701 *	(1612,1611)
SAVI	**0.675** *	(413,934)	0.662 *	(1358,1393)	0.638 *	(1611,1612)
NDVI	**0.664** *	(840,843)	0.599 *	(1043,1045)	0.574 *	(844,839)
TVI	**0.689** *	(694,616)	0.661 *	(1353,680)	0.594 *	(1129,487)
mSR	**0.671** *	(840,843)	0.597 *	(1043,1045)	0.584 *	(844,839)
mNDI	**0.670** *	(840,841)	0.597 *	(1043,1045)	0.584 *	(844,839)
PI	**0.676** *	(856,854)	0.642 *	(753,1356)	0.593 *	(1046,1047)
SI	**0.686** *	(783,1368)	0.663 *	(840,1368)	0.579 *	(759,1129)
VI6	**0.732** *	(841,842)	0.658 *	(972,981)	0.639 *	(840,843)

3.3. Construction of Soybean LNC Estimation Model at a Multi-Spatial Vertical Scale

In the preceding section, we computed the correlation coefficients between the spectral parameters and soybean LNCs at various layers. We selected different types of spectral parameters as inputs for the machine learning model, namely, Combination 1 (IPVI, OSAVI, NDNI, AVI, $D_{678/500}$, $D_{800/550}$, $D_{800/680}$, $D_{833/658}$, DVI_{MSS}, and DDI), Combination 2 (D_b, D_y, D_r, R_g, S_b, and S_r), Combination 3 (RI, DI, SAVI, NDVI, TVI, mSR, mNDI, PI, SI, and VI6), and Combination 4 (IPVI, OSAVI, NDNI, AVI, $D_{678/500}$, $D_{800/550}$, $D_{800/680}$, $D_{833/658}$, DVI_{MSS}, D_D, D_b, D_y, D_r, R_g, S_b, S_r, RI, DI, SAVI, NDVI, TVI, mSR, mNDI, PI, SI, and VI6). The selected spectral parameters for each combination were consistent across different leaf layers. Subsequently, using the aforementioned combinations as independent variables and the soybean LNCs at various layers as the response variables, we employed PLSR, RF, and BPNN to construct soybean LNC estimation models. Model accuracy was comprehensively evaluated based on R^2. The prediction results for the soybean leaf areas by different modeling methods are illustrated in Figures 5–7. The results indicate that, with the increase in the vertical height of soybean, the accuracy of the estimation model gradually improved. The N_{cl} exhibited the best estimation performance, with a validation set R^2 higher by 9.9% to 16.0% compared to other layers under similar conditions. Analyzing the estimation model for N_{cl}, random forest (RF) demonstrated the highest accuracy in estimating LNC. The validation set R^2 was 6.2% to 8.8% higher than other models, RMSE was 2.1% to 7.0% lower, and MRE was 4.7% to 5.6% lower than other models. Considering different input combinations, Combination 4 as the input achieved the highest accuracy, with a validation set R^2 higher by 2.3% to 13.7% compared to other models. In summary, using Combination 4 as the input, the RF model produced the optimal estimation results for N_{CL}, with a validation set R^2 of 0.856, RMSE of 0.551, and MRE of 10.405%.

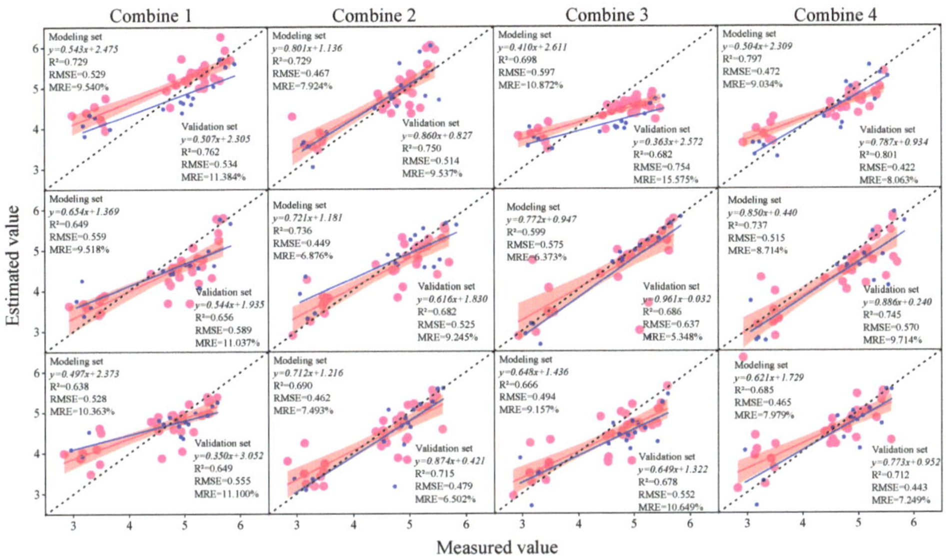

Figure 5. The modeling set and validation sets of the BPNN estimation models with different input variables and leaf layers. The red dots and red lines represent the modeling sets and the modeling set fitted curves, the blue dots and blue lines represent the verification sets and the verification sets fitted curve, and the dotted lines represent the 1:1 lines.

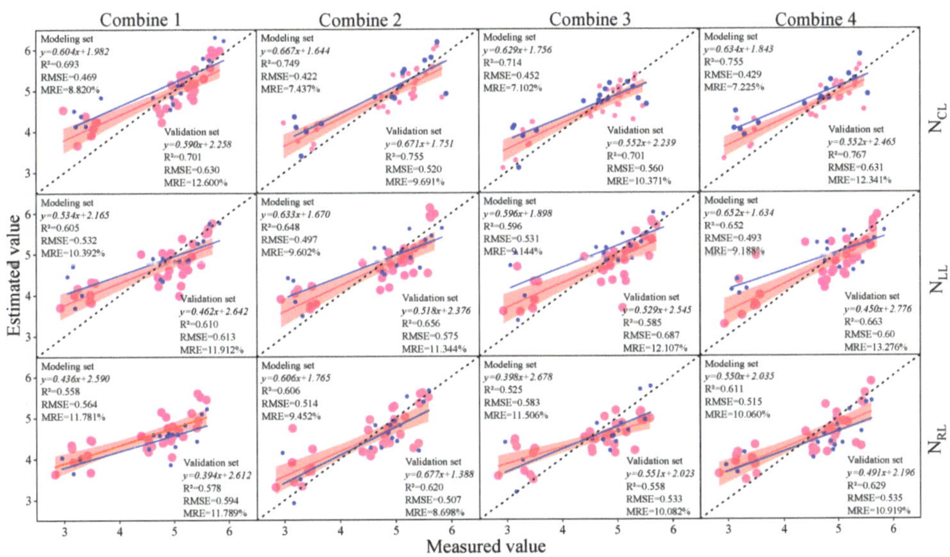

Figure 6. The modeling sets and validation sets of the PLSR estimation models with different input variables and leaf layers. The red dots and red lines represent the modeling sets and the modeling set fitted curves, the blue dots and blue lines represent the verification sets and the verification set fitted curves, and the dotted lines represent the 1:1 lines.

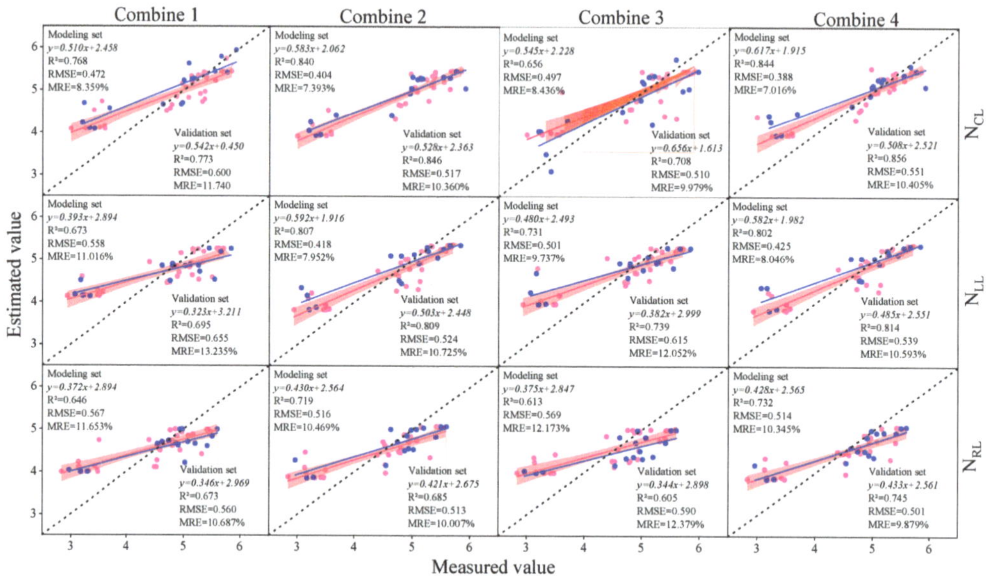

Figure 7. The modeling sets and validation sets of RF estimation models with different input variables and leaf layers. The red dots and red lines represent the modeling sets and the modeling set fitted curves, the blue dots and blue lines represent the verification sets and the verification set fitted curves, and the dotted lines represent the 1:1 lines.

4. Discussion

Spectral information serves as the foundational data for the optical remote sensing of spectral traits [46,47], and it is determined by various factors such as canopy structure and biochemical composition. Due to the uneven distribution of nutrients along the vertical scale of crops, it is often manifested on the leaves [29]. Studying the vertical variation in LNC can better elucidate changes in plant physiological processes and adjust the photosynthesis and nutrient absorption of crops, thereby optimizing agricultural management practices more effectively.

This study revealed that with the deepening of vertical spatial scale (closer to the root), LNC gradually decreased. This may be attributed to the fact that the growth point of crops is usually located at the top of the plant or at the top of its branches, which is the primary region for new tissue development. New tissues have a higher demand for nitrogen as it is a crucial component in building proteins and other biomolecules [48]. Simultaneously, upper leaves typically receive more sunlight, making them more exposed to sunlight [49]. Photosynthesis, the process through which plants synthesize organic compounds using sunlight, has a high demand for nitrogen [50]. Therefore, to meet the needs of photosynthesis, upper leaves may have a relatively higher nitrogen concentration to support the development of new tissues. This study also found that when predicting crop leaf LNCs on the vertical scale using spectral parameters, the prediction performance was relatively better for upper leaves, while the prediction for middle and lower leaves was poorer. This is because the ability of light to penetrate through crop plants may significantly differ at different vertical levels. Upper leaves are usually directly exposed to sunlight, making them easier to identify for spectral sensors. In contrast, middle and lower leaves may be shaded by upper leaves, resulting in poor light transmission [51]. This may reduce the quality of observations for these leaves, and the spectral signals at different vertical levels may mix together, especially when the sensor resolution is relatively low. This can make it challenging to differentiate the spectral signals at different vertical levels, and in general, upper leaves usually have more sensitive spectral signals, making them easier to detect and interpret, while the signal mixing at the middle and lower layers may complicate the estimation of LNC [29]. We observed that the increase in nitrogen application led to a corresponding rise in LNC. This is attributed to the heightened nitrogen application, which enhances soybean's capacity to absorb nitrogen, consequently elevating LNC [7]. Simultaneously, our findings indicate that inoculation with rhizobia enhanced soybean LNC. This is attributed to the symbiotic relationship between rhizobia and soybeans, wherein rhizobia facilitate nitrogen fixation, thereby promoting nitrogen absorption [52].

Through the correlation analysis of different spectral parameters with LNC at different leaf layers, it was found that randomly selected two-band spectral indices had higher correlation coefficients with the LNC at different leaf layers compared to empirical spectral indices and three-side parameters. This is because, for different study subjects, due to variations in growth environments, growth stages, and other factors, the physiological information of crops may differ, leading to different spectral characteristics. In such cases, using empirically calculated spectral parameters with the same wavelengths may underutilize spectral data [26–28]. Additionally, three-side spectral parameters may not sufficiently extract information related to nitrogen concentration in plant leaves. Conversely, randomly selected spectral indices may more effectively capture this information by searching for correlations between randomly chosen wavelengths [53].

The accuracy of the soybean LNC estimation models varied significantly based on the different model inputs and machine learning methods. When constructing soybean LNC estimation models using different input combinations, the soybean LNC estimation model based on random forest (RF) had the highest accuracy. This suggests that RF is more advantageous in estimating soybean LNC compared to other models, which is consistent with previous results for crop physiological growth indicators [28,53]. This is because RF is an ensemble learning method based on decision trees, which can effectively capture nonlinear relationships. The relationship between the LNC and spectral indices may be a complex nonlinear relationship, and RF may be more suitable for handling such nonlinear relationships [54]. Additionally, random forests typically exhibit good resistance to overfitting, which is helpful when dealing with small sample data or noise in the data. RF introduces randomness by building multiple decision trees and combining their results, reducing the risk of overfitting [53]. In contrast, when there are a large number of highly correlated predictor variables, partial least squares regression (PLSR) adapts to fewer components by considering the dependent variable, reducing the estimation accuracy of the model [55]. However, the backpropagation neural network (BPNN) has limitations due to the selection of model parameters, and too many feature parameters can lead to computational redundancy and insufficient timeliness, resulting in reduced accuracy [56]. Notably, when the machine learning models were the same, we found that the accuracy of the model input with Combination 2 (randomly selected two-band spectral indices) was lower than that with Combination 4 (various spectral parameters selected). This finding can be attributed to Combination 4 incorporating features from Combinations 1–3, maximizing the extraction of hyperspectral information. The lower accuracy of the models built with Combinations 1 to 3 as inputs was due to their inherent limitations as empirical parameters, as explained earlier. Similarly, it is not difficult to explain that the estimation of N_{cl} is better under the same conditions, as hyperspectral collection is more sensitive to the upper canopy.

This study aimed to estimate soybean LNC in vertical layers using spectral parameters calculated from drone multispectral image information. As soybean field management has significant practical implications, further research is needed. Future studies could consider developing more precise and accurate new indices rather than relying solely on conventional indices. Additionally, the use of drone hyperspectral imaging and thermal infrared imaging will be considered. Furthermore, spectral sensors will be adjusted multiple times in terms of angles and heights to capture more spectral information of soybean plants in three-dimensional spaces, achieving precise monitoring of physiological data for lower leaves and ultimately improving the estimation accuracy of soybean LNC and enhancing soybean yield.

5. Conclusions

This study, based on plot experiments and hyperspectral measured data, estimated the nitrogen concentration values in different leaf layers of soybeans. This was achieved by constructing spectral parameters, including empirical spectral indices, spectral indices selected from any two bands, and "three-edge" parameters. Three machine learning models, namely PLSR, RF, and BPNN, were employed for estimation. The conclusions drawn were the following: the majority of spectral parameters showed significant correlations with soybean leaf nitrogen concentration (LNC) at a significance level of $p < 0.05$. As soybean height increased, the correlation coefficients between LNC and spectral parameters also increased. Among them, VI6 exhibited the highest correlation with upper-layer leaf nitrogen concentration, reaching 0.732, with a wavelength combination of 841 nm and 842 nm. In summary, using Combination 4 (spectral parameters significantly correlated ($p < 0.05$) with leaf LNC) as the input, the RF model yielded the optimal estimation results for the upper-layer LNC, with a validation set R^2 of 0.856, RMSE of 0.551, and MRE of 10.405%.

Author Contributions: Data curation, T.S. and W.X.; investigation, Z.W., Z.Z. (Zhiheng Zhu), Y.Z., S.C., G.C. and W.Y.; methodology, T.S. and Y.L.; project administration, T.S.; resources, F.Z. and Z.L.; software, T.S. and W.X.; supervision, Z.Z. (Zhitao Zhang); visualization, Z.L.; writing—original draft, T.S.; writing—review and editing, T.S. and Z.L. All authors have read and agreed to the published version of the manuscript.

Funding: This study was supported by the National Natural Science Foundation of China (No. 52279047).

Data Availability Statement: Data are contained within the article.

Conflicts of Interest: The authors declare no conflicts of interest.

References

1. Buezo, J.; Sanz-Saez, Á.; Moran, J.F.; Soba, D.; Aranjuelo, I.; Esteban, R. Drought tolerance response of high-yielding soybean varieties to mild drought: Physiological and photochemical adjustments. *Physiol. Plant* **2019**, *166*, 88–104. [CrossRef] [PubMed]
2. Guo, S.; Zhang, Z.; Guo, E.; Fu, Z.; Gong, J.; Yang, X. Historical and projected impacts of climate change and technology on soybean yield in China. *Agric. Syst.* **2022**, *203*, 103522. [CrossRef]
3. Sun, Y.; Chen, X.; Shan, J.; Xian, J.; Cao, D.; Luo, Y.; Yao, R.; Zhang, X. Nitrogen mitigates salt stress and promotes wheat growth in the Yellow River Delta, China. *Water* **2022**, *14*, 3819. [CrossRef]
4. Sharma, S.K.; Sharma, P.K.; Mandeewal, R.L.; Sharma, V.; Chaudhary, R.; Pandey, R.; Gupta, S. Effect of foliar application of nano-urea under different nitrogen levels on growth and nutrient content of pearl millet (*Pennisetum glaucum* L.). *Int. J. Plant Sci.* **2022**, *34*, 149–155. [CrossRef]
5. Olveira, A.L.; Rozas, H.S.; CastroFranco, M.; Carciochi, W.; Nieto, L.; Balzarini, M.; Ciampitti, I.; Reussi Calvo, N. Monitoring corn nitrogen concentration from radar (C-SAR), optical, and sensor satellite data fusion. *Remote Sens.* **2023**, *15*, 824. [CrossRef]
6. Li, H.; Li, M.-R.; Guo, Y.-L.; Wei, Y.; Zhang, H.; Song, X.; Feng, W.; Guo, T. Angular effect of algorithms for monitoring leaf nitrogen concentration of wheat using multi-angle remote sensing data. *Comput. Electron. Agric.* **2022**, *195*, 106815. [CrossRef]
7. Wang, X.; Li, W.; An, J.; Shi, H.; Tang, Z.; Zhao, X.; Guo, J.; Jin, L.; Xiang, Y.; Li, Z. Effects of Nitrogen Supply on Dry Matter Accumulation, Water-Nitrogen Use Efficiency and Grain Yield of Soybean (*Glycine max* L.) under Different Mulching Methods. *Agronomy* **2023**, *13*, 606. [CrossRef]
8. Wang, X.; Xiang, Y.; Guo, J.; Tang, Z.; Zhao, S.; Wang, H.; Li, Z.; Zhang, F. Coupling effect analysis of drip irrigation and mixed slow-release nitrogen fertilizer on yield and physiological characteristics of winter wheat in Guanzhong area. *Field Crop Res.* **2023**, *302*, 109103. [CrossRef]
9. Gao, W.; Shou, N.; Jiang, C.; Ma, R.; Yang, X. Optimizing N Application for Forage Sorghum to Maximize Yield, Quality, and N Use Efficiency While Reducing Environmental Costs. *Agronomy* **2022**, *12*, 2969. [CrossRef]
10. Koppensteiner, L.J.; Kaul, H.P.; Piepho, H.P.; Barta, N.; Euteneuer, P.; Bernas, J.; Klimek-Kopyra, A.; Gronauer, A.; Neugschwandtner, R.W. Yield and yield components of facultative wheat are affected by sowing time, nitrogen fertilization and environment. *Eur. J. Agron.* **2022**, *140*, 126591. [CrossRef]
11. Sun, J.; Shi, S.; Gong, W.; Yang, J.; Du, L.; Song, S.; Chen, B.; Zhang, Z. Evaluation of hyperspectral LiDAR for monitoring rice leaf nitrogen by comparison with multispectral LiDAR and passive spectrometer. *Sci. Rep.* **2017**, *7*, 40362. [CrossRef] [PubMed]

12. Xiao, S.; Xu, D.; He, L.; Feng, W.; Wang, Y.; Wang, Z.; Coburn, C.A.; Guo, T. Using multi-angle hyperspectral data to monitor canopy leaf nitrogen content of wheat. *Precis. Agric.* **2016**, *17*, 721–736. [CrossRef]
13. Cheng, M.; He, J.; Wang, H.; Fan, J.; Xiang, Y.; Liu, X.; Liao, Z.; Tang, Z.; Abdelghany, A.E.; Zhang, F. Establishing critical nitrogen dilution curves based on leaf area index and aboveground biomass for greenhouse cherry tomato: A Bayesian analysis. *Eur. J. Agron.* **2022**, *141*, 126615. [CrossRef]
14. Guo, J.; Fan, J.; Xiang, Y.; Zhang, F.; Zheng, J.; Yan, S.; Yan, F.; Hou, X.; Li, Y.; Yang, L. Effects of nitrogen type on rainfed maize nutrient uptake and grain yield. *Agron. J.* **2021**, *113*, 5454–5471. [CrossRef]
15. Li, D.; Chen, J.M.; Yan, Y.; Zheng, H.; Yao, X.; Zhu, Y.; Cao, W.; Cheng, T. Estimating leaf nitrogen content by coupling a nitrogen allocation model with canopy reflectance. *Remote Sens. Environ.* **2022**, *283*, 113314. [CrossRef]
16. Schlemmer, M.; Gitelson, A.; Schepers, J.; Ferguson, R.; Peng, Y.; Shanahan, J.; Rundquist, D. Remote estimation of nitrogen and chlorophyll contents in maize at leaf and canopy levels. *Int. J. Appl. Earth Obs.* **2013**, *25*, 47–54. [CrossRef]
17. Moreno-Martínez, Á.; Camps-Valls, G.; Kattge, J.; Robinson, N.; Reichstein, M.; van Bodegom, P.; Kramer, K.; Cornelissen, J.H.C.; Reich, P.; Bahn, M.; et al. A methodology to derive global maps of leaf traits using remote sensing and climate data. *Remote Sens. Environ.* **2018**, *218*, 69–88. [CrossRef]
18. Wang, Y.; Suarez, L.; Poblete, T.; Gonzalez-Dugo, V.; Ryu, D.; Zarco-Tejada, P.J. Evaluating the role of solar-induced fluorescence (SIF) and plant physiological traits for leaf nitrogen assessment in almond using airborne hyperspectral imagery. *Remote Sens. Environ.* **2022**, *279*, 113141. [CrossRef]
19. Wan, L.; Zhou, W.; He, Y.; Wanger, T.C.; Cen, H. Combining transfer learning and hyperspectral reflectance analysis to assess leaf nitrogen concentration across different plant species datasets. *Remote Sens. Environ.* **2022**, *269*, 112826. [CrossRef]
20. Raj, R.; Walker, J.P.; Pingale, R.; Banoth, B.N.; Jagarlapudi, A. Leaf nitrogen content estimation using top-of-canopy airborne hyperspectral data. *Int. J. Appl. Earth Obs.* **2021**, *104*, 102584. [CrossRef]
21. Leo, S.; De Antoni Migliorati, M.; Nguyen, T.H.; Grace, P.R. Combining remote sensing-derived management zones and an auto-calibrated crop simulation model to determine optimal nitrogen fertilizer rates. *Agric. Syst.* **2023**, *205*, 103559. [CrossRef]
22. Wang, D.; Li, R.; Liu, T.; Liu, S.; Sun, C.; Guo, W. Combining vegetation, color, and texture indices with hyperspectral parameters using machine-learning methods to estimate nitrogen concentration in rice stems and leaves. *Field Crop Res.* **2023**, *304*, 109175. [CrossRef]
23. Fan, L.; Zhao, J.; Xu, X.; Liang, D.; Yang, G.; Feng, H.; Yang, H.; Wang, Y.; Chen, G.; Wei, P. Hyperspectral-Based Estimation of Leaf Nitrogen Content in Corn Using Optimal Selection of Multiple Spectral Variables. *Sensors* **2019**, *19*, 2898. [CrossRef] [PubMed]
24. Zhao, Y.; Wang, J.; Chen, L.; Fu, Y.; Zhu, H.; Feng, H.; Xu, X.; Li, Z. An entirely new approach based on remote sensing data to calculate the nitrogen nutrition index of winter wheat. *J. Integr. Agric.* **2021**, *20*, 2535–2551. [CrossRef]
25. Shu, M.; Zhu, J.; Yang, X.; Gu, X.; Li, B.; Ma, Y. A spectral decomposition method for estimating the leaf nitrogen status of maize by UAV-based hyperspectral imaging. *Comput. Electron. Agric.* **2023**, *212*, 108100. [CrossRef]
26. Tang, Z.; Guo, J.; Xiang, Y.; Lu, X.; Wang, Q.; Wang, H.; Cheng, M.; Wang, H.; Wang, X.; An, J.; et al. Estimation of Leaf Area Index and Above-Ground Biomass of Winter Wheat Based on Optimal Spectral Index. *Agronomy* **2022**, *12*, 1729. [CrossRef]
27. Shi, H.; Guo, J.; An, J.; Tang, Z.; Wang, X.; Li, W.; Zhao, X.; Jin, L.; Xiang, Y.; Li, Z.; et al. Estimation of Chlorophyll Content in Soybean Crop at Different Growth Stages Based on Optimal Spectral Index. *Agronomy* **2023**, *13*, 663. [CrossRef]
28. Zhang, W.; Li, Z.; Pu, Y.; Zhang, Y.; Tang, Z.; Fu, J.; Xu, W.; Xiang, Y.; Zhang, F. Estimation of the Leaf Area Index of Winter Rapeseed Based on Hyperspectral and Machine Learning. *Sustainability* **2023**, *15*, 12930. [CrossRef]
29. Ji, J.; Li, N.; Cui, H.; Li, Y.; Zhao, X.; Zhang, H.; Ma, H. Study on Monitoring SPAD Values for Multispatial Spatial Vertical Scales of Summer Maize Based on UAV Multispectral Remote Sensing. *Agriculture* **2023**, *13*, 1004. [CrossRef]
30. *GB20287-2006*; Microbial Inoculants in Agriculture. Ministry of Agriculture: Beijing, China, 2006.
31. Zhang, C.; Cai, H.; Li, Z. Estimation of Fraction of Absorbed Photosynthetically Active Radiation for Winter Wheat Based on Hyperspectral Characteristic Parameters. *Spectrosc. Spectr. Anal.* **2015**, *35*, 2644–2649.
32. Zheng, J.; Li, F.; Du, X. Using Red Edge Position Shift to Monitor Grassland Grazing Intensity in Inner Mongolia. *J. Indian Soc. Remote* **2018**, *46*, 81–88. [CrossRef]
33. Blackburn, G.A. Hyperspectral Remote Sensing of Plant Pigments. *J. Exp. Bot.* **2007**, *58*, 855–867. [CrossRef] [PubMed]
34. Horler, D.N.H.; Barber, J.; Darch, J.P.; Ferns, D.C.; Barringer, A.R. Approaches to Detection of Geochemical Stress in Vegetation. *Adv. Space Res.* **1983**, *3*, 175–179. [CrossRef]
35. Song, G.; Wang, Q. Coupling Effective Variable Selection with Machine Learning Techniques for Better Estimating Leaf Photosynthetic Capacity in a Tree Species (Fagus crenata Blume) from Hyperspectral Reflectance. *Agric. For. Meteorol.* **2023**, *338*, 109528. [CrossRef]
36. Rondeaux, G.; Steven, M.; Baret, F. Optimization of Soil-Adjusted Vegetation Indices. *Remote Sens. Environ.* **1996**, *55*, 95–107. [CrossRef]
37. Barnes, E.M.; Clarke, T.; Richards, S.E.; Colaizzi, P.D.; Haberland, J.; Kostrzewski, M.; Waller, P.M.; Choi, C.Y.; Riley, E.; Thompson, T.L.; et al. Coincident Detection of Crop Water Stress, Nitrogen Status and Canopy Density Using Ground-Based Multispectral Data. 2000. Available online: https://api.semanticscholar.org/CorpusID:128773162 (accessed on 19 July 2000).
38. Ashburn, P. The vegetative index number and crop identification NASA. In Proceedings of the Technical description of the Large Area Crop Inventory Experiment, Houston, TX, USA, 23–26 October 1978; Volumes 1 and 2.

39. Merzlyak, M.N.; Gitelson, A.A.; Chivkunova, O.B.; Rakitin, V.Y.U. Non-destructive optical detection of pigment changes during leaf senescence and fruit ripening. *Physiol. Plantarum.* **1999**, *106*, 135–141. [CrossRef]
40. Buschmann, C.; Nagel, E. In vivo spectroscopy and internal optics of leaves as basis for remote sensing of vegetation. *Int. J. Remote Sens.* **1993**, *14*, 711–722. [CrossRef]
41. Jordan, C.F. Derivation of Leaf-Area Index from Quality of Light on the Forest Floor. *Ecology* **1969**, *50*, 663–666. [CrossRef]
42. Shibayama, M.; Salli, A.; HÃ¤me, T.; Iso-Iivari, L.; Heino, S.; Alanen, M.; Morinaga, S.; Inoue, Y.; Akiyama, T. Detecting Phenophases of Subarctic Shrub Canopies by Using Automated Reflectance Measurements. *Remote Sens. Environ.* **1999**, *67*, 160–180. [CrossRef]
43. Bannari, A.; Morin, D.; Bonn, F.; Huete, A.R. A Review of Vegetation Indices. *Remote Sens. Rev.* **1995**, *13*, 95–120. [CrossRef]
44. Main, R.; Cho, M.A.; Mathieu, R.; O'Kennedy, M.M.; Ramoelo, A.; Koch, S. An investigation into robust spectral indices for leaf chlorophyll estimation. *Isprs J. Photogramm.* **2011**, *66*, 751–761. [CrossRef]
45. Hong, Y.; Guo, L.; Chen, S.; Linderman, M.; Mouazen, A.M.; Yu, L.; Chen, Y.; Liu, Y.; Liu, Y.; Cheng, H.; et al. Exploring the potential of airborne hyperspectral image for estimating topsoil organic carbon: Effects of fractional-order derivative and optimal band combination algorithm. *Geoderma* **2020**, *365*, 114228. [CrossRef]
46. Ren, K.; Dong, Y.; Huang, W.; Guo, A.; Jing, X. Monitoring of winter wheat stripe rust by collaborating canopy SIF with wavelet energy coefficients. *Comput. Electron. Agric.* **2023**, *215*, 108366. [CrossRef]
47. Deng, J.; Zhang, X.; Yang, Z.; Zhou, C.; Wang, R.; Zhang, K.; Lv, X.; Yang, L.; Wang, Z.; Li, P.; et al. Pixel-level regression for UAV hyperspectral images: Deep learning-based quantitative inverse of wheat stripe rust disease index. *Comput. Electron. Agric.* **2023**, *215*, 108434. [CrossRef]
48. Department of Plant Nutrition, China Agricultural University; Li, W.; Li, C. Comparison of N uptake and internal use efficiency in two tobacco varieties. *Crop J.* **2015**, *3*, 80–86. [CrossRef]
49. Monadjemi, S.; El Roz, M.; Richard, C.; Ter Halle, A. Photoreduction of chlorothalonil fungicide on plant leaf models. *Environ. Sci. Technol.* **2011**, *45*, 9582–9589. [CrossRef] [PubMed]
50. Ge, Z.; Ma, Y.; Xing, W.; Wu, Y.; Peng, S.; Mao, L.; Miao, Z. Inorganic Nitrogen-Containing Aerosol Deposition Caused "Excessive Photosynthesis" of Herbs, Resulting in Increased Nitrogen Demand. *Plants* **2022**, *11*, 2225. [CrossRef]
51. Raza, M.A.; Feng, L.Y.; Khalid, M.H.B.; Iqbal, N.; Meraj, T.A.; Hassan, M.J.; Ahmed, S.; Chen, Y.K.; Feng, Y.; Yang, W. Optimum leaf excision increases the biomass accumulation and seed yield of maize plants under different planting patterns. *Ann. Appl. Biol.* **2019**, *175*, 54–68. [CrossRef]
52. Su, Y.; Wang, Y.; Liu, G.; Zhang, Z.; Li, X.; Chen, G.; Guo, Z.; Gao, Q. Nitrogen (N) supplementation, slow release, and retention strategy improves N use efficiency via the synergistic effect of biochar, nitrogen-fixing bacteria, and dicyandiamide. *Sci. Total Environ.* **2023**, *908*, 168518. [CrossRef]
53. Yang, B.; Zhang, H.; Lu, X.; Wan, H.; Zhang, Y.; Zhang, J.; Jin, Z. Inversion of Leaf Water Content of Cinnamomum camphora Based on Preferred Spectral Index and Machine Learning Algorithm. *Forests* **2023**, *14*, 2285. [CrossRef]
54. Xiang, Y.; Wang, X.; An, J.; Tang, Z.; Li, W.; Shi, H. Estimation of Leaf Area Index of Soybean Based on Fractional OrderDifferentiation and Optimal Spectral Index. *Trans. Chin. Soc. Agric. Mach.* **2023**, *54*, 329–342.
55. Tang, Z.; Xiang, Y.; Wang, X.; An, J.; Guo, J.; Wang, H.; Li, Z.; Zhang, F. Comparison of SPAD Value and LAI Spectral Estimation of Soybean Leaves Based on Different Analysis Models. *Soybean Sci.* **2023**, *42*, 55–63.
56. Wang, Q.; Lu, X.; Zhang, H.; Yang, B.; Gong, R.; Zhang, J.; Jin, Z.; Xie, R.; Xia, J.; Zhao, J. Comparison of Machine Learning Methods for Estimating Leaf Area Index and Aboveground Biomass of Cinnamomum camphora Based on UAV Multispectral Remote Sensing Data. *Forests* **2023**, *14*, 1688. [CrossRef]

plants

MDPI

Article

Soybean (*Glycine max* L.) Leaf Moisture Estimation Based on Multisource Unmanned Aerial Vehicle Image Feature Fusion

Wanli Yang [1,2], Zhijun Li [1,2,*], Guofu Chen [1,2], Shihao Cui [1,2], Yue Wu [1,2], Xiaochi Liu [1,2], Wen Meng [1,2], Yucheng Liu [1,2], Jinyao He [1,2], Danmao Liu [1,2], Yifan Zhou [1,2], Zijun Tang [1,2], Youzhen Xiang [1,2] and Fucang Zhang [1,2]

1 Key Laboratory of Agricultural Soil and Water Engineering in Arid and Semiarid Areas of Ministry of Education, Northwest A&F University, Xianyang 712100, China; 2022012387@nwsuaf.edu.cn (W.Y.); 2022012367@nwsuaf.edu.cn (G.C.); 2022012353@nwsuaf.edu.cn (S.C.); 2022012363@nwsuaf.edu.cn (Y.W.); 2023055903@nwafu.edu.cn (X.L.); mengwen@nwsuaf.edu.cn (W.M.); 2022012268@nwsuaf.edu.cn (Y.L.); 2022012263@nwsuaf.edu.cn (J.H.); 2022012351@nwsuaf.edu.cn (D.L.); 2022012368@nwsuaf.edu.cn (Y.Z.); tangzijun@nwsuaf.edu.cn (Z.T.); youzhenxiang@nwsuaf.edu.cn (Y.X.); zhangfc@nwsuaf.edu.cn (F.Z.)

2 Institute of Water-Saving Agriculture in Arid Areas of China, Northwest A&F University, Xianyang 712100, China

* Correspondence: lizhij@nwsuaf.edu.cn

Abstract: Efficient acquisition of crop leaf moisture information holds significant importance for agricultural production. This information provides farmers with accurate data foundations, enabling them to implement timely and effective irrigation management strategies, thereby maximizing crop growth efficiency and yield. In this study, unmanned aerial vehicle (UAV) multispectral technology was employed. Through two consecutive years of field experiments (2021–2022), soybean (*Glycine max* L.) leaf moisture data and corresponding UAV multispectral images were collected. Vegetation indices, canopy texture features, and randomly extracted texture indices in combination, which exhibited strong correlations with previous studies and crop parameters, were established. By analyzing the correlation between these parameters and soybean leaf moisture, parameters with significantly correlated coefficients ($p < 0.05$) were selected as input variables for the model (combination 1: vegetation indices; combination 2: texture features; combination 3: randomly extracted texture indices in combination; combination 4: combination of vegetation indices, texture features, and randomly extracted texture indices). Subsequently, extreme learning machine (ELM), extreme gradient boosting (XGBoost), and back propagation neural network (BPNN) were utilized to model the leaf moisture content. The results indicated that most vegetation indices exhibited higher correlation coefficients with soybean leaf moisture compared with texture features, while randomly extracted texture indices could enhance the correlation with soybean leaf moisture to some extent. RDTI, the random combination texture index, showed the highest correlation coefficient with leaf moisture at 0.683, with the texture combination being Variance1 and Correlation5. When combination 4 (combination of vegetation indices, texture features, and randomly extracted texture indices) was utilized as the input and the XGBoost model was employed for soybean leaf moisture monitoring, the highest level was achieved in this study. The coefficient of determination (R^2) of the estimation model validation set reached 0.816, with a root-mean-square error (RMSE) of 1.404 and a mean relative error (MRE) of 1.934%. This study provides a foundation for UAV multispectral monitoring of soybean leaf moisture, offering valuable insights for rapid assessment of crop growth.

Keywords: leaf moisture content; multispectral; soil moisture content; soybean; texture features; vegetation indices

Citation: Yang, W.; Li, Z.; Chen, G.; Cui, S.; Wu, Y.; Liu, X.; Meng, W.; Liu, Y.; He, J.; Liu, D.; et al. Soybean (*Glycine max* L.) Leaf Moisture Estimation Based on Multisource Unmanned Aerial Vehicle Image Feature Fusion. *Plants* **2024**, *13*, 1498. https://doi.org/10.3390/plants13111498

Academic Editor: Stefano Martellos

Received: 6 May 2024
Revised: 23 May 2024
Accepted: 27 May 2024
Published: 29 May 2024

1. Introduction

Soybean (*Glycine max* L.), as one of the major leguminous crops globally, plays a crucial role in global food security and sustainable agriculture [1]. In arid and semi-arid regions, soybean cultivation faces multiple challenges, often associated with limited water resources and irregular precipitation patterns [2]. Being a water-consuming crop, soybean requires adequate water for normal growth [3]; however, water scarcity in dry areas frequently leads to water stress, constraining soybean growth and resulting in yield reduction [4]. Therefore, timely monitoring of soybean leaf moisture is essential for identifying plant moisture status, adjusting irrigation strategies, and enhancing yield.

Plant leaf moisture is influenced by multiple factors such as sunlight, soil moisture, and air temperature, making it difficult to measure accurately and rapidly [5]. Currently, commonly used methods for measuring plant leaf moisture include oven drying, Karl Fischer titration, and capacitance methods; however, these methods are often time-consuming, labor-intensive, and limited in applicability, failing to provide timely and accurate field monitoring data [6]. Hence, rapid acquisition of plant leaf moisture status and timely adjustment of soil moisture management strategies remain a challenge for large-scale agricultural operations.

Remote sensing technology has been widely applied to qualitatively and quantitatively analyze the water and nutrient status of large-scale plants nondestructively [7]. Among them, multispectral remote sensing technology can simultaneously obtain data from multiple bands, covering various information during the crop growth cycle [8]. In contrast, traditional field measurement methods require measurements at different times and locations, which are inefficient and make it difficult to achieve comprehensive monitoring [9]. Compared with hyperspectral remote sensing, multispectral remote sensing has fewer bands but still covers key bands related to crop growth and health conditions, with more flexible band selection according to specific application requirements, making UAV multispectral technology more widely applicable in field management [10].

Spectra can focus on the internal optical response of crops [11], while images capture external morphological information of crops [12]. Multispectral data provide reflectance of crops in different bands, which correlates with crop leaf moisture to some extent [13], thus enabling indirect inference of crop moisture status through multispectral data analysis. Vegetation indices, computed based on multispectral data, directly reflect the growth status of crops [14]. On the other hand, vegetation canopy texture features, obtained through image data analysis, reflect the spatial distribution and structural characteristics of crops, including leaf morphology, density, and arrangement, which also indicate the moisture status of crops [15]. Researchers have conducted relevant studies on monitoring physiological growth indicators of crops based on vegetation indices; however, constrained by crop types and meteorological factors, using fixed formulas to calculate vegetation indices for monitoring physiological growth indicators of crops can limit prediction accuracy. Some studies have shown that combining texture features with vegetation indices can improve the inversion accuracy of physiological growth indicators of crops (biomass [16], leaf area index [17], chlorophyll content [18], etc.). Constructing inversion models based on multiple input variables has higher accuracy compared with single input variables.

Machine learning methods have been proven effective in solving complex nonlinear problems with multiple factors. Some results indicate that back propagation neural networks (BPNN) have higher accuracy in monitoring physiological growth indicators [19,20], while other studies suggest that extreme gradient boosting (XGBoost) may be more suitable [21,22]. Overall, there is uncertainty in the existing research regarding the optimal feature extraction and modeling methods for monitoring physiological growth indicators. Therefore, this study further explores monitoring soybean leaf moisture.

In this study, our aim was to determine the relationship between soybean leaf moisture and vegetation indices and texture features. To achieve this, we employed machine learning algorithms such as ELM, XGBoost, and BPNN to explore the optimal combination of these features and the best monitoring depth for monitoring soybean physiological growth indicators, aiming to provide rapid and efficient theoretical support for field water management.

2. Materials and Methods

2.1. Research Area and Test Design

The experiment was conducted during 2021–2022 at the Institute of Water-saving Agriculture in Arid Areas, Northwest A&F University, located in the southern part of the Loess Plateau in northwest China (34°14′ N, 108°10′ E). The experimental area is a typical dryland agricultural region with an average annual precipitation of 632 mm and an evaporation of 1500 mm. Daily temperature and rainfall data for the two growing seasons were carefully recorded by an automatic weather station located within the experimental fields, as shown in Figure 1. During the period from June to October 2021, the annual average maximum and minimum temperatures were 30.3 °C and 20.0 °C, respectively, while in 2022, they were 31.3 °C and 21.2 °C, respectively. The precipitation during the soybean growing seasons in 2021 and 2022 was 432.6 mm (from 18 June 2021 to 30 September 2021) and 279.5 mm (from 10 June 2022 to 20 September 2022), respectively. For basic terrain and meteorological information of the experimental site, please refer to reference [1].

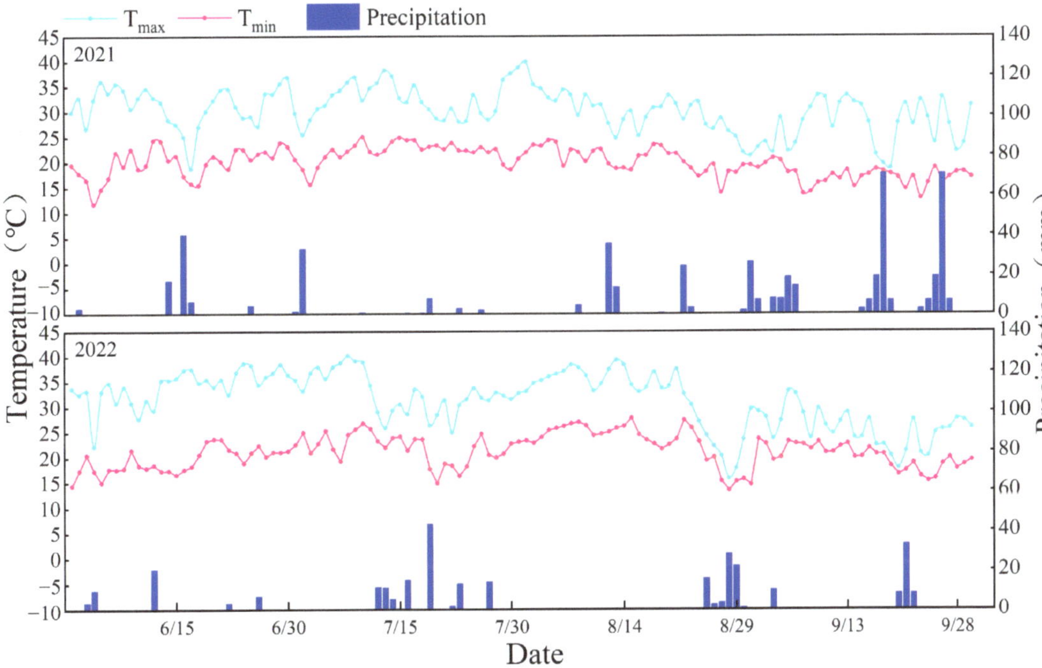

Figure 1. The daily temperature and precipitation of soybean growing season at the Yangling experimental station in China in 2021 and 2022.

In this experiment, a split-plot design with two factors was employed, including different cover treatments and supplementary irrigation strategies. The cover treatments consisted of three types: straw mulch (SM), ridge-film mulch (FM), and no mulch (NM). Additionally, three supplementary irrigation treatments were included: W1 (irrigation during branching stage, V4), W2 (irrigation during podding stage, R2), and W3 (irrigation during both V4 and R2 stages simultaneously). This resulted in a total of nine treatments, each replicated three times, comprising 27 experimental plots. Each irrigation event applied 40 mm of water. The detailed information of the experiment is presented in Table 1. Each plot had an area of 24 m^2 (4 m × 6 m) and was arranged randomly, with a 2 m buffer zone around each plot.

Before sowing, each plot received phosphorus and potassium fertilizers at a rate of 30 kg ha^{-1} and nitrogen fertilizer at a rate of 120 kg ha^{-1}. The nitrogen fertilizer used in the experiment was urea (46% N), the phosphorus fertilizer was calcium superphosphate (16% P_2O_5), and the potassium fertilizer was potassium chloride (62% K_2O).

For the FM treatment, a ridge (50 cm wide, 30 cm high) was used, and seeds were sown in furrows covered by the ridge. The ridge–furrow ratio was 1:1. Before sowing, two rows of soybeans were planted at the bottom of each ridge. The straw mulch rate was 9000 kg ha^{-1}, and wheat straw was used to cover the soil within 7 days after sowing. The planting density of soybeans was 300,000 plants ha^{-1}, with row spacing of 50 cm and plant spacing of 10 cm. Soybeans were sown on 18 June 2021 and 10 June 2022 and harvested on 30 September 2021 and 20 September 2022, respectively.

Additionally, to ensure proper germination, approximately 20 mm of water was applied to each plot after sowing. Other field management practices, including spraying and weeding, remained consistent with local practices.

2.2. *Data Collection and Preprocessing*

2.2.1. Drone Data Acquisition

This study utilized a DJI Matrice M300 RTK quadcopter equipped with an MS600 Pro multispectral camera platform to acquire multispectral remote sensing data. The camera platform comprised six spectral channels and was equipped with six CMOS image sensors, with a pixel resolution of 1.2×10^6. The sensors covered the following spectral bands: blue band (center wavelength 450 nm, band 1), green band (center wavelength 555 nm, band 2), red band (center wavelength, band 3), red edge band 1 (center wavelength 720 nm, band 4), red edge band 2 (center wavelength 750 nm, band 5), and near-infrared band (center wavelength 840 nm, band 6). Data were collected during the soybean flowering period (5 August 2021 and 10 August 2022) at noon under clear-sky conditions. Flight routes were planned for the study area, and whiteboard calibration was conducted. The flight altitude was set at 30 m, with a speed of 2.5 m per second and a pixel resolution of 4.09 cm. The forward and lateral overlap ratios were set at 75% and 65%, respectively. Figure 2 shows the aerial photos of some residential areas in the test area.

Figure 2. UAV photo of soybean plots in this experimental area.

2.2.2. Obtaining Leaf Moisture Content

Simultaneously with the collection of multispectral information by drones, the soybean leaf moisture content was determined using the drying method. Five average growing soybeans plants were selected from each plot. Fresh leaves were harvested from various directions and heights of each plant, totaling 100 g, using an analytic balance. These leaves were then placed in parchment bags, labeled, and subjected to dehydration in a drying oven at 105 °C for 0.5 h, followed by drying at 80 °C until a constant mass was achieved.

The dry mass, after subtracting the mass of the parchment bags, represented the moisture content. The average moisture content of the five soybean plants was considered indicative of the entire plot's soybean leaf moisture content.

2.2.3. Multispectral Image Processing

In scientific research, precise handling and analysis of remote sensing data are crucial. In this study, we employed the Yusense Map V2.2.2 software to process multispectral imagery collected by unmanned aerial vehicles (UAVs). Initially, the software was used for image mosaicking to ensure continuity and integrity of all images within the study area. To enhance the accuracy of subsequent analyses, geometric correction was applied to the mosaicked images, eliminating distortions caused by variations in UAV flight altitude and angle. This was followed by radiometric preprocessing to mitigate the impact of sensor sensitivity, solar radiation intensity, and atmospheric conditions on the imagery, ensuring that the images accurately reflected the ground truth. The preprocessed UAV multispectral image information was then imported into ENVI 5.3 software. ENVI is a widely used remote sensing image processing software that supports a variety of data analysis and image processing functions. Within this software, we extracted spectral reflectance, a key metric for measuring the reflection of solar energy by surface objects. To focus on the study area, we clipped corresponding spectral images centered around each experimental plot from the imagery. During the clipping process, special attention was given to exclude areas with soil and film shadows as these could affect the purity of spectral data. Subsequently, regions of interest (ROIs) were defined within each experimental plot, and the average reflectance spectra of the soybean leaf samples were extracted from these areas. This average reflectance spectrum represented the spectral reflectance within the plot, providing us with valuable information about the growth conditions of the soybeans. Ultimately, we obtained spectral reflectance data across different bands, which will be used for further analysis, such as assessing crop health, monitoring vegetation cover changes, or estimating biophysical parameters. Through these detailed spectral data, researchers can gain a deeper understanding of the environmental conditions affecting crop growth, offering a scientific basis for precision agriculture. Figure 3 shows the reflectivity performance of each experimental treatment in each band.

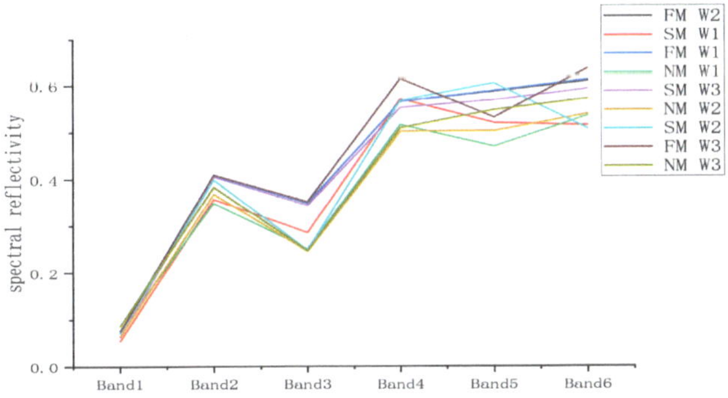

Figure 3. Spectral reflectance of soybean under different field treatments in each band.

2.3. Selection and Construction of Vegetation Index and Texture Features

Crop growth and nutritional status can be effectively reflected by vegetation indices [23]. In this study, based on existing research, ten classic vegetation indices were selected for investigation, with calculation formulas and references provided in Table 1. The texture is a visual feature reflecting homogeneity phenomena in images, indicating the arrangement properties of surface structures with slow or periodic changes. In this paper, ENVI 5.3 software was employed to extract texture features (TFs) based on second-order statistical filtering (co-occurrence measures). Eight TFs were extracted from the near-infrared band: mean (MEA), variance (VAR), homogeneity (HOM), contrast (CON), dissimilarity (DIS), entropy (ENT), second moment (SEM), and correlation (COR). A window size of 7×7 and default spatial offset values of 1 were used for texture analysis. To explore the potential applications of texture features in estimating soybean leaf moisture content from UAV multispectral images, randomly combined texture features were extracted in this study. Subsequently, based on previous research experience and formulas, seven types of texture indices (TIs) [24] were constructed, including normalized difference texture index (NDTI), ratio texture index (RTI), difference texture index (DTI), additive texture index (ATI), reciprocal difference texture index (RDTI), and reciprocal additive texture index (RATI). The specific calculation formulas are as follows:

$$\mathrm{RTI} = \mathrm{T_i}/\mathrm{T_j} \tag{1}$$

$$\mathrm{DTI} = \mathrm{T_i} - \mathrm{T_j} \tag{2}$$

$$\mathrm{ATI} = \mathrm{T_i} + \mathrm{T_j} \tag{3}$$

$$\mathrm{NDTI} = \left(\mathrm{T_i} - \mathrm{T_j}\right)/\left(\mathrm{T_i} + \mathrm{T_j}\right) \tag{4}$$

$$\mathrm{RDTI} = 1/\mathrm{T_i} - 1/\mathrm{T_j} \tag{5}$$

$$\mathrm{RATI} = 1/\mathrm{T_i} + 1/\mathrm{T_j} \tag{6}$$

Table 1. Vegetation index and its calculation formula.

Selected Spectra Parameters	Calculation Formula	Reference
Soil-adjusted vegetation index (SAVI)	$(1+0.5)(R_{NIR} - R_{RED})/(R_{NIR} + R_{RED} + 0.5)$	[25]
Enhanced vegetation index (EVI)	$2.5 \times (R_{NIR} - R_{RED})/((R_{NIR} + 6 \times R_{RED} - 7.5 \times R_B) + 1)$	[26]
Modified simple ratio vegetation index (MSR)	$\left(\frac{R_{NIR}}{R_{RED}} - 1\right)\left(\frac{R_{NIR}}{R_{RED}} + 1\right)^{-0.5}$	[27]
Optimized soil-adjusted vegetation index (OSAVI)	$(1+0.16)(R_{NIR} - R_G)/(R_{NIR} + R_G + 0.16)$	[25]
Renormalized difference vegetation index (RDVI)	$(R_{NIR} - R_{RED}/R_{NIR} + R_{RED})^{(0.5)}$	[25]
Modified soil-adjusted vegetation index (MSAVI)	$0.5((2R_{NIR} - 1) + ((2R_{NIR} + 1)^2 - 8(R_{NIR} - R_{RED})^2)^{0.5})$	[27]
Atmospheric resistance vegetation index (ARVI)	$(R_{NIR} - 2R_{RED} + R_B)/(R_{NIR} + 2R_{RED} - R_B)$	[28]
Green normalized difference vegetation index (GNDVI)	$R_{NIR} - R_G/R_{NIR} + R_G$	[26]
Meris terrestrial chlorophyll index (MTCI)	$R_{NIR} - R_{RE}/R_{RE} - R_{RED}$	[29]
Chlorophyll index (CI)	$R_{NIR}/R_{RE} - 1$	[30]

Note: R_RED, R_G, R_B, R_NIR, and R_RE represent the reflectance of the red, green, blue, near-infrared, and red-edge bands, respectively.

2.4. Sample Set Partitioning, Model Methods, and Model Evaluation

During the soybean flowering stage, a total of 54 valid samples were collected. Two-thirds of the samples were randomly selected as the training set, while the remaining one-third was reserved as the validation set. Figure 4 presents the sample counts and statistical characteristics of both the training and validation sets.

First, the correlation between vegetation indices, texture features, and soybean leaf moisture content was analyzed. Parameters significantly correlated with soybean leaf moisture content ($p < 0.05$) were selected as input variables for the model. These include combination 1, vegetation indices; combination 2, texture features; combination 3, texture indices extracted from random combinations; and combination 4, vegetation indices, texture features, and randomly extracted texture indices combined. Subsequently, ELM, XGBoost, and BPNN were employed to model the leaf moisture content. Detailed descriptions of these machine learning models can be found in references [7,31].

For the ELM model, a sigmoid function was utilized, and parameters $(a_i, b_i)_{i=1}^{L}$ for the hidden layer were randomly generated within the range [−1, 1]. The number of hidden layer nodes was set to 1000 [7], and the number of neurons started at 15, incrementing by 15 until reaching 120. Each model was run 50 times to select the optimal training result, and the final number of neurons was determined to be 60.

For the XGBoost algorithm, the optimal parameters were refined through a grid search, setting 100 weak learners (n_estimators), a learning rate of 0.03, and a maximum tree depth (max_depth) of 5 [31].

In BPNN, the transfer function for the hidden layer was set as TANSIG, and the Levenberg–Marquardt algorithm based on numerical optimization theory (Train-LM) was used as the network training function. After multiple training iterations, the number of neurons in the middle layer was determined to be 15 [31]. Figure 5 shows the process of UAV multispectral data processing, the acquisition of vegetation index and texture features, and the construction process of soybean leaf moisture content model.

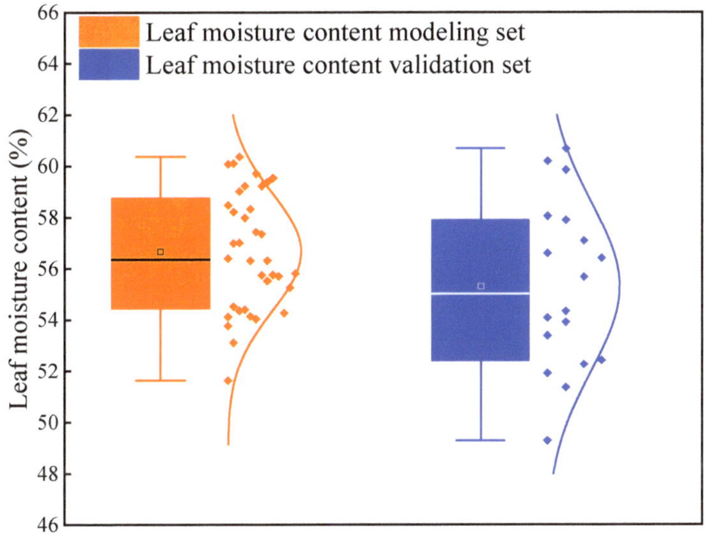

Figure 4. Descriptive statistics of soybean leaf moisture content. The horizontal line in the box line diagram represents the median, and the white box represents the average value.

To validate the model's prediction accuracy and capability, this study selected three evaluation metrics: the coefficient of determination (R^2), root mean square error (RMSE), and mean relative error (MRE). These metrics were used to assess the model's precision [32]. A higher R^2 value closer to 1 and lower RMSE and MRE values closer to 0 indicated better model fitting. The formulas for these metrics were as follows:

$$R^2 = \frac{\sum_{i=1}^{n}(\hat{y}_i - \bar{y})^2}{\sum_{i=1}^{n}(y_i - \bar{y})^2} \tag{7}$$

$$RMSE = \sqrt{\frac{\sum_{i=1}^{n}(\hat{y}_i - y_i)^2}{n}} \tag{8}$$

$$MRE = \frac{1}{n}\sum_{i=1}^{n}\frac{|\hat{y}_i - y_i|}{y_i} \times 100\% \tag{9}$$

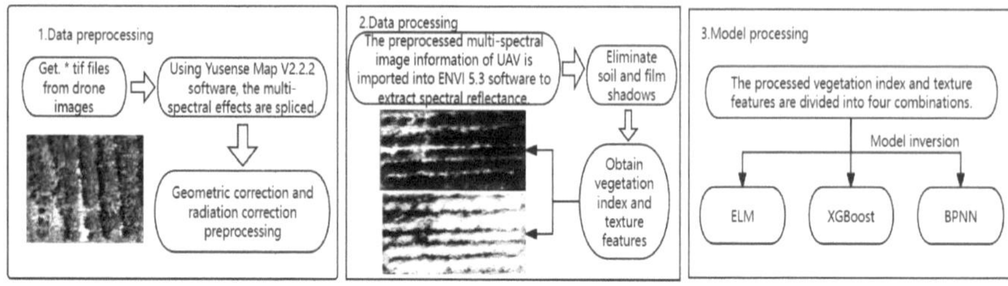

Figure 5. The process of UAV multispectral data processing, the acquisition of vegetation index and texture features, and the construction process of the soybean leaf moisture content model. * on behalf of each small piece of UAV image stitching.

3. Results and Analysis

3.1. Correlation Analysis between Vegetation Index, Texture Features, and Leaf Moisture Content

The correlation analysis between the vegetation indices and the soybean leaf moisture content is presented in Table 2, the correlation analysis between the texture feature and the soybean leaf moisture content is presented in Table 3. The results indicated that the majority of vegetation indices and texture features exhibited significant correlations with the soybean leaf moisture content ($p < 0.05$). Among these, the vegetation index with the highest correlation coefficient was MSR, with a value of 0.649. Additionally, the texture feature with the highest correlation coefficient to leaf moisture content was the mean in band 2, with a coefficient of 0.644. Subsequently, an analysis of randomly extracted texture indices was conducted (Table 4 and Figure 6). It was found that the randomly extracted texture indices, after screening, also demonstrated significant correlations with soybean leaf moisture content ($p < 0.05$). Among these, RDTI stood out as the combination with the highest correlation coefficient to leaf moisture content, with a coefficient of 0.683. The texture combination comprised Variance1 and Correlation5.

Table 2. The calculation results of vegetation index and correlation coefficient with soybean leaf moisture content (* significant at $p < 0.05$).

Vegetation Index	Correlation Coefficient
Soil-adjusted vegetation index (SAVI)	0.511 *
Enhanced vegetation index (EVI)	0.517 *
Modified simple ratio vegetation index (MSR)	0.649 *
Optimized soil-adjusted vegetation index (OSAVI)	0.636 *
Renormalized difference vegetation index (RDVI)	0.506 *
Modified soil-adjusted vegetation index (MSAVI)	0.619 *
Atmospheric resistance vegetation index (ARVI)	0.643 *
Green normalized difference vegetation index (GNDVI)	0.189
Meris terrestrial chlorophyll index (MTCI)	0.249
Chlorophyll index (CI)	0.171

From this, we selected four combinations to serve as input for the model: combination 1 (SAVI, EVI, MSR, OSAVI, RDVI, MSAVI, and ARVI), combination 2 (Mean1, Variance1, Homogeneity1, Contrast1, Dissimilarity1, Entropy1, Second Moment1, Correlation1, Mean2, Variance2, Homogeneity2, Contrast2, Second Moment2, Correlation2, Mean3, Variance3, Homogeneity3, Contrast3, Second Moment3, Correlation3, Mean4, Variance4, Contrast4, Second Moment4, Correlation4, Variance5, Homogeneity5, Contrast5, Dissimilarity5, Second Moment5, Correlation5, Variance6, Homogeneity6, Contrast6, Dissimilarity6, Second Moment6, and Correlation6), combination 3 (RTI, DTI, ATI, NDTI, RATI, and RDTI), combination 4 (SAVI, EVI, MSR, OSAVI, RDVI, MSAVI, ARVI, Mean1, Variance1, Homogeneity1, Contrast1, Dissimilarity1, Entropy1, Second Moment1, Correlation1, Mean2, Variance2, Homogeneity2, Contrast2, Second Moment2, Correlation2, Mean3, Variance3,

Homogeneity3, Contrast3, Second Moment3, Correlation3, Mean4, Variance4, Contrast4, Second Moment4, Correlation4, Variance5, Homogeneity5, Contrast5, Dissimilarity5, Second Moment5, Correlation5, Variance6, Homogeneity6, Contrast6, Dissimilarity6, Second Moment6, Correlation6, RTI, DTI, ATI, NDTI, RATI, and RDTI).

Table 3. Texture features and calculation results of correlation coefficient with soybean leaf moisture content (* significant at $p < 0.05$).

Texture Features	Correlation Coefficients					
	Band 1	Band 2	Band 3	Band 4	Band 5	Band 6
Mean	0.597 *	0.644 *	0.618 *	0.606 *	0.249	0.259
Variance	0.578 *	0.559 *	0.398 *	0.513 *	0.485 *	0.505 *
Homogeneity	0.554 *	0.328 *	0.268 *	0.164	0.429 *	0.424 *
Contrast	0.585 *	0.581 *	0.439 *	0.531 *	0.492 *	0.506 *
Dissimilarity	0.581 *	0.223	0.046	0.170	0.505 *	0.523 *
Entropy	0.389 *	0.210	0.229	0.138	0.251	0.248
Second moment	0.457 *	0.285 *	0.399 *	0.358 *	0.398 *	0.396 *
Correlation	0.471 *	0.593 *	0.599 *	0.588 *	0.667 *	0.654 *

Table 4. The calculation results of texture index extracted by random combination and correlation coefficient with soybean leaf moisture content (* significant at $p < 0.05$).

Texture Features Extracted by Random Combination	Maximum Correlation Coefficient	
	Correlation Coefficient	Texture Feature Combination
RTI	0.663 *	Homogeneity3, Mean2
DTI	0.645 *	Dissimilarity1, Mean2
ATI	0.653 *	Mean2, Second Moment3
NDTI	0.647 *	Correlation5, Mean4
RATI	0.670 *	Variance5, Correlation5
RDTI	0.683 *	Variance1, Correlation5

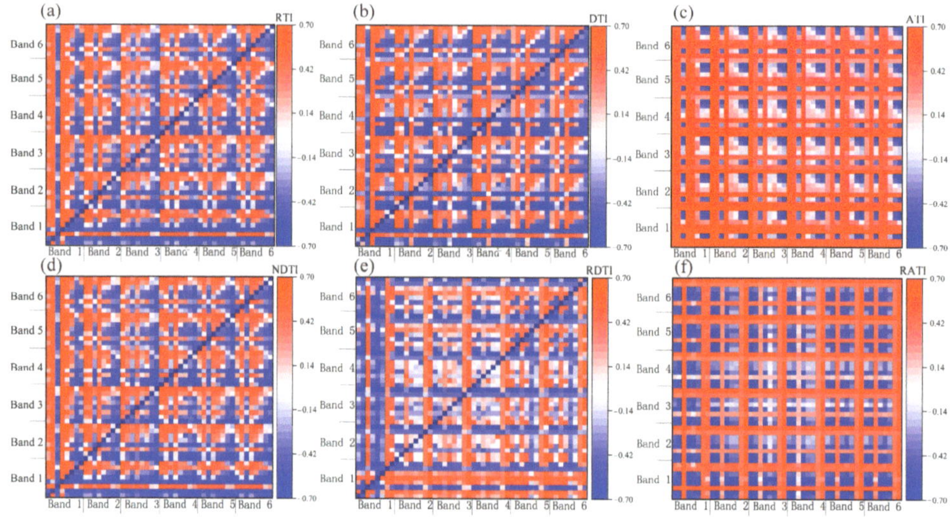

Figure 6. The correlation coefficients between the moisture content and the texture index of soybean leaves were (**a**) RTI, (**b**) DTI, (**c**) ATI, (**d**) NDTI, (**e**) RDTI, and (**f**) RATI. Any point in the figure represents the correlation coefficient between the texture index and the moisture content of soybean leaves. The texture index is calculated by the two texture eigenvalues corresponding to the horizontal and vertical coordinates of the point. Band 1 in the image consists of the following parameters from start to finish: Mean1, Variance1, Homogeneity1, Contrast1, Dissimilarity1, Entropy1, Second Moment1, and Correlation1; band 2 consists of Mean2, Variance2, Homogeneity2, Contrast2, Dissimilarity2, Entropy2, Second Moment2, and Correlation2; band 3 consists of Mean3, Variance3, Homogeneity3, Contrast3, Dissimilarity3, Entropy3, Second Moment3, and Correlation3; band 4 consists of Mean4, Variance4, Homogeneity4, Contrast4, Dissimilarity4, Entropy4, Second Moment4, and Correlation4; band 5 consists of Mean5, Variance5, Homogeneity5, Contrast5, Dissimilarity5, Entropy5, Second Moment5, and Correlation5; and band 6 consists of Mean6, Variance6, Homogeneity6, Contrast6, Dissimilarity6, Entropy6, Second Moment6, and Correlation6.

3.2. *Construction of a Monitoring Model for Soybean Leaf Moisture Content*

The four selected combinations from Section 3.1 were utilized as inputs for modeling using ELM, XGBoost, and BPNN. The model results are illustrated in Figure 7. The findings revealed that when the machine learning models were consistent, the combination of vegetation indices and texture features (combination 4) yielded the highest estimation accuracy for soybean leaf moisture content. This was evidenced by the highest R^2, and the lowest RMSE and MRE values in the validation set. Furthermore, among models with the same input combinations, XGBoost demonstrated the optimal capability for monitoring the soybean leaf moisture content.

In summary, in this study, the combination of vegetation indices and texture features (combination 4) as input, combined with the XGBoost model, achieved the highest level of soybean leaf moisture content monitoring. The estimated R^2 for the validation set was 0.816, with an RMSE of 1.404 and an MRE of 1.934%.

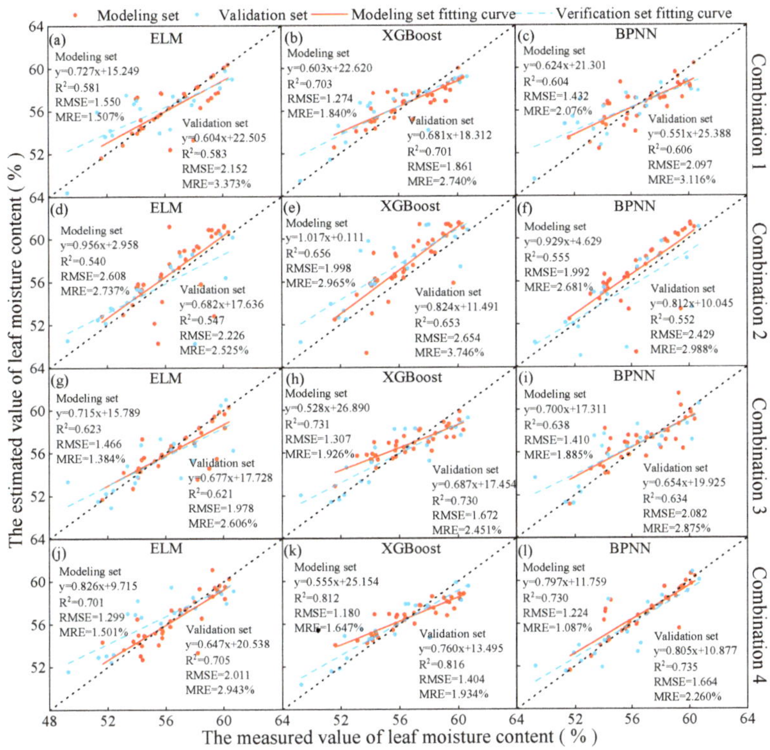

Figure 7. The prediction results of soybean leaf moisture content based on ELM, XGBoost, and BPNN. The red dots in the figure are the modeling set, and the blue dots are the verification set. The predicted results of the soybean leaf moisture content inversion models using different input variables and modeling methods are presented for both the modeling and validation datasets. (**a**–**c**) The prediction models constructed using combination 1 with ELM, XGBoost, and BPNN as the methods. (**d**–**f**) The prediction models constructed using combination 2 with ELM, XGBoost, and BPNN as the methods. (**g**–**i**) The prediction models constructed using combination 3 with ELM, XGBoost, and BPNN as the methods. (**j**–**l**) The prediction models constructed using combination 4 with ELM, XGBoost, and BPNN as the methods.

4. Discussion

Water is a crucial element for photosynthesis and nutrient transport in plants. The water content in plant leaves directly influences the plant's growth process [5]. Therefore, a profound understanding of variations in the leaf moisture content is essential for grasping the plant's growth status, assessing its water condition, and effectively managing plant water resources. This awareness has garnered widespread attention in fields such as agriculture, forestry, and horticulture [6]. By analyzing vegetation indices and texture features in unmanned aerial vehicle multispectral data, the crop water status can be inferred more accurately, providing a scientific basis for field management and irrigation decisions [12].

In this study, to ensure the accuracy and reliability of experimental outcomes, a stringent control variable approach was adopted. Apart from water management, all other agricultural practices were kept uniform, including but not limited to soil fertility, planting density, and pest control. By doing so, we were able to eliminate the interference of these potential factors on the experimental results, ensuring that the observed outcomes could be attributed to the varying treatments. This research aimed to elucidate the direct impact of water management on soybean growth and yield by precisely controlling all management measures except for water, thereby offering a scientific basis for sustainable agricultural practices.

This study found that the correlation coefficient between vegetation indices and leaf moisture content was generally higher than that of texture features. This is because texture features typically reflect the spatial distribution and structural characteristics of vegetation (such as density, height, coverage, etc.). However, when measuring canopy texture features, the shadowing effect may affect the sensor's observations, leading to less accurate extraction of canopy texture features [24]. In contrast, vegetation indices are more capable of comprehensively reflecting the growth status of vegetation; therefore, they may be more reliable in terms of correlation with leaf moisture content [33].

The study also found that when using the same machine learning model for modeling, the accuracy of the combined model using combination 4 for leaf moisture content monitoring was higher than combinations 1, 2, and 3. This may be because combination 4, which included vegetation indices, texture features, and texture indices, provided richer and more diverse information. Compared with using vegetation indices or texture features alone, the combination of these two in combination 4 could offer a more comprehensive feature description. Vegetation indices typically reflect the growth status and photosynthetic activity of vegetation [34], while texture features provide information about the structure and spatial distribution of vegetation canopies [35]. By combining these two features, the model could more accurately capture the complex relationship between leaf moisture content and vegetation growth status. Additionally, vegetation indices and texture features often have different sensitivities and feature expression capabilities. That is, vegetation indices may be more suitable for reflecting vegetation growth status, while texture features can better describe the spatial distribution and structural characteristics of vegetation canopies [36]. Therefore, combining these two features can complement each other's shortcomings and improve the model's accuracy in estimating the leaf moisture content.

In the process of constructing leaf moisture content models, when the input combinations were the same, it was found that the XGBoost model had higher accuracy compared with the SVM and BPNN models. This may be because XGBoost performed well in handling nonlinear relationships and complex data patterns, enabling better fitting of the complex relationship between leaf moisture content and input features [31]. In contrast, ELM and BPNN models may be less flexible in handling high-dimensional, nonlinear data, resulting in relatively lower accuracy [7,31]. Additionally, XGBoost improved model generalization by optimizing the loss function, which enabled it to perform well beyond the training dataset. This means that the XGBoost model could better adapt to new datasets and exhibit more stable performance on the test set, thereby enhancing model accuracy [37]. Finally, in terms of feature selection and processing, the XGBoost model had unique advantages [22]. It could automatically select the most important features and had good capabilities in handling missing values and outliers, helping to reduce the model's sensitivity to noise and unnecessary features and thereby improving model accuracy [21].

Currently, monitoring the leaf moisture content based on remote sensing data and machine learning models still has certain limitations. Despite using various feature combinations and machine learning models for modeling, there still exists a certain degree of error and uncertainty. First, vegetation water status is influenced by multiple factors, including climatic conditions, soil types, vegetation types, etc., and existing models may not fully consider the complex relationships among these factors. Additionally, current research mainly focuses on monitoring and evaluating vegetation water status, while challenges remain in translating these monitoring results into effective agricultural management and irrigation decisions in practical applications. Further research and development are needed to establish intelligent agricultural decision support systems based on monitoring results, integrating multiple data sources such as meteorological data, soil information, etc., to achieve precise management and optimal utilization of farmland water resources.

To address these issues, future research will consider in-depth exploration of the water requirements and response patterns of different vegetation types at different growth stages, optimizing vegetation water monitoring models to improve the accuracy and reliability of vegetation water status assessment. Additionally, combining meteorological data, soil information, and other multiple data sources, conducting correlation analyses between vegetation water content and factors such as climatic conditions and soil types, will deepen the understanding of influencing factors and variation patterns of vegetation water status.

5. Conclusions

This study employed plot experiments and multispectral data obtained from drones, combined with vegetation indices and texture features, and utilized three machine learning models, extreme learning machine (ELM), extreme gradient boosting tree (XGBoost), and back propagation neural network (BPNN), to estimate soybean leaf moisture content. The results indicated that most vegetation indices and texture features were significantly correlated with the soybean leaf moisture content ($p < 0.05$). Among them, the vegetation index with the highest correlation coefficient was MSR, at 0.649, while the texture feature with the highest correlation coefficient with leaf moisture content was the mean in band 2, at 0.644. All texture indices were significantly correlated with the soybean leaf moisture content ($p < 0.05$), with RATI being the randomly combined texture feature with the highest correlation coefficient, at 0.683. The texture combination was Variance1 and Correlation5, and the prediction model's fitting accuracy for leaf moisture content was ranked as follows: XGBoost > BPNN > ELM. Furthermore, using the XGBoost model, combination 4 (vegetation indices, texture features, and randomly combined texture features) provided the best monitoring effect for leaf moisture content, with an R^2 of 0.816, RMSE of 1.404, and MRE of 1.934% on the model validation set. These results provide important references for establishing a nondestructive, rapid, and efficient model for monitoring crop leaf moisture content.

In the research on three machine learning models based on vegetation indices and texture features, there are still some issues to be addressed. For example, this study, along with the majority of researchers, primarily focused on a single growth period of a single plant species as the experimental subject. The feasibility of applying these research findings to the entire growth period of vegetation requires further investigation. Therefore, achieving a higher level of universality and accuracy in simulating leaf moisture content for both individual growth periods and the entire growth period of most plants still requires further research and practical exploration.

Author Contributions: W.Y., conceptualization, data curation, methodology, investigation, roles, and writing—original draft. G.C., S.C., Y.W., X.L., W.M., Y.L., J.H., D.L. and Y.Z., data curation, formal analysis, investigation, and methodology. Z.L., Y.X., Z.T. and F.Z., conceptualization, writing—review and editing, supervision, formal analysis, and funding support. All authors have read and agreed to the published version of the manuscript.

Funding: This study was supported by the National Natural Science Foundation of China (No. 52179045), Innovative Experimental Project for College Students (202400860A9).

Data Availability Statement: Data are contained within the article.

Conflicts of Interest: The authors declare no conflicts of interest.

References

1. Tang, Z.J.; Wang, X.; Xiang, Y.Z.; Liang, J.P.; Guo, J.J.; Li, W.Y.; Lu, J.S.; Du, R.Q.; Li, Z.J.; Zhang, F.C. Application of hyperspectral technology for leaf function monitoring and nitrogen nutrient diagnosis in soybean (*Glycine max* L.) production systems on the Loess Plateau of China. *Eur. J. Agron.* **2024**, *154*, 127098. [CrossRef]
2. Tang, Z.J.; Lu, J.S.; Xiang, Y.Z.; Shi, H.Z.; Sun, T.; Zhang, W.; Wang, H.; Zhang, X.Y.; Li, Z.J.; Zhang, F.C. Farmland mulching and optimized irrigation increase water productivity and seed yield by regulating functional parameters of soybean (*Glycine max* L.) leaves. *Agric. Water Manag.* **2024**, *298*, 108875. [CrossRef]

3. Zhang, P.; Li, L.; Fu, Q.; Zhang, Z.; Li, H.; Zhao, L.; Liu, W.; Wang, Y.; Li, M.; Yang, A. Sustainable Management of Water, Nitrogen and Biochar Resources for Soybean Growth Considering Economic, Environmental and Resource Use Efficiency Aspects: An Integrated Modeling Approach. *J. Clean. Prod.* **2023**, *428*, 139236. [CrossRef]
4. Ketema, Z.; Claas, N. Yield response and water productivity of soybean (*Glycine max* L.) to deficit irrigation and sowing time in south-eastern Australia. *Agric. Water. Manag.* **2024**, *296*, 108815. [CrossRef]
5. Junttila, S.; Hölttä, T.; Saarinen, N.; Kankare, V.; Yrttimaa, T.; Hyyppä, J.; Vastaranta, M. Close-range hyperspectral spectroscopy reveals leaf water content dynamics. *Remote Sens. Environ.* **2022**, *277*, 113071. [CrossRef]
6. Serrano-Finetti, E.; Castillo, E.; Alejos, S.; León Hilario, L.M. Toward noninvasive monitoring of plant leaf water content by electrical impedance spectroscopy. *Comput. Electron. Agric.* **2023**, *210*, 107907. [CrossRef]
7. Tang, Z.J.; Zhang, W.; Xiang, Y.Z.; Liu, X.C.; Wang, X.; Shi, H.Z.; Li, Z.J.; Zhang, F.C. Monitoring of Soil Moisture Content of Winter Oilseed Rape (Brassica napus L.) Based on Hyperspectral and Machine Learning Models. *J. Soil Sci. Plant Nutr.* **2024**, *24*, 1250–1260. [CrossRef]
8. Elshikha, D.E.M.; Hunsaker, D.J.; Waller, P.M.; Thorp, K.R.; Dierig, D.; Wang, G.Y.; Cruz, V.M.V.; Katterman, M.E.; Bronson, K.F.; Wall, G.W.; et al. Estimation of direct-seeded guayule cover, crop coefficient, and yield using UAS-based multispectral and RGB data. *Agric. Water Manag.* **2022**, *265*, 107540. [CrossRef]
9. Tang, Z.J.; Guo, J.J.; Xiang, Y.Z.; Lu, X.H.; Wang, Q.; Wang, H.D.; Cheng, M.H.; Wang, H.; Wang, X.; An, J.Q.; et al. Estimation of Leaf Area Index and Above-Ground Biomass of Winter Wheat Based on Optimal Spectral Index. *Agronomy* **2022**, *12*, 1729. [CrossRef]
10. Blekanov, I.; Molin, A.; Zhang, D.; Mitrofanov, E.; Mitrofanova, O.; Li, Y. Monitoring of grain crops nitrogen status from uav multispectral images coupled with deep learning approaches. *Comput. Electron. Agric.* **2023**, *212*, 108047. [CrossRef]
11. Choudhury, B.U.; Narzari, R.; Zafar, M.; Singh, N.; Mishra, V.K.; Prabhakar, M. Spectral library of crops and discrimination of major vegetables grown in the eastern Himalayan ecosystem: A proximal hyperspectral remote sensing approach. *Ecol. Inform.* **2023**, *77*, 102263. [CrossRef]
12. Liu, Y.; An, L.L.; Wang, N.; Tang, W.J.; Liu, M.J.; Liu, G.H.; Sun, H.; Li, M.Z.; Ma, Y.T. Leaf area index estimation under wheat powdery mildew stress by integrating UAV-based spectral, textural and structural features. *Comput. Electron. Agric.* **2023**, *213*, 108169. [CrossRef]
13. Yang, N.; Zhang, Z.T.; Zhang, J.R.; Guo, Y.H.; Yang, X.Z.; Yu, G.D.; Bai, X.Q.; Chen, J.Y.; Chen, Y.W.; Shi, L.S.; et al. Improving estimation of maize leaf area index by combining of UAV-based multispectral and thermal infrared data: The potential of new texture index. *Comput. Electron. Agric.* **2023**, *214*, 108294. [CrossRef]
14. Berger, A.; Ettlin, G.; Quincke, C.; Rodríguez-Bocca, P. Predicting the Normalized Difference Vegetation Index (NDVI) by training a crop growth model with historical data. *Comput. Electron. Agric.* **2019**, *161*, 305–311. [CrossRef]
15. Liu, Y.; Fan, Y.G.; Feng, H.K.; Chen, R.Q.; Bian, M.B.; Ma, Y.P.; Yue, J.B.; Yang, G.J. Estimating potato above-ground biomass based on vegetation indices and texture features constructed from sensitive bands of UAV hyperspectral imagery. *Comput. Electron. Agric.* **2024**, *220*, 108918. [CrossRef]
16. Tang, Z.J.; Xiang, Y.Z.; Zhang, W.; Wang, X.; Zhang, F.C.; Chen, J.Y. Research on potato (*Solanum tuberosum* L.) nitrogen nutrition diagnosis based on hyperspectral data. *Agron. J.* **2024**, *116*, 531–541. [CrossRef]
17. Wang, X.X.; Yan, S.C.; Wang, W.T.; Yin, L.B.; Li, M.; Yu, Z.; Chang, S.H.; Hou, F.J. Monitoring leaf area index of the sown mixture pasture through UAV multispectral image and texture characteristics. *Comput. Electron. Agric.* **2023**, *214*, 108333. [CrossRef]
18. Qiao, L.; Tang, W.J.; Gao, D.H.; Zhao, R.M.; An, L.L.; Li, M.Z.; Sun, H.; Song, D. UAV-based chlorophyll content estimation by evaluating vegetation index responses under different crop coverages. *Comput. Electron. Agric.* **2022**, *196*, 106775. [CrossRef]
19. Wang, L.; Wang, P.X.; Liang, S.L.; Zhu, Y.C.; Khan, J.; Fang, S.B. Monitoring maize growth on the North China Plain using a hybrid genetic algorithm-based back-propagation neural network model. *Comput. Electron. Agric.* **2020**, *170*, 105238. [CrossRef]
20. Tang, X.P.; Liu, H.J.; Feng, D.X.; Zhang, W.J.; Chang, J.; Li, L.; Yang, L. Prediction of field winter wheat yield using fewer parameters at middle growth stage by linear regression and the BP neural network method. *Eur. J. Agron.* **2022**, *141*, 126621. [CrossRef]
21. Han, L.; Yang, G.J.; Yang, X.D.; Song, X.Y.; Xu, B.; Li, Z.H.; Wu, J.T.; Yang, H.; Wu, J.W. An explainable XGBoost model improved by SMOTE-ENN technique for maize lodging detection based on multi-source unmanned aerial vehicle images. *Comput. Electron. Agric.* **2022**, *194*, 106804. [CrossRef]
22. Joshi, A.; Pradhan, B.; Chakraborty, S.; Behera, M.D. Winter wheat yield prediction in the conterminous United States using solar-induced chlorophyll fluorescence data and XGBoost and random forest algorithm. *Ecol. Inform.* **2023**, *77*, 102194. [CrossRef]
23. Mandal, D.; Kumar, V.; Ratha, D.; Dey, S.; Bhattacharya, A.; Lopez-Sanchez, J.M.; McNairn, H.; Rao, Y.S. Dual polarimetric radar vegetation index for crop growth monitoring using sentinel-1 SAR data. *Remote Sens. Environ.* **2020**, *247*, 111954. [CrossRef]
24. Yang, Z.Y.; Yu, Z.R.; Wang, X.Y.; Yan, W.G.; Sun, S.J.; Feng, M.C.; Sun, J.J.; Su, P.Y.; Sun, X.K.; Wang, Z.G.; et al. Estimation of Millet Aboveground Biomass Utilizing Multi-Source UAV Image Feature Fusion. *Agronomy* **2024**, *14*, 701. [CrossRef]
25. Li, T.S.; Wang, H.J.; Cui, J.; Wang, W.J.; Li, W.R.Y.; Jiang, M.H.; Shi, X.Y.; Song, J.H.; Wang, J.G.; Lv, X.; et al. Improving the accuracy of cotton seedling emergence rate estimation by fusing UAV-based multispectral vegetation indices. *Front. Plant Sci.* **2024**, *15*, 1333089. [CrossRef] [PubMed]

26. Zhou, W.Q.; Song, C.; Liu, C.L.; Fu, Q.; An, T.H.; Wang, Y.J.; Sun, X.B.; Wen, N.; Tang, H.; Wang, Q. A Prediction Model of Maize Field Yield Based on the Fusion of Multitemporal and Multimodal UAV Data: A Case Study in Northeast China. *Remote Sens.* **2023**, *15*, 3483. [CrossRef]
27. Pilaš, I.; Gašparović, M.; Novkinić, A.; Klobučar, D. Mapping of the Canopy Openings in Mixed Beech–Fir Forest at Sentinel-2 Subpixel Level Using UAV and Machine Learning Approach. *Remote Sens.* **2020**, *12*, 3925. [CrossRef]
28. Duque, A.F.; Patino, D.; Colorado, J.D.; Petro, E.; Rebolledo, M.C.; Mondragon, I.F.; Espinosa, N.; Amezquita, N.; Puentes, O.D.; Mendez, D.; et al. Characterization of Rice Yield Based on Biomass and SPAD-Based Leaf Nitrogen for Large Genotype Plots. *Sensors* **2023**, *23*, 5917. [CrossRef] [PubMed]
29. Kapari, M.; Sibanda, M.; Magidi, J.; Mabhaudhi, T.; Nhamo, L.; Mpandeli, S. Comparing Machine Learning Algorithms for Estimating the Maize Crop Water Stress Index (CWSI) Using UAV-Acquired Remotely Sensed Data in Smallholder Croplands. *Drones* **2024**, *8*, 61. [CrossRef]
30. Musungu, K.; Dube, T.; Smit, J.; Shoko, M. Using UAV multispectral photography to discriminate plant species in a seep wetland of the Fynbos Biome. *Wetl. Ecol. Manag.* **2024**, *32*, 207–227. [CrossRef]
31. Wang, Q.; Lu, X.H.; Zhang, H.N.; Yang, B.C.; Gong, R.X.; Zhang, J.; Jin, Z.N.; Xie, R.X.; Xia, J.W.; Zhao, J.M. Comparison of Machine Learning Methods for Estimating Leaf Area Index and Aboveground Biomass of Cinnamomum camphora Based on UAV Multispectral Remote Sensing Data. *Forests* **2023**, *14*, 1688. [CrossRef]
32. Yang, B.C.; Zhang, H.N.; Lu, X.H.; Wan, H.L.; Zhang, Y.; Zhang, J.; Jin, Z.N. Inversion of Leaf Water Content of Cinnamomum camphora Based on Preferred Spectral Index and Machine Learning Algorithm. *Forests* **2023**, *14*, 2285. [CrossRef]
33. Hu, J.Y.; Feng, H.; Wang, Q.L.; Shen, J.N.; Wang, J.; Liu, Y.; Feng, H.K.; Yang, H.; Guo, W.; Qiao, H.B.; et al. Pretrained Deep Learning Networks and Multispectral Imagery Enhance Maize LCC, FVC, and Maturity Estimation. *Remote Sens.* **2024**, *16*, 784. [CrossRef]
34. Xie, J.X.; Wang, J.X.; Chen, Y.F.; Gao, P.; Yin, H.L.; Chen, S.Y.; Sun, D.Z.; Wang, W.X.; Mo, H.D.; Shen, J.Y.; et al. Estimating the SPAD of Litchi in the Growth Period and Autumn Shoot Period Based on UAV Multi-Spectrum. *Remote Sens.* **2023**, *15*, 5767. [CrossRef]
35. Zhao, X.D.; Hu, W.M.; Han, J.; Wei, W.; Xu, J.X. Urban Above-Ground Biomass Estimation Using GEDI Laser Data and Optical Remote Sensing Images. *Remote Sens.* **2024**, *16*, 1229. [CrossRef]
36. Ma, R.; Zhang, N.N.; Zhang, X.; Bai, T.C.; Yuan, X.T.; Bao, H.; He, D.D.; Sun, W.J.; He, Y. Cotton Verticillium wilt monitoring based on UAV multispectral-visible multi-source feature fusion. *Comput. Electron. Agric.* **2024**, *217*, 108628. [CrossRef]
37. Li, Y.C.; Zeng, H.W.; Zhang, M.; Wu, B.F.; Zhao, Y.; Yao, X.; Cheng, T.; Qin, X.L.; Wu, F.M. A county-level soybean yield prediction framework coupled with XGBoost and multidimensional feature engineering. *Int. J. Appl. Earth Obs.* **2023**, *118*, 103269. [CrossRef]

plants

MDPI

Article

Monitoring of Chlorophyll Content of Potato in Northern Shaanxi Based on Different Spectral Parameters

Hongzhao Shi [1,2], Xingxing Lu [1,3], Tao Sun [1,2], Xiaochi Liu [1,2], Xiangyang Huang [1,2], Zijun Tang [1,2], Zhijun Li [1,2,*], Youzhen Xiang [1,2,*], Fucang Zhang [1,2] and Jingbo Zhen [1,2]

1 Key Laboratory of Agricultural Soil and Water Engineering in Arid and Semiarid Areas of Ministry of Education, Northwest A&F University, Xianyang 712100, China; shihongzhao7@nwsuaf.edu.cn (H.S.); luxingxing625@163.com (X.L.); 2021050986@nwsuaf.edu.cn (T.S.); 2023055903@nwafu.edu.cn (X.L.); huangxiangyang2023@163.com (X.H.); tangzijun@nwsuaf.edu.cn (Z.T.); zhangfc@nwsuaf.edu.cn (F.Z.); zhenjingbo@nwsuaf.edu.cn (J.Z.)

2 Institute of Water-Saving Agriculture in Arid Areas of China, Northwest A&F University, Xianyang 712100, China

3 Department of Mechanical Engineering, College of Mechanical and Electrical Engineering, Yangling Vocational & Technical College, Xianyang 712100, China

* Correspondence: lizhij@nwsuaf.edu.cn (Z.L.); youzhenxiang@nwsuaf.edu.cn (Y.X.)

Abstract: Leaf chlorophyll content (LCC) is an important physiological index to evaluate the photosynthetic capacity and growth health of crops. In this investigation, the focus was placed on the chlorophyll content per unit of leaf area (LCC_A) and the chlorophyll content per unit of fresh weight (LCC_W) during the tuber formation phase of potatoes in Northern Shaanxi. Ground-based hyperspectral data were acquired for this purpose to formulate the vegetation index. The correlation coefficient method was used to obtain the "trilateral" parameters with the best correlation between potato LCC_A and LCC_W, empirical vegetation index, any two-band vegetation index constructed after 0–2 fractional differential transformation (step size 0.5), and the parameters with the highest correlation among the three spectral parameters, which were divided into four combinations as model inputs. The prediction models of potato LCC_A and LCC_W were constructed using the support vector machine (SVM), random forest (RF) and back propagation neural network (BPNN) algorithms. The results showed that, compared with the "trilateral" parameter and the empirical vegetation index, the spectral index constructed by the hyperspectral reflectance after differential transformation had a stronger correlation with potato LCC_A and LCC_W. Compared with no treatment, the correlation between spectral index and potato LCC and the prediction accuracy of the model showed a trend of decreasing after initial growth with the increase in differential order. The highest correlation index after 0–2 order differential treatment is DI, and the maximum correlation coefficients are 0.787, 0.798, 0.792, 0.788 and 0.756, respectively. The maximum value of the spectral index correlation coefficient after each order differential treatment corresponds to the red edge or near-infrared band. A comprehensive comparison shows that in the LCC_A and LCC_W estimation models, the RF model has the highest accuracy when combination 3 is used as the input variable. Therefore, it is more recommended to use the LCC_A to estimate the chlorophyll content of crop leaves in the agricultural practices of the potato industry. The results of this study can enhance the scientific understanding and accurate simulation of potato canopy spectral information, provide a theoretical basis for the remote sensing inversion of crop growth, and promote the development of modern precision agriculture.

Keywords: potato; hyperspectral; chlorophyll content; machine learning

Citation: Shi, H.; Lu, X.; Sun, T.; Liu, X.; Huang, X.; Tang, Z.; Li, Z.; Xiang, Y.; Zhang, F.; Zhen, J. Monitoring of Chlorophyll Content of Potato in Northern Shaanxi Based on Different Spectral Parameters. *Plants* **2024**, *13*, 1314. https://doi.org/10.3390/plants13101314

Academic Editor: Georgia Ouzounidou

Received: 10 April 2024
Revised: 5 May 2024
Accepted: 8 May 2024
Published: 10 May 2024

1. Introduction

Potato, the fourth largest staple crop in the world, is widely distributed and exhibits strong adaptability, high yield, and rich nutritional content. It is suitable for storage as both food and industrial raw material, playing a crucial role in improving people's living standards and ensuring food security [1]. Shaanbei, as one of the major potato-producing regions in China, possesses soil, temperature, and light conditions favorable for the growth and development of potatoes. However, outdated irrigation and fertilization techniques in this region have led to soil fertility degradation and environmental pollution, severely hindering the development of its potato industry [2]. Therefore, addressing the issues of unstable potato yields and inconsistent quality in this area is imperative. Leaf chlorophyll content (LCC) serves as a vital indicator for measuring crop growth, reflecting the growth status and health of crops. Monitoring its content changes aids in distinguishing the physiological characteristics of crops [3]. In recent years, with the rapid development of intelligent agriculture, the rapid and non-destructive estimation of chlorophyll content has been realized, which is of great significance for evaluating and managing crop canopy photosynthetic capacity.

Traditional methods for determining chlorophyll content mainly rely on ethanol extraction, which is time-consuming and cumbersome [4–6]. In recent years, commonly used units for chlorophyll content include chlorophyll content per unit leaf area (LCC_A) and chlorophyll content per unit fresh weight (LCC_W) [7]. Expressing LCC_A is not affected by changes in crop plant internal water content, resulting in more stable outcomes. Meanwhile, LCC_W is widely used in agricultural research to describe chlorophyll content [8,9]. Thus, clarifying the chlorophyll content in different measurement units is of significant importance for reflecting the actual value of crop chlorophyll. Traditional measurement methods are destructive and yield unstable results. Utilizing hyperspectral remote sensing technology provides a new approach for monitoring dynamic changes in crop leaf chlorophyll and offers technical means for selecting the most representative measurement units of crop leaf chlorophyll. Modern information technology provides a new method for intelligent agriculture. With the rapid development and integration of modern information technologies such as remote sensing, big data, machine learning and cloud computing, other technologies such as intelligent identification, accurate measurement, model construction, information collection are becoming more and more mature. It provides a new method for monitoring crop growth parameters by remote sensing, which is of great significance for crop water and fertilizer management and agricultural decision-making [10]. Liu et al. (2021) collected SPAD and remote sensing information of soybean leaves and successfully monitored the chlorophyll content using mathematical models [11]. Based on feature optimization, Zhao et al. (2022) used a variety of machine learning methods to invert farmland surface soil moisture. The experimental results show that the random forest model has higher inversion accuracy and the best fitting effect, and the inversion accuracy is greatly improved after feature optimization [12]. Existing studies mostly construct spectral indices from original canopy hyperspectral reflectance to infer crop growth physiological indicators, but the prediction accuracy and results are not satisfactory. Introducing differential transformation methods can reduce noise interference, enhance model applicability, and optimize fitting effects. Shi et al. (2023) selected the optimal spectral indices and established models using first-order differentially processed hyperspectral reflectance, which significantly improved model accuracy [13]. Zhao et al. (2022) used five methods to process the original spectrum, and found that FOD achieved good results regardless of the modeling method [14]. Currently, chlorophyll content determination often involves averaging chlorophyll content at the individual plant level [15]. Although this method is simple and easy to implement, it fails to accurately reflect the overall level of LCC. Spectral indices are linear or nonlinear combinations of different sensitive bands, closely related to the reflection, absorption, and growth of different plants in different spectral bands. Constructing a prediction model requires appropriate band combinations to enhance model accuracy [16]. When plants are subjected to disease stress, chlorophyll digestion, water content reduction and coverage reduction often accompany plant growth [17], leading to the degree of reflection of canopy spectral information on plant physiological growth indicators to decrease significantly [18]. In such cases, the use of spectral indices related to relevant bands may fail to extract all spectral information, resulting in poor model fitting [19]. A correlation matrix analysis is commonly used in crop growth parameter and spectral index correlation analysis. By selecting the optimal bands highly correlated with crop growth physiological indicators across the full spectrum, it greatly enhances the utilization of spectral information and optimizes model performance [20].

This study utilized spectral data and employed the correlation coefficient method to select three sensitive parameters for potato LCC_A and LCC_W. Additionally, empirical vegetation indices, vegetation indices obtained from the differentiation of spectral bands from 0 to 2 (with a step size of 0.5), and the most highly correlated parameters among these three spectral parameters were identified. These parameters were divided into four combinations and used as inputs for model construction. Support Vector Machine (SVM), Random Forest (RF), and Back Propagation Neural Network (BPNN) were employed to build prediction models for potato LCC_A and LCC_W. The study aimed to identify the most effective method for reflecting crop chlorophyll content to enhance the scientific understanding and accurate simulation of potato canopy spectral information, providing a theoretical basis for the remote sensing inversion of crop growth.

2. Materials and Methods

2.1. Research Area and Test Design

This experiment was conducted at the Potato Experimental Demonstration Station of Northwest A&F University in Yulin City (Figure 1), Shaanxi Province, China (38°23′ N, 109°43′ E) during the months of May to October in both 2022 and 2023. The experimental variety used was the local main cultivar, 'Qingshu 9′. Planting took place on 5 May 2022, and 1 May 2023, respectively. In 2022, the average temperature during the entire potato growing period was 22 °C, and the total rainfall was 482.20 mm. In 2023, the average temperature during the entire potato growing period was also 22 °C, while the total rainfall was 212.10 mm. The soil was sandy loam, with the following physical and chemical properties: the bulk density of the cultivation layer (0–40 cm) was 1.73 g/cm^3, the ammonium nitrogen content was 6.35 mg/kg, the nitrate nitrogen content was 11.45 mg/kg, the available phosphorus content was 4.43 mg/kg, the available potassium content was 107 mg/kg, the pH value was 8.1 (H_2O was used to determine soil pH in the experiment), and the organic matter content was 4.31 g/kg. The experiment encompassed five nitrogen application levels: N0 (0 kg N/hm^2), N1 (90 kg N/hm^2), N2 (180 kg N/hm^2), N3 (270 kg N/hm^2), and N4 (360 kg N/hm^2). Additionally, two biochar application levels were implemented: B0 (0 t/hm^2) and B1 (30 t/hm^2), resulting in a total of 10 experimental treatments. Phosphorus and potassium fertilizers were applied once before sowing, and nitrogen fertilizers were applied together using the water and fertilizer integration facilities during irrigation. The test fertilizers were urea (N—46%), diammonium phosphate (N—18%, P_2O_5—46%) and potassium nitrate (N—13.5%, K_2O—46%). Each treatment was replicated three times, yielding a total of 30 plots. The plot dimensions were 4 m × 12 m, equivalent to 48 m^2, and the plots were arranged randomly with a protective strip of 3 m surrounding the experimental area. The potatoes were planted by artificial sowing, with a row spacing of 0.9 m, a plant spacing of 25 cm, and a sowing depth of 8~10 cm. Before potato planting, biochar was evenly incorporated into the top 20 cm of the soil and mixed evenly, and other field treatments were consistent with the locale.

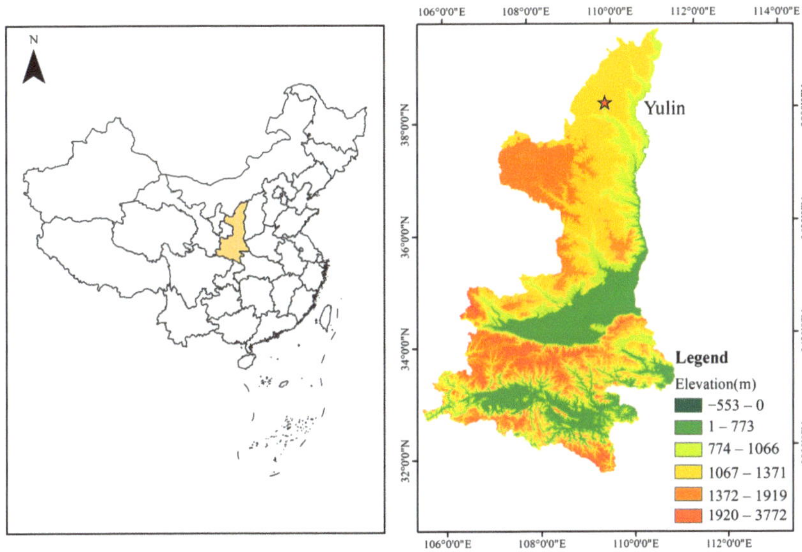

Figure 1. Geographic location of study area.

2.2. Data Collection and Preprocessing

2.2.1. Acquisition of Spectral Data

During the tuber formation stage of the potato, spectral data were collected on days with clear weather and no cloud cover. The spectral reflectance was measured using an ASD Field-Spec 3 portable spectrometer, following the method described in reference [20]. Spectral data were collected on 7 July 2022, and 8 July 2023, between 11:00 and 13:00. There are 60 groups of samples in this study.

2.2.2. Acquisition of Agronomic Parameters

The Leaf Chlorophyll Content (LCC) was determined using the 100% ethanol extraction method. Potato leaves corresponding to the hyperspectral measurement plots were collected. After removing the leaf veins, leaf disks were obtained using a hole punch method. Nine leaf disks with a diameter of 1 cm were collected and thoroughly ground. Additionally, 0.1 g of the remaining crushed leaves was weighed. A total of 10 mL of 100% ethanol was added, soaking and extracting the chlorophyll in potato leaves in a dark place at room temperature for approximately 3 days. Periodic shaking during soaking can shorten the duration, until the leaves become colorless or white. After all the chlorophyll in the crushed leaves was extracted into the ethanol solution (adjusted to a total volume of 25 mL), the absorbance at wavelengths of 663 nm and 645 nm was measured. The LCC_A and LCC_W were calculated using the following formulas [8,9]:

$$\text{Chlorophyll a content} = (12.7D_{663\text{nm}} - 2.69D_{645\text{nm}}) \times \frac{1}{40 \times m} \tag{1}$$

$$\text{Chlorophyll b content} = (22.9D_{663\text{nm}} - 4.68D_{645\text{nm}}) \times \frac{1}{40 \times m} \tag{2}$$

$$\text{Total chlorophyll content} = (20.21D_{645\text{nm}} + 8.02D_{663\text{nm}}) \times \frac{1}{40 \times m} \tag{3}$$

In the equations, $D_{663\text{nm}}$ and $D_{645\text{nm}}$ represent the absorbance at 663 nm and 645 nm, respectively. m denotes the fresh weight (g) or leaf area (dm^2). When m represents the fresh weight of the leaf, it yields the LCC_W. When m represents the leaf area, it yields the LCC_A.

Specific Leaf Weight (SLW) refers to the weight of leaf per unit leaf area (fresh weight). In this study, Specific Leaf Weight (g/dm^2) is calculated as the LCC_A divided by the LCC_W.

2.2.3. Spectral Data Processing

The original spectra of 60 samples in this study were obtained using View Spec Pro Version 6.2 software. In this study, 0–2 order fractional differential (FD) processing was performed on the spectral data after SG (Savitzky–Golay) smoothing pretreatment [20,21]. SG smoothing was implemented in The Unscrambler X 10.4 software.

The preprocessing of spectral data and the calculation of vegetation indices were conducted using MATLAB 2022 (MathWorks, Inc., Natick, MA, USA). The drawing charts were created using Origin 2024 (OriginLab Corp., Northampton, MA, USA).

2.3. Model Construction and Validation

Three different spectral indices were selected to more accurately screen for the wavelength combinations with the highest correlation with LCC_A and LCC_W:

(1) Previous research has demonstrated better correlations between empirical vegetation indices and crop parameters; therefore, this study also selected some empirical vegetation indices.
(2) The "trilateral" spectral parameters, which encompass the regions in the blue edge, yellow edge, and red edge spectra, are derived by extracting the peak value, valley value, area, or a combination of different bands from the blue edge, yellow edge, and red edge.
(3) The inversion of agricultural parameters can be effectively achieved by selecting any two-band vegetation index as the input parameter for the model. In this study, three arbitrary dual-band indices were initially chosen and then subjected to a 0–2 order fractional differential operation. Within the range of its spectral measurement wavelength, the combination index of the optimal order and the best vegetation index were selected.

Then, the two spectral indices with the highest correlation to potato LCC_A or LCC_W were further selected, constituting the optimal combination indices. The detailed calculation formulas are provided in Table 1.

Table 1. The empirical vegetation index selected for the study.

Selected Spectra Parameters	Calculation Formula	Reference
CARI	$(R_{700} - R_{670}) - 0.2 \times (R_{700} + R_{670})$	[22]
GRVI	R_{800}/R_{550}	[22]
PRI	$(R_{570} - R_{530})/(R_{570} + R_{530})$	[22]
IPVI	$R_{800} \times (R_{800} + R_{670})$	[22]
PRI1	$(R_{531} - R_{570})/(R_{531} + R_{570})$	[23]
SR1	R_{750}/R_{700}	[24]
SR3	R_{750}/R_{550}	[24]
SR705	R_{750}/R_{705}	[25]
SR680	R_{800}/R_{680}	[25]
SIPI	$(R_{800} - R_{445})/(R_{800} - R_{680})$	[25]
D_b	The highest value of the blue edge band (490–530 nm) in the 1-FD order spectral.	[26]
Dy	The highest value of the yellow edge band (462–642 nm) after the 1-FD order treatment.	[26]
Dr	The highest value of the red edge band (670–760 nm) after the 1-FD order treatment.	[27]
R_g	The highest value of the green edge band (510~560 nm).	[27]
R_r	The lowest value of the red edge band (650~690 nm).	[27]
SD_b	The sum of the blue edge wavelength range in the spectral reflectance after the 1-FD order treatment.	[28]

Table 1. *Cont.*

Selected Spectra Parameters	Calculation Formula	Reference
SD_y	The sum of the yellow edge wavelength range after the 1-FD order treatment	[28]
SD_r	The sum of the red edge wavelength range after the 1-FD order treatment.	[28]
SDr-SDb	/	[29]
SDr/SDy	/	[29]
Difference Index (DI)	$R_i - R_j$	[13]
Soil-Adjusted Vegetation Index (SAVI)	$(1+0.16)\frac{R_i - R_j}{R_i + R_j + 0.16}$	[13]

Notes: R_i (i = 1, 2, 3) is any value of the wavelength reflectivity in the measurement range (350~1830 nm), 1-FD is the first-order differential, and R_{number} is the spectral reflectivity of the digital band.

2.4. Model Approach

From the empirical spectral indices, "trilateral" spectral parameters, fractional order differentiation processed spectral indices within 0–2 order, and all spectral indices, the spectral index with the best correlation with LCC_A and LCC_W was selected as the model input. Subsequently, SVM, RF and BPNN models were separately employed to model LCC_A and LCC_W. For SVM, both Gaussian kernel and polynomial kernel were used as base kernel functions. The model parameters C and γ are 20 and 0.02, respectively [30]. RF belongs to the bagging algorithm in Ensemble Learning. The CART tree model is used as the base learner, the number of decision trees is 100 [31]. In BPNN, through data forward propagation and error back propagation, the input has undergone multiple iterations and repeated training [32]. The final fitted result is the average of multiple predictions from the machine learning model.

2.5. Model Evaluation Index

The model fitting results are evaluated using R^2, RMSE, and MRE. A higher R^2 signifies improved predictive accuracy, whereas smaller RMSE and MRE values indicate greater model stability and more focused prediction outcomes [21].

3. Results

3.1. LCC_A, LCC_W, SLW and Yield (GY)

In Figure 2, the trends of LCC_A, LCC_W, SLW, and GY under different treatments are illustrated. When the application rate of biochar is constant, LCC_A, LCC_W, SLW, and GY initially increase and then decrease with the increase in nitrogen fertilizer. Among them, the highest values of LCC_A, LCC_W, SLW, and GY are observed at N3. When the nitrogen fertilizer application rate is constant, the values of LCC_A, LCC_W, SLW, and GY in treatment B1 are higher than those in treatment B0, with increases of 5.44%, 7.61%, 1.32%, and 4.82%, respectively, compared to B0. Treatment B1N3 maximally enhances LCC_A, LCC_W, SLW, and GY of the crops.

The significant analysis of the effects of different biochar types and nitrogen application rates on LCC_A, LCC_W, SLW and yield (GY) is presented in Table 2. Different nitrogen fertilizer application rates significantly affect LCC_A, SLW, and GY ($p < 0.05$). Different biochar application rates significantly affect LCC_A and GY, and the interaction between nitrogen fertilizer and biochar application rates significantly influences LCC_A, SLW, and GY ($p < 0.05$). The effects of nitrogen fertilizer application rates, biochar application rates, and their interaction on LCC_W are not significant.

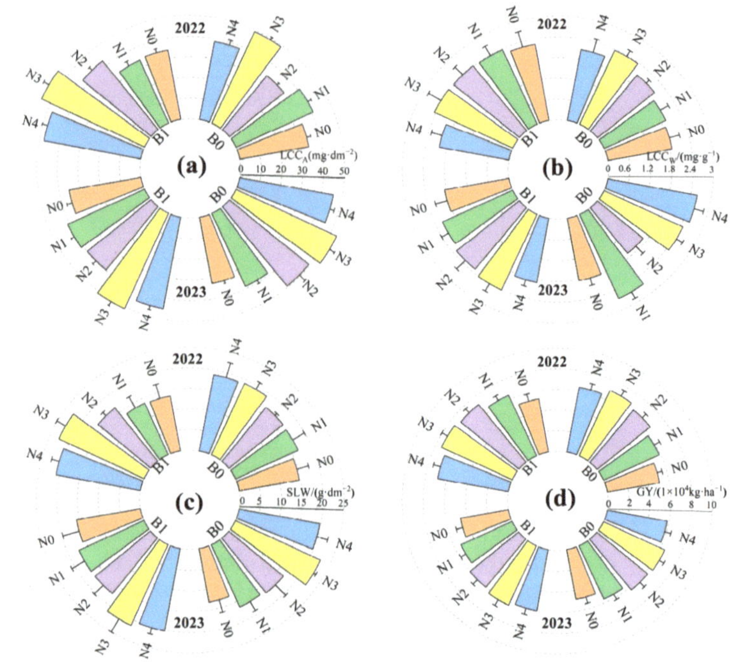

Figure 2. LCC_A (**a**), LCC_W (**b**), SLW (**c**), and GY (**d**) under different treatments.

Table 2. Effects of different biochar and nitrogen application rates on LCC_A, LCC_W, SLW and GY.

Year	Treatment		LCC_A	LCC_W	SLW	GY
			$mg \cdot dm^{-2}$	$mg \cdot g^{-1}$	$g \cdot dm^{-2}$	$kg \cdot ha^{-1}$
2022	B0	N0	33.60 hi	2.07 ab	16.54 bcd	50,520.34 f
		N1	33.87 hi	2.16 ab	19.00 abcd	58,533.81 e
		N2	38.44 fgh	2.34 ab	14.63 cd	65,618.02 d
		N3	49.00 bcd	2.64 ab	18.72 bcd	69,750.80 bc
		N4	40.78 ef	2.33 ab	16.80 bcd	63,574.08 d
	B1	N0	33.55 hi	2.24 ab	13.43 cd	52,084.25 f
		N1	34.09 hi	2.65 ab	13.65 cd	64,221.69 d
		N2	42.74 ef	2.70 ab	16.53 bcd	71,766.71 ab
		N3	54.95 cde	2.73 ab	22.73 ab	74,203.79 a
		N4	44.77 a	2.50 ab	20.73 abcd	68,307.19 c
2023	B0	N0	35.22 ghi	2.07 ab	15.66 bcd	46,397.08 e
		N1	39.75 fg	2.45 ab	17.32 bcd	53,743.41 d
		N2	40.87 ef	2.57 ab	17.33 bcd	60,027.95 b
		N3	50.91 abc	2.69 ab	21.41 abc	63,212.80 a
		N4	45.11 de	2.11 ab	20.17 abcd	60,036.28 b
	B1	N0	32.56 i	2.04 ab	13.14 d	48,711.15 e
		N1	38.81 fgh	2.89 a	16.48 bcd	56,542.28 c
		N2	49.38 bcd	1.88 b	17.01 bcd	62,935.39 a
		N3	53.35 ab	2.97 a	26.52 a	64,432.88 a
		N4	45.79 cde	2.76 ab	19.74 abcd	58,184.26 bc
			Significant level			
	B		**	ns	ns	**
	N		**	ns	**	**
	B×N		**	ns	*	*

Notes: The letters after the values of each column indicated that there were significant differences between treatments ($p < 0.05$), and * ($p < 0.05$) and ** ($p < 0.01$) indicated that there were significant differences in different degrees, ns means no significant difference.

3.2. Correlation Analysis between LCC_A, LCC_W and Spectral Index

The correlation analysis between various spectral indices and potato LCC_A and LCC_W was conducted to select the optimal vegetation index as the model input variable. Table 3 displays the correlation coefficients between empirical spectral indices, "trilateral" parameters, and potato LCC_A and LCC_W. The correlation analysis between empirical spectral indices and potato LCC_A indicates that the top seven indices with the highest correlation coefficients are IPVI, SR1, SR705, SR3, SR680, GRVI, and CARI, ranging from 0.4 to 0.8. Among them, IPVI exhibits the highest correlation coefficient of 0.771. In contrast, the top seven indices with the optimal correlation between empirical spectral indices and potato LCC_W are IPVI, CARI, SR1, SR3, SR705, SR680, and SIPI, ranging from 0.3 to 0.7. The Integrated Phenotypic Vegetation Index (IPVI) exhibits a peak correlation of 0.695. When assessed against LCC_W, empirical spectral indices have demonstrated a higher degree of correlation with LCC_A. Furthermore, the so-called "trilateral" parameters have shown a consistently strong correlation with potato LCC_A and LCC_W. The top seven parameters with the optimal correlation with potato LCC_A are SDr-SDb, SDr, Dr, Dy, Db, SDb, and Rg, ranging from 0.5 to 0.8. Among them, SDr-SDb exhibits the optimal correlation of 0.717. Similarly, the top seven parameters with the optimal correlation with potato LCC_W are SDr, SDr-SDb, Dr, Dy, Db, Rg, and SDb, ranging from 0.4 to 0.7. SDr shows the highest correlation coefficient of 0.613. Two arbitrary two-band spectral indices were constructed based on the spectral reflectance after 0–2 order (step size 0.5) differential processing, and their correlation with LCC_A and LCC_W were analyzed (Table 4). A graphical representation of the correlation matrix, referred to as Figures 3 and 4, was constructed. In this visualization, a color gradient ranging from yellow to green is utilized to depict the degree of correlation between various two-band spectral indices and the concentration of LCC_A or LCC_W. The gradient indicates a spectrum of correlation values, transitioning from strongly negative to strongly positive. The correlation analysis between spectral indices and LCC_A indicates that the spectral indices constructed from spectra processed with 0.5, 1, and 1.5 order differentials exhibit significantly improved correlation coefficients with potato LCC_A, with the highest correlation coefficient observed for DI constructed from spectra processed with a 0.5 order differential, reaching a maximum value of 0.798, with corresponding wavelength positions at (755,697). In contrast, the spectral indices constructed from spectra processed with a 2 order differential show a decrease in correlation coefficients. The order of correlation coefficients in terms of order is: 0.5 order > 1 order > 1.5 order > 0 order > 2 order. Similarly, the spectral index with the optimal correlation in the correlation analysis between spectral indices and LCC_W is DI processed with a 1.5 order differential, with a value of 0.737 and corresponding wavelength combination of (726,680). The order of correlation coefficients in terms of order is: 1.5 order > 1 order > 0.5 order > 2 order > 0 order. Compared to spectral indices established from the original reflectance, the correlation coefficients of spectral indices calculated from fractional order differentials significantly improved with LCC_A or LCC_W.

Table 3. Empirical spectral index and 'trilateral' parameters and potato LCC_A and LCC_W correlation coefficients.

Index	Spectral Index Category	Spectral Index	r
LCC_A	Empirical spectral index	CARI	0.496
		GRVI	0.404
		PRI	0.317
		IPVI	0.771
		PRI1	0.338
		SR1	0.669
		SR3	0.533
		SR705	0.658
		SR680	0.504
		SIPI	0.372
		Db	0.567
	"trilateral" parameters	Dy	0.568
		Dr	0.673
		Rg	0.536
		Rr	−0.087
		SDb	0.565
		SDy	−0.262
		SDr	0.711
		SDr-SDb	0.717
		SDr/SDy	0.432
LCC_W	Empirical spectral index	CARI	0.563
		GRVI	0.133
		PRI	−0.106
		IPVI	0.695
		PRI1	0.123
		SR1	0.515
		SR3	0.473
		SR705	0.394
		SR680	0.383
		SIPI	0.302
		Db	0.531
	"trilateral" parameters	Dy	0.532
		Dr	0.560
		Rg	0.548
		Rr	0.106
		SDb	0.481
		SDy	−0.064
		SDr	0.613
		SDr-SDb	0.612
		SDr/SDy	0.398

Table 4. Optimal spectral index wavelength combinations under different differential orders.

Index	Spectral Index	Differential Order	r_{max}	Position of Wavelength (i, j)/(nm)
LCC_A	DI	0	0.787	740,733
		0.5	0.798	755,697
		1	0.792	737,758
		1.5	0.788	736,748
		2	0.756	702,753
	SAVI	0	0.700	708,756
		0.5	0.787	694,755
		1	0.792	754,745
		1.5	0.785	748,736
		2	0.756	753,702
LCC_W	DI	0	0.684	757,724
		0.5	0.723	756,671
		1	0.723	739,670
		1.5	0.737	726,680
		2	0.702	694,751
	SAVI	0	0.612	674,678
		0.5	0.706	671,756
		1	0.723	670,739
		1.5	0.736	751,731
		2	0.702	751,694

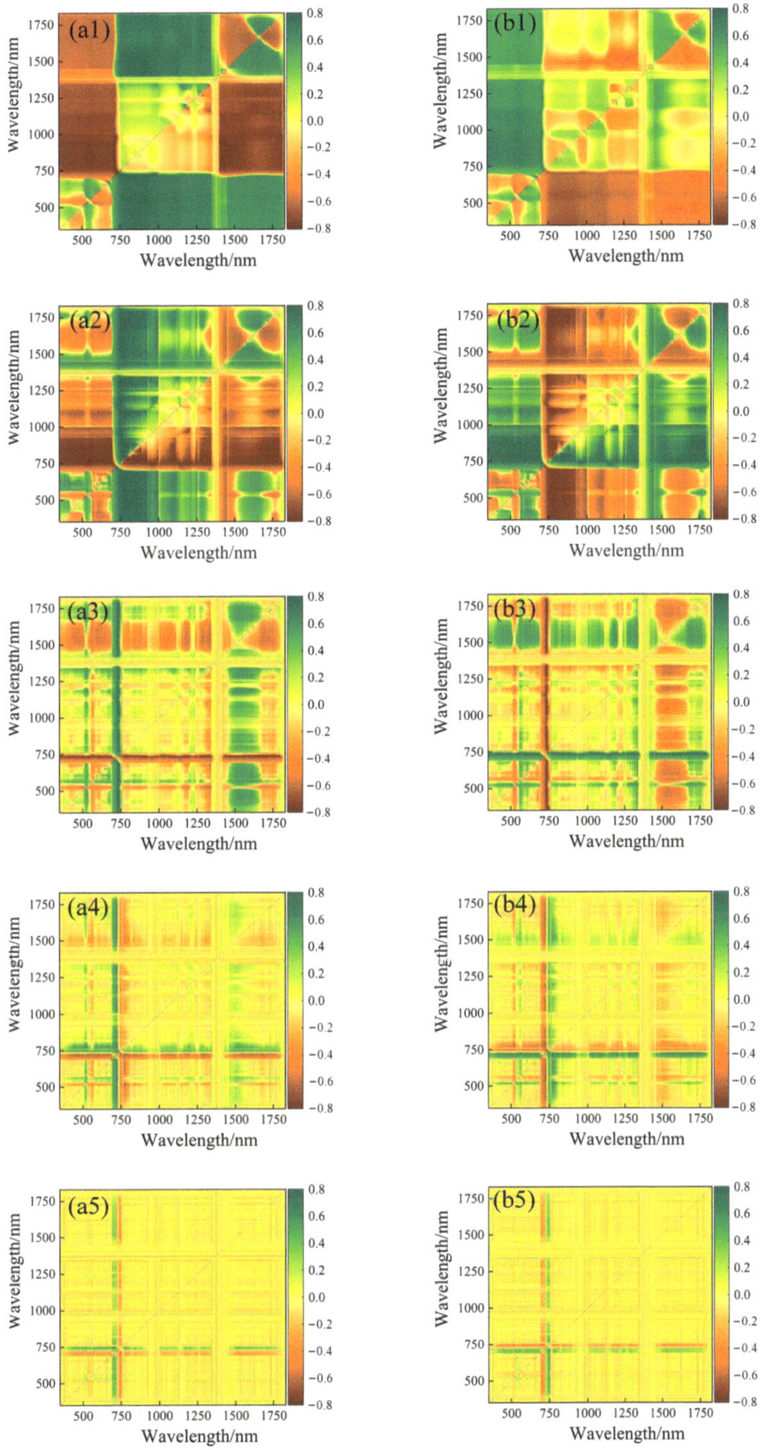

Figure 3. Correlation matrix of DI, SAVI with LCC_A (**a1–a5**,**b1–b5**).

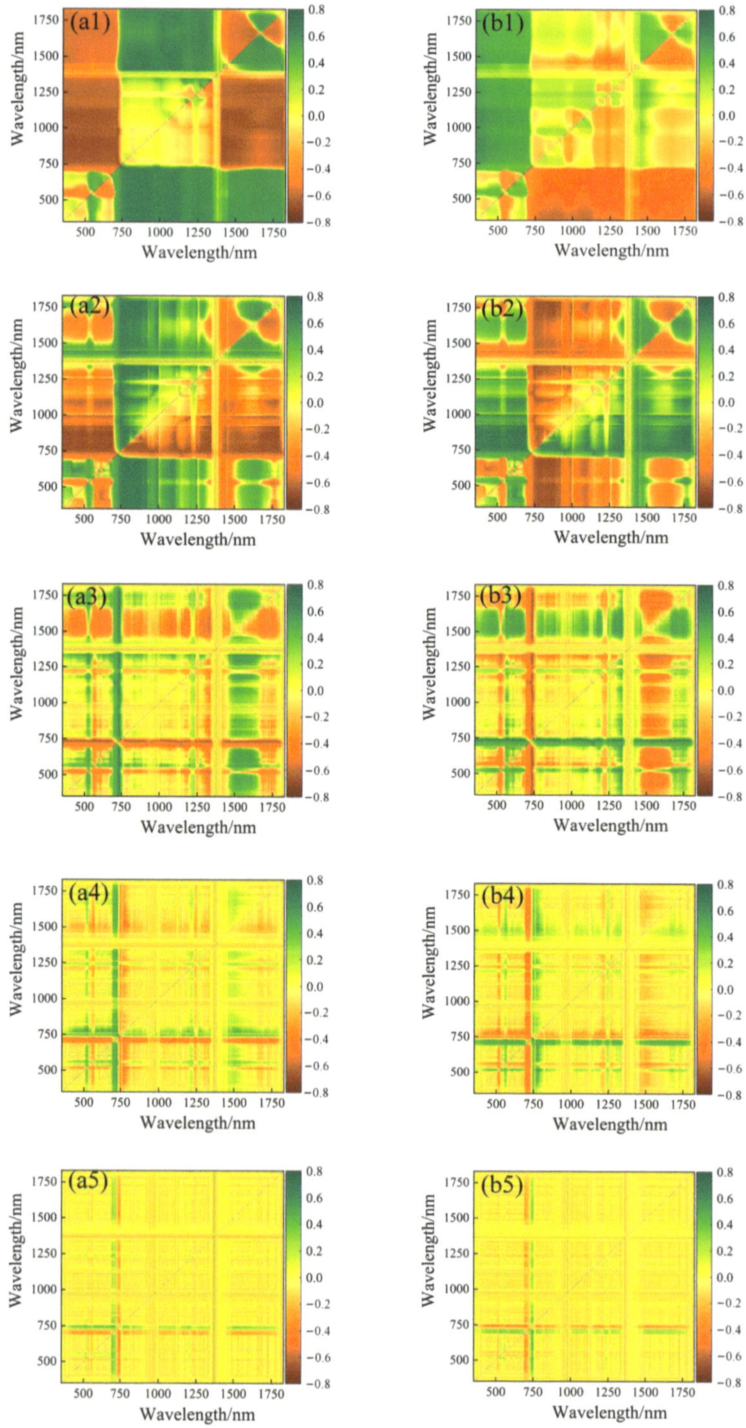

Figure 4. Correlation matrix of DI, SAVI with LCC_W (**a1–a5**,**b1–b5**).

3.3. Establishment of Estimation Model of LCC_A and LCC_W Based on Optimal Spectral Index

Section 2.3 introduces empirical spectral indices, "trilateral" spectral parameters, and arbitrary two-band vegetation indices. The parameters with the optimal correlation in these three types of spectral parameters are divided into four combinations for correlation analysis. Then, the top seven spectral indices with the optimal correlation with potato LCC_A or LCC_W in each combination are chosen as the input for the model. Potato LCC_A or LCC_W is used as the response variable, and SVM, RF, and BPNN are used to construct prediction models for potato tuber formation period LCC_A and LCC_W. The performance and fitting effect of the models are comprehensively evaluated based on three indicators: R^2, RMSE, and MRE (Figures 5 and 6).

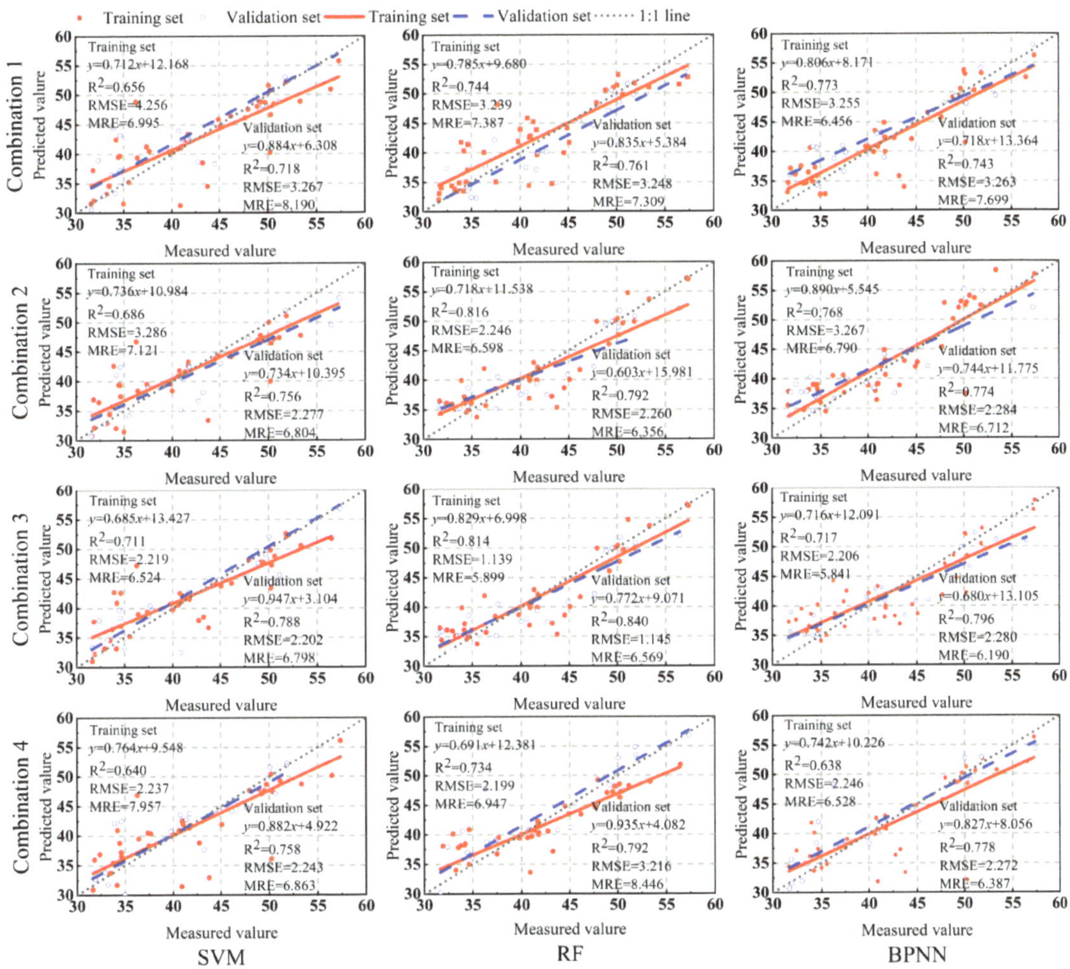

Figure 5. Precision evaluation of potato LCC_A model under different input variables and different model combinations.

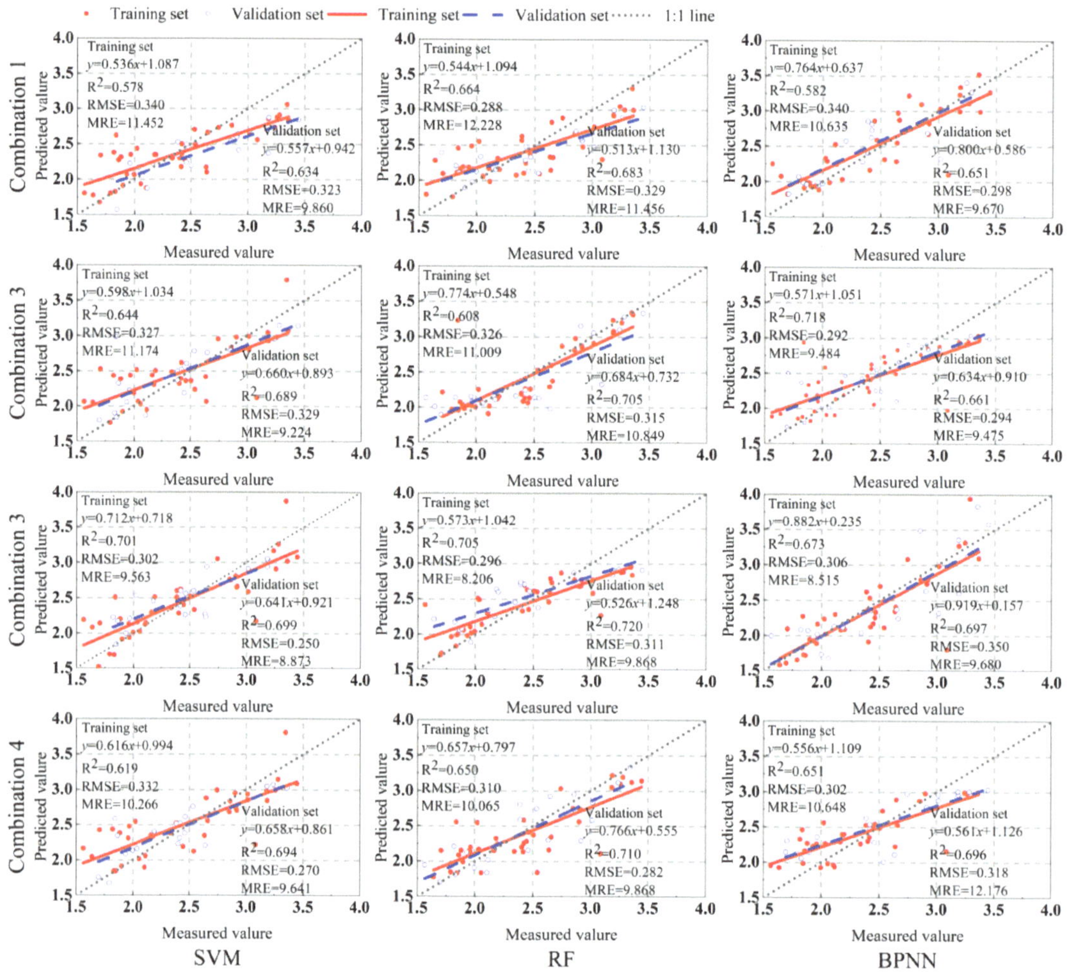

Figure 6. Precision evaluation of potato LCC_W model under different input variables and different model combinations.

In a parallel comparison of the three models, the model accuracy for estimating potato LCC_A and LCC_W is as follows: RF > BPNN > SVM. In the potato LCC_A estimation model, both RF and BPNN have R^2 values higher than 0.7, with RMSE and MRE maintained at relatively low levels, indicating good model performance and fitting effect. In the potato LCC_A estimation model, when the input variables are different combinations, the validation set R^2 values are all higher than 0.7, indicating good model performance and fitting effect. In contrast, in the potato LCC_W estimation model, the SVM and BPNN models have R^2 values ranging from 0.5 to 0.7 when the input variables are combination 1, indicating a lower fitting accuracy. However, for combination 3, both the modeling set and validation set have the highest R^2 values, with lower RMSE and MRE, specifically showing: combination 3 > combination 4 > combination 2 > combination 1. Overall, the R^2 of the LCC_A model is higher than that of the LCC_W model, and the MRE shows lower values, indicating higher model accuracy and better performance and fitting effect. When the input variables and modeling methods are combination 3 and RF, the optimal potato LCC_A and LCC_W prediction models can be constructed. The R^2 values of the validation set are 0.840 and 0.720, RMSE values are 1.145 and 0.311, and MRE values are 6.569% and 11.868%, respectively.

4. Discussion

In recent years, with the rapid development and integration of modern information technologies such as remote sensing, big data, machine learning and cloud computing, there have been numerous applications in monitoring crop growth or pest and disease infestations in agriculture. Hyperspectral imaging, due to its wide spectral range and nearly continuous spectral information of objects, can accurately record multidimensional information and component data [33]. It has been widely used in monitoring crop parameters such as leaf area index (LAI) [21], LCC [13], above-ground biomass (AGB) [34], soil moisture content [35], and surface parameters. Chlorophyll content directly determines the photosynthetic activity of crops and is an important physiological indicator for evaluating crop growth status. Combining hyperspectral remote sensing to estimate crop LCC is beneficial for accurately assessing its estimation capability and comprehensively evaluating crop growth status [36].

The construction of three types of spectral indices or parameters, including empirical spectral indices, "trilateral" spectral parameters, and arbitrary two-band spectral indices, revealed that the selection of arbitrary two-band spectral indices showed the highest correlation with potato LCC. This is because the arbitrary two-band spectral indices created by combining two bands utilize hyperspectral reflectance data processed through 0–2 order differentials, which helps reduce the basic background noise of the original spectral reflectance data and highlights their detailed spectral features [37]. As the order of differentiation increases, both the correlation between spectral indices and potato LCC and the predictive fitting performance of the model initially increase, but then decrease. When fractional order differentials (such as 0.5 order and 1.5 order) are used, the correlation between arbitrary two-band spectral indices and potato LCC exceeds that of the integer-order differentials (such as 1 order and 2 order). This is mainly because fractional order differentials can capture gradient information missed by integer-order differentials [38]. Most empirical vegetation indices based on fixed bands tend to saturate. When the crop canopy coverage is high, empirical vegetation indices tend to saturate, leading to decreased sensitivity to the reflecting of chlorophyll content and thus a decrease in correlation [39]. In sparse canopy conditions, where soil reflectance dominates, the effectiveness of empirical vegetation indices in reflecting vegetation growth parameters is often poor [40]. Additionally, due to the influence of the crop growth stage, environment, and pests and diseases, different spectral information may be generated, resulting in the phenomenon of "same object with different spectra" or "different objects with the same spectrum". In such cases, the use of empirical vegetation indices and "trilateral" parameters based on correlated bands may not fully utilize spectral information, leading to reduced correlation [41].

Our study utilized the rich spectral information contained in hyperspectral data to construct various spectral indices combined with different machine learning methods. We aimed to establish models for predicting LCC_A and LCC_W in potatoes, and to explore the rationality and applicability of these two different units of chlorophyll content. The results indicated that the LCC_A model exhibited higher R^2 compared to LCC_W, with a lower Mean Relative Error (MRE), indicating higher accuracy and better fitting of the model. This suggests that using hyperspectral data to extract information about crop LCC_A is richer and more correlated compared to LCC_W. This is attributed to the instability of chlorophyll content represented per unit fresh weight under different water and fertilizer supply conditions and growth environments. When chlorophyll content is expressed per unit leaf area, its representation per unit fresh weight is greatly affected by leaf water content variations, resulting in significant variability [42]. Therefore, expressing LCC_A can effectively avoid the interference of crop leaf water content, better approximate the true value of crop leaf chlorophyll content, and make full use of spectral information to accurately monitor crop photosynthetic capacity and understand crop growth conditions in a timely manner. Among the three models used in this study, the LCC_A and LCC_W prediction models-based RF method showed the optimal fitting performance, attributed to RF's good noise and interference resistance and its resistance to overfitting [43], whereas Back Propagation Neural Network (BPNN) suffers from slow convergence and is prone to local minima during training [44]. Support Vector Machine (SVM) exhibited the lowest accuracy, possibly due to its sensitivity to model parameters such as the kernel function and penalty factor, which hindered its predictive ability [45]. Therefore, the RF model is considered the optimal method for predicting crop LCC, and expressing LCC_A is recommended for estimating crop leaf chlorophyll content in the agricultural practices of the potato industry.

This study mainly focuses on the model inversion results of LCC_W and LCC_A, which shows that LCC_A has a strong ability to represent the real chlorophyll content of crops. These strategies can efficiently and non-destructively monitor the chlorophyll content of crops, grasp the real-time growth status of crops, and formulate corresponding solutions. Under the premise of paying attention to the efficient and rational use of water resources and fertilizers, they can effectively guide precision fertilization, scientific irrigation, and integrated pest control. These models not only save agricultural water and fertilizer, but also make an important contribution to the sustainable utilization of agricultural resources and the protection of the ecological environment. In the future, multi-source remote sensing data (hyperspectral, multispectral, thermal infrared, etc.) will be used as model input variables, and other types of models will be tried. The field measured data (different varieties, regions, time and space) will be compared and verified at a larger scale, in order to strengthen the real-time monitoring of crop physiological growth and promote the development of intelligent agriculture.

5. Conclusions

In this study, four combinations of model input variables were constructed, including empirical spectral indices, "trilateral" spectral parameters, any two-band vegetation indices, and the most highly correlated parameters among these three spectral parameters. SVM, RF and BPNN machine learning methods were employed to construct models for predicting LCC_A and LCC_W during the tuber differentiation stage of potatoes. The conclusions drawn from the study are as follows:

(1) Compared to "trilateral" parameters and empirical vegetation indices, any two-band vegetation indices constructed from hyperspectral reflectance after fractional order differentiation processing exhibit stronger correlations with potato LCC. As the order of differentiation increases, both the correlation between spectral indices and potato LCC and the predictive accuracy of the models initially increase but then decrease. When employing fractional order differentiations (e.g., 0.5th order and 1.5th order), the correlation between any two-band spectral indices and potato LCC exceeds that obtained when using integer-order differentiations (e.g., 1st order and 2nd order). Among them, the maximum correlation coefficients of the DI with the highest correlation after 0–2 order differentiation processing are: 0.787, 0.798, 0.792, 0.788, and 0.756, respectively.

(2) In the constructed LCC_A and LCC_W models, the performance and fitting effects are as follows: RF > BPNN > SVM, with the input combinations ranked as follows: combination 3 > combination 4 > combination 2 > combination 1. The RF method consistently demonstrates the highest accuracy and best fitting performance in model construction. The optimal input variables and modeling method for both LCC_A and LCC_W models are combination 3 and RF method. Therefore, expressing LCC_A is recommended for estimating crop leaf chlorophyll content in agricultural practice.

Author Contributions: Data curation: H.S., Investigation: Y.X. and F.Z., Methodology: Z.L. and J.Z., Project administration: Z.T. and H.S., Resources: Y.X. and F.Z., Software: X.L. (Xingxing Lu), T.S., X.H. and X.L. (Xiaochi Liu), Supervision: Z.T., Visualization: H.S., Writing—original draft: H.S.,

Writing—review and editing: H.S. All authors have read and agreed to the published version of the manuscript.

Funding: This study was supported by the Natural Science Foundation of Basic Research Project of Shaanxi Province (2023-JC-QN-0377).

Data Availability Statement: Data are contained within the article.

Conflicts of Interest: The authors declare no conflicts of interest.

References

1. Scott, G.J.; Petsakos, A.; Juarez, H. Climate Change, Food Security, and Future Scenarios for Potato Production in India to 2030. *Food Secur.* **2019**, *11*, 43–56. [CrossRef]
2. Xing, Y.; Wang, N.; Niu, X.; Jiang, W.; Wang, X. Assessment of Potato Farmland Soil Nutrient Based on MDS-SQI Model in the Loess Plateau. *Sustainability* **2021**, *13*, 3957. [CrossRef]
3. Gitelson, A.; Viña, A.; Ciganda, V.; Rundquist, D.; Arkebauer, T. Remote estimation of canopy chlorophyll content in crops. *Geophys. Res. lett.* **2005**, *32*. [CrossRef]
4. Croft, H.; Chen, J.M.; Luo, X.; Bartlett, P.; Chen, B.; Staebler, R.M. Leaf chlorophyll content as a proxy for leaf photosynthetic capacity. *Glob. Chang. Biol.* **2017**, *23*, 3513–3524. [CrossRef] [PubMed]
5. Ali, J.; Jan, I.; Ullah, H.; Fahad, S.; Saud, S.; Adnan, M.; Ali, B.; Liu, K.; Harrison, M.T.; Hassan, S.; et al. Biochemical Response of Okra (*Abelmoschus esculentus* L.) to Selenium (Se) under Drought Stress. *Sustainability* **2023**, *15*, 5694. [CrossRef]
6. Lu, J.; Wang, D.; Liu, K.; Chu, G.; Huang, L.; Tian, X.; Zhang, Y. Inbred varieties outperformed hybrid rice varieties under dense planting with reducing nitrogen. *Sci. Rep.* **2020**, *10*, 8769. [CrossRef]
7. Xu, D. Determination of Chlorophyll Content and Several Problems in Its Application. *Plant Physiol. Newsl.* **2009**, *45*, 896–898, (In Chinese with English Abstract). [CrossRef]
8. Li, J.; Zhou, X.; Zhou, J.; Shang, R.; Wang, Y.; Jing, P. Comparative study on several determination methods of chlorophyll content in plants. *IOP Conf. Ser. Mater. Sci. Eng.* **2020**, *730*, 012066. [CrossRef]
9. Ritchie, R.J. Consistent sets of spectrophotometric chlorophyll equations for acetone, methanol and ethanol solvents. *Photosynth. Res.* **2006**, *89*, 27–41. [CrossRef]
10. Tang, Z.; Zhang, W.; Huang, X.; Zhang, F.; Chen, J. Estimation Model of Soybean Yield Based on Ground Hyperspectral Remote Sensing. *Trans. Chin. Soc. Agric. Mach.* **2024**, *55*, 145–153+240. (In Chinese with English Abstract). [CrossRef]
11. Liu, S.; Yu, H.; Zhang, J.; Zhou, H.; Kong, L.; Huang, L.; Sui, Y. Research on Inversion Model of Soybean Leaf Chlorophyll Content Based on Optimal Spectral Index. *Spectrosc. Spect. Anal.* **2021**, *41*, 1912–1919, (In Chinese with English Abstract).
12. Zhao, J.; Zhang, C.; Min, L.; Guo, Z.; Li, N. Retrieval of Farmland Surface Soil Moisture Based on Feature Optimization and Machine Learning. *Remote Sens.* **2022**, *14*, 5102. [CrossRef]
13. Shi, H.; Guo, J.; An, J.; Tang, Z.; Wang, X.; Li, W.; Zhao, X.; Jin, L.; Xiang, Y.; Li, Z.; et al. Estimation of Chlorophyll Content in Soybean Crop at Different Growth Stages Based on Optimal Spectral Index. *Agronomy* **2023**, *13*, 663. [CrossRef]
14. Zhao, H.; Gan, S.; Yuan, X.; Hu, L.; Wang, J.; Liu, S. Application of a Fractional Order Differential to the Hyperspectral Inversion of Soil Iron Oxide. *Agriculture* **2022**, *12*, 1163. [CrossRef]
15. Nehela, Y.; Mazrou, Y.S.A.; Taha, N.A.; Elzaawely, A.A.; Xuan, T.D.; Makhlouf, A.H.; El-Nagar, A. Hydroxylated Cinnamates Enhance Tomato Resilience to Alternaria alternated, the Causal Agent of Early Blight Disease, and Stimulate Growth and Yield Traits. *Plants* **2023**, *12*, 1775. [CrossRef]
16. Li, S.; Peng, B.; Fang, L.; Li, Q. Hyperspectral Band Selection via Optimal Combination Strategy. *Remote Sens.* **2022**, *14*, 2858. [CrossRef]
17. Martinelli, F.; Scalenghe, R.; Davino, S.; Scuderi, G.; Ruisi, P.; Dandekar, A.M. Advanced Methods of Plant Disease Detection. A Review. *Agro. Sustain. Dev.* **2015**, *35*, 1–25. [CrossRef]
18. Feng, W.; Shen, W.; He, L.; Duan, J.; Guo, B.; Li, Y.; Guo, T. Improved Remote Sensing Detection of Wheat Powdery Mildew Using Dual-Green Vegetation Indices. *Precis. Agric.* **2016**, *17*, 608–627. [CrossRef]
19. Han, X.; Jiang, Z.; Liu, Y.; Zhao, J.; Sun, Q.; Li, Y. A Spatial–Spectral Combination Method for Hyperspectral Band Selection. *Remote Sens.* **2022**, *14*, 3217. [CrossRef]
20. Pang, L.; Xiao, J.; Ma, J.; Yan, L. Hyperspectral imaging technology to detect the vigor of thermal-damaged Quercus variabilis seeds. *J. Forestry Res.* **2021**, *32*, 461–469. [CrossRef]
21. Xiang, Y.; Wang, X.; An, J.; Tang, Z.; Li, W.; Shi, H. Estimation of Soybean Leaf Area Index Based on Fractional Differential and Optimal Spectral Index. *Trans. Chin. Soc. Agric. Mach.* **2023**, *54*, 329–342. (In Chinese with English Abstract)
22. Song, G.; Wang, Q. Coupling Effective Variable Selection with Machine Learning Techniques for Better Estimating Leaf Photosynthetic Capacity in a Tree Species (Fagus crenata Blume) from Hyperspectral Reflectance. *Agr. Forest Meteorol.* **2023**, *338*, 109528. [CrossRef]
23. Gamon, J.; Penuelas, J.; Field, C. A Narrow-Waveband Spectral Index that Tracks Diurnal Changes in Photosynthetic Efficiency. *Remote Sens. Environ.* **1992**, *41*, 35–44. [CrossRef]

24. Gitelson, A.; Merzlyak, M. Remote Estimation of Chlorophyll Content in Higher Plant Leaves. *J. Indian Soc. Remote.* **1997**, *18*, 2691–2697. [CrossRef]
25. Sims, D.A.; Gamon, J.A. Relationships Between Leaf Pigment Content and Spectral Reflectance Across a Wide Range of Species, Leaf Structures and Developmental Stages. *Remote Sens. Environ.* **2002**, *81*, 337–354. [CrossRef]
26. Tao, X.; Liang, S.; He, T.; Jin, H. Estimation of Fraction of Absorbed Photosynthetically Active Radiation from Multiple Satellite Data: Model Development and Validation. *Remote Sens. Environ.* **2016**, *184*, 539–557. [CrossRef]
27. Zheng, J.; Li, F.; Du, X. Using Red Edge Position Shift to Monitor Grassland Grazing Intensity in Inner Mongolia. *J. Indian Soc. Remote.* **2018**, *46*, 81–88. [CrossRef]
28. Blackburn, G.A. Hyperspectral Remote Sensing of Plant Pigments. *J. Exp. Bot.* **2007**, *58*, 855–867. [CrossRef]
29. Liu, S.; Li, L.; Fan, H.; Guo, X.; Wang, S.; Lu, J.W. Real-time and Multi-stage Recommendations for Nitrogen Fertilizer Topdressing Rates in Winter Oilseed Rape Based on Canopy Hyperspectral Data. *Ind. Crop. Prod.* **2020**, *154*, 112699. [CrossRef]
30. Cherkassky, V.; Ma, Y. Practical selection of SVM parameters and noise estimation for SVM regression. *Neural networks* **2004**, *17*, 113–126. [CrossRef]
31. Tang, Z.; Xiang, Y.; Wang, X.; An, J.; Guo, J.; Wang, H.; Zhang, F. Comparison of SPAD Value and LAI Spectral Estimation of Soybean Leaves Based on Different Analysis Models. *Soyb. Sci.* **2023**, *42*, 55–63. (In Chinese with English Abstract) [CrossRef]
32. Tang, Z.; Wang, X.; Xiang, Y.; Liang, J.; Guo, J.; Li, W.; Zhang, F. Application of Hyperspectral Technology for Leaf Function Monitoring and Nitrogen Nutrient Diagnosis in Soybean (*Glycine max* L.) Production Systems on the Loess Plateau of China. *Eur. J. Agron.* **2024**, *154*, 127098. [CrossRef]
33. Blanes, I.; Serra-Sagrista, J.; Marcellin, M.W.; Bartrina-Rapesta, J. Divide-and-Conquer Strategies for Hyperspectral Image Processing: A Review of Their Benefits and Advantages. *IEEE Signal Proc. Mag.* **2012**, *29*, 71–81. [CrossRef]
34. Yang, C.; Xu, J.; Feng, M.; Bai, J.; Sun, H.; Song, L.; Wang, C.; Yang, W.; Xiao, L.; Zhang, M. Evaluation of Hyperspectral Monitoring Model for Aboveground Dry Biomass of Winter Wheat by Using Multiple Factors. *Agronomy* **2023**, *13*, 983. [CrossRef]
35. Tang, Z.; Zhang, W.; Xiang, Y.; Li, Z.; Zhang, F.; Chen, J. Monitoring of Soil Moisture Content of Winter Wheat Based on Hyperspectral and Machine Learning Models. *Trans. Chin. Soc. Agric. Mach.* **2023**, *54*, 350–358, (In Chinese with English Abstract).
36. Sonobe, R.; Sugimoto, Y.; Kondo, R.; Seki, H.; Sugiyama, E.; Kiriiwa, Y.; Suzuki, K. Hyperspectral Wavelength Selection for Estimating Chlorophyll Content of Muskmelon Leaves. *Eur. J. Remote Sens.* **2021**, *54*, 513–524. [CrossRef]
37. Liu, J.; Li, Y.; Zhao, F.; Liu, Y. Hyperspectral Remote Sensing Images Feature Extraction Based on Spectral Fractional Differentiation. *Remote Sens.* **2023**, *15*, 2879. [CrossRef]
38. Li, C.; Li, X.; Meng, X.; Xiao, Z.; Wu, X.; Wang, X.; Ren, L.; Li, Y.; Zhao, C.; Yang, C. Hyperspectral Estimation of Nitrogen Content in Wheat Based on Fractional Difference and Continuous Wavelet Transform. *Agriculture* **2023**, *13*, 1017. [CrossRef]
39. Qian, B.; Ye, H.; Huang, W.; Xie, Q.; Pan, Y.; Xing, N.; Lan, Y. A Sentinel-2-Based Triangular Vegetation Index for Chlorophyll Content Estimation. *Agr. Forest Meteorol.* **2022**, *322*, 109000. [CrossRef]
40. Zhang, J.; Tian, H.; Wang, D.; Li, H.; Mouazen, A.M. A Novel Spectral Index for Estimation of Relative Chlorophyll Content of Sugar Beet. *Comput. Electron. Agric.* **2021**, *184*, 106088. [CrossRef]
41. Van der Werff, H.M.A.; Van der Meer, F.D. Shape-Based Classification of Spectrally Identical Objects. *ISPRS J Photogramm.* **2008**, *63*, 251–258. [CrossRef]
42. Inoue, Y.; Moran, M.S.; Horie, T. Analysis of spectral measurements in paddy field for predicting rice growth and yield based on a simple crop simulation model. *Plant Prod. Sci.* **1998**, *1*, 269–279. [CrossRef]
43. Yuan, H.; Yang, G.; Li, C.; Wang, Y.; Liu, J.; Yu, H.; Feng, H.; Yang, X. Retrieving soybean leaf area index from unmanned aerial vehicle hyperspectral remote sensing: Analysis of RF, ANN, and SVM regression models. *Remote Sens.* **2017**, *9*, 309. [CrossRef]
44. Zhao, L.; Hu, Y.; Zhou, W.; Liu, Z.; Pan, Y.; Shi, Z.; Wang, L.; Wang, G. Estimation Methods for Soil Mercury Content Using Hyperspectral Remote Sensing. *Sustainability* **2018**, *10*, 2474. [CrossRef]
45. Yang, H.; Hu, Y.; Zheng, Z.; Qiao, Y.; Zhang, K.; Guo, T.; Chen, J. Estimation of Potato Chlorophyll Content from UAV Multispectral Images with Stacking Ensemble Algorithm. *Agronomy* **2022**, *12*, 2318. [CrossRef]

plants

MDPI

Review

L-Band Synthetic Aperture Radar and Its Application for Forest Parameter Estimation, 1972 to 2024: A Review

Zilin Ye [1,2,3], Jiangping Long [1,2,3], Tingchen Zhang [1,2,3], Bingbing Lin [1,2,3] and Hui Lin [1,2,3,*]

1 Research Center of Forestry Remote Sensing & Information Engineering, Central South University of Forestry and Technology, Changsha 410004, China; loyzer@163.com (Z.Y.); longjiangping@csuft.edu.cn (J.L.); 20220100015@csuft.edu.cn (T.Z.); lbb15623238233@outlook.com (B.L.)
2 Key Laboratory of Forestry Remote Sensing Based Big Data & Ecological Security for Hunan Province, Changsha 410004, China
3 Key Laboratory of State Forestry Administration on Forest Resources Management and Monitoring in Southern Area, Changsha 410004, China
* Correspondence: t19911090@csuft.edu.cn; Tel.: +86-0731-8562-3848

Abstract: Optical remote sensing can effectively capture 2-dimensional (2D) forest information, such as woodland area and percentage forest cover. However, accurately estimating forest vertical-structure relevant parameters such as height using optical images remains challenging, which leads to low accuracy of estimating forest stocks like biomass and carbon stocks. Thus, accurately obtaining vertical structure information of forests has become a significant bottleneck in the application of optical remote sensing to forestry. Microwave remote sensing such as synthetic aperture radar (SAR) and polarimetric SAR provides the capability to penetrate forest canopies with the L-band signal, and is particularly adept at capturing the vertical structure information of forests, which is an alternative ideal remote-sensing data source to overcome the aforementioned limitation. This paper utilizes the Citexs data analysis platform, along with the CNKI and PubMed databases, to investigate the advancements of applying L-band SAR technology to forest canopy penetration and structure-parameter estimation, and provides a comprehensive review based on 58 relevant articles from 1978 to 2024 in the PubMed database. The metrics, including annual publication numbers, countries/regions from which the publications come, institutions, and first authors, with the visualization of results, were utilized to identify development trends. The paper summarizes the state of the art and effectiveness of L-band SAR in addressing the estimation of forest height, moisture, and forest stocks, and also examines the penetration depth of the L-band in forests and highlights key influencing factors. This review identifies existing limitations and suggests research directions in the future and the potential of using L-band SAR technology for forest parameter estimation.

Keywords: L band SAR; Citexs data; forest canopy penetration

Citation: Ye, Z.; Long, J.; Zhang, T.; Lin, B.; Lin, H. L-Band Synthetic Aperture Radar and Its Application for Forest Parameter Estimation, 1972 to 2024: A Review. *Plants* **2024**, *13*, 2511. https://doi.org/10.3390/plants13172511

Academic Editor: Stefano Martellos

Received: 28 June 2024
Revised: 16 August 2024
Accepted: 22 August 2024
Published: 7 September 2024

1. Introduction

The rapid advancement in technologies of satellites [1], unmanned aerial vehicles (UAVs) [2] and radar [3] has facilitated the acquisition of fast and precise ground information for forest resource surveys. Consequently, the reliance on remote sensing technologies in forestry has significantly increased. Remote sensing technologies now enable accurate extraction of two-dimensional (2D) forest information such as woodland area and percentage forest cover, along with other parameters. However, due to the limited penetration ability of electromagnetic waves and the complexity of forest canopies, obtaining accurate information of vertical structures remains challenging. Estimating forest canopy height and related parameters such as stock, biomass, and carbon storage continues to pose difficulties. Therefore, acquiring comprehensive information of forest vertical structures represents a major obstacle that hampers the widespread application of remote sensing technologies in forestry.

Spaceborne microwave signals used by Synthetic Aperture Radar (SAR) and polarimetric SAR possess the capability to penetrate forest canopies. Particularly, L-band SAR signals exhibit strong forest canopy-penetration ability and sensitivity [4] to forest vertical structures. Thus, the L-band SAR offers a valuable means of acquiring extensive forest vertical-structure information across multiple frequencies and serves as an optimal remote sensing data source for addressing this "bottleneck". However, current research findings indicate that the estimation results of forest height using L-band SAR data, combined with InSAR or PolInSAR technology, are still not satisfied. This is primarily attributed to the unclear understanding of the penetration process and response mechanism of the L-band SAR signal in forests, as well as the lack of clarity regarding its penetration depth in different types of forests. The quantitative description of transmittance and behavior patterns exhibited by the L-band SAR signal within forest canopies remains elusive, impeding the establishment of a reliable model for determining the L-band SAR signal's penetration depth [5]. Therefore, it becomes imperative to characterize the attenuation process and penetration depth of the L-band SAR signal based on information of forest canopies obtained from air-ground cooperative microwave radiometers and LiDAR data. Additionally, interpreting the forest penetration mechanism associated with the L-band SAR signal will enable us to develop a theoretical framework and methodology for retrieving accurate estimates pertaining to forest penetration depth of the L-band, and thus effectively overcome the existing limitations related to extracting accurate information of forests through remote sensing technologies.

2. Literature Analysis of Forest Penetration of L-Band SAR Signal

2.1. Introduction of L-Band SAR Satellites

In 1978, the United States launched the first Ocean Satellite (Seasat) with the L-band SAR sensor. Then, Japan launched the JELLS-1 satellite in 1992 and the Advanced Land Observing Satellite (ALOS) in January 2006 with L-band SAR sensors and a maximum positioning accuracy of 10 m. In November 2009, the European Space Agency launched the Soil Moisture and Ocean Salinity Satellite (SMOS), which can emit L-band radiation energy to the ground and provide 9 km spatial resolution products. In January 2015, the United States launched the Active and Passive Soil Moisture Monitoring Satellite (SMAP) with an L-band SAR sensor, which can provide global moisture products with a resolution of 40 km × 40 km. In October 2018, Argentina launched the SAOCOM 1A satellite with L-band and a spatial resolution of 10 m × 10 m to 100 m × 100 m. In January 2022, China launched the Land Probe 1 Group 01 A/B satellite (LT-1A/1B), also known as the L-band differential Interferometric SAR satellite, with spatial resolutions of 3 m × 3 m, 6 m × 6 m, 12 m × 12 m, 20 m × 20 m and 30 m × 30 m. A summary of the launched satellites with L-band is shown in Table 1.

Table 1. SAR satellites with L band in the world.

Serial Number	Launch Time	Country or Region	Satellite Name
1	February 1978	Global Positioning System	NAVSTAR GPS
2	June 1978	American Seasat satellite	Seasat
3	February 1992	Japan Earth Resource Satellite	JERS-1
4	January 2006	Japan Advanced Land Observing Satellite	ALOS/ALOS-2
5	November 2009	European Space Agency Soil Moisture and Ocean Salinity Satellite	SMOS
6	January 2015	United States Active and Passive Soil Moisture Monitoring Satellite	SMAP
7	October 2018	Argentine Microwave Observation Satellite 1A	SAOCOM-1A
8	September 2019	China Yunhai-1 02 satellite	Yunhai-1 02
9	August 2020	Argentine Microwave Observation Satellite 1B	SAOCOM-1B
10	January 2022	China Landexplorer-1 Group 01A satellite	LT-1A
11	January 2022	China Landexplorer-1 Group 01 B satellite	LT-1B

Compared with the number of other remote sensing satellites, L-band SAR satellites are very few, and even fewer are in orbit. Thus, it is relatively difficult to obtain L-band data. The emergence of LT-1 has started to change this situation.

2.2. Trend in Annual Publications

Based on the Citexs comprehensive literature database, this paper adopted the bibliometrics method to carry out mining of the literature big database by selecting year, country, institution, author, journal, and so on. Moreover, analyzing and visualizing the general trend and distribution in this field was conducted. Using L band SAR and forest penetration as keywords, we identified 0 papers from January 1978 to April 2024 in the CNKI database and 58 papers from January 1978 to April 2024 in the PubMed database. The average annual number of the papers published was three, and the annual numbers of the published papers are shown in Figure 1.

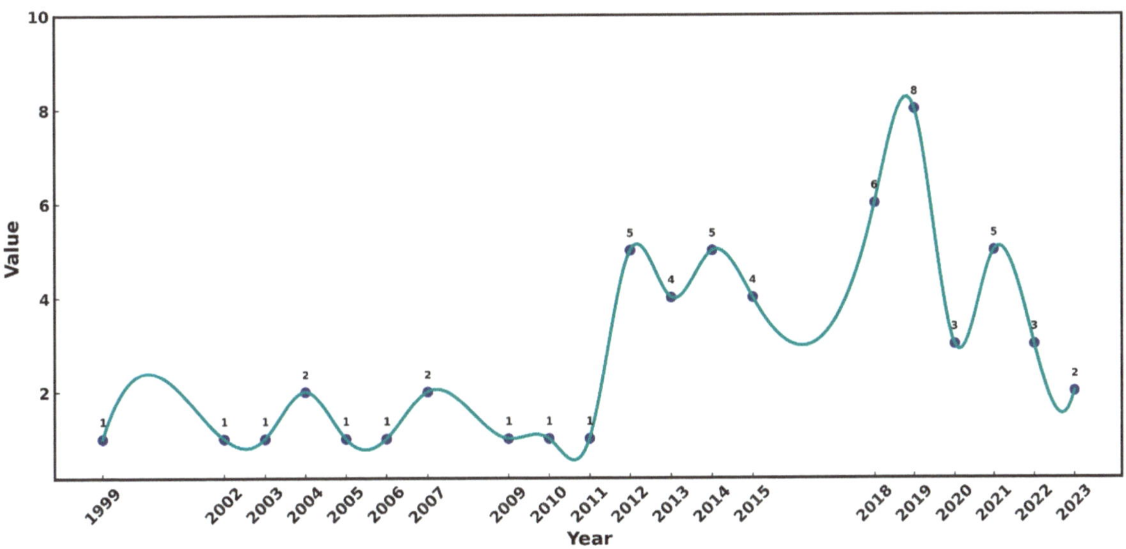

Figure 1. SCI published papers from 1978 to 2024 dealing with L-band SAR data and related to forest penetration.

Figure 1 shows that the publications related to the use of L-band SAR data in the field of forest canopy penetration first appeared in 1999, indicating a relatively late start. From 1999 to 2011, there is only one SCI paper published each year, which indicates the stage of slow development. From 2012 to 2023, the annual number of the published papers increased to more than two, reaching the peak of eight in 2019. The fastest growth happened from 2012 to 2019, indicating that the research in this field was in a rising stage of rapid development, and then a decreased trend was found.

2.3. Publications by Country and Region

From January 1978 to April 2024, the distribution of the top 23 countries/regions in the studies of L band SAR data and forest canopy penetration in the world is shown in Figure 2. The countries/regions in which there is the largest number of publications were China (13 papers, 22.41%), then Germany (7 papers, 12.07%) and France (7 papers, 12.07%).

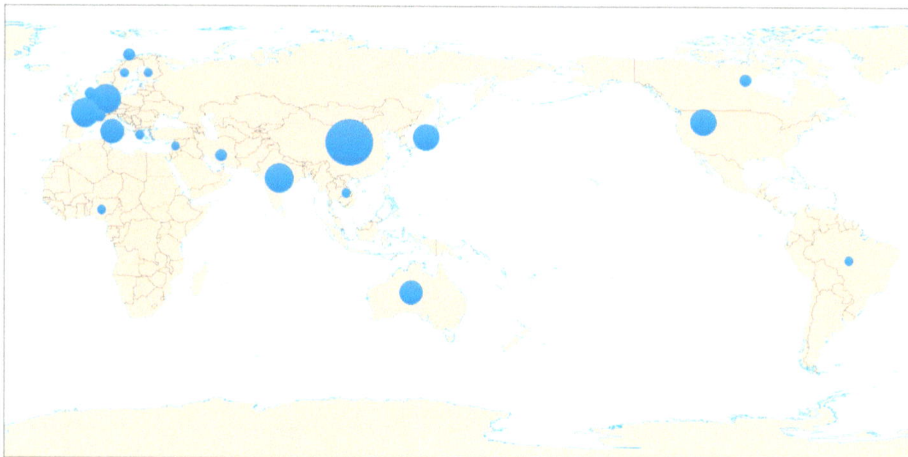

Figure 2. The distribution of the SCI publications related to the use of L-band SAR data and forest canopy penetration.

2.4. Publications by Research Institution

The top-20 national research institutions in which the authors published their papers in the field of L band SAR data applications and forest canopy penetration from January 1978 to April 2024 are shown in Figure 3. The institutions can be divided into three groups. The top group includes the Chinese Academy of Sciences, the German Aerospace Center, the Japan Aerospace Exploration Agency and the Indian Institute of Remote Sensing, in which there are three papers in this field for each institute. The second group consists of 12 research institutions, each publishing two articles in this field. The last group is composed of four institutes in which there is only one publication each.

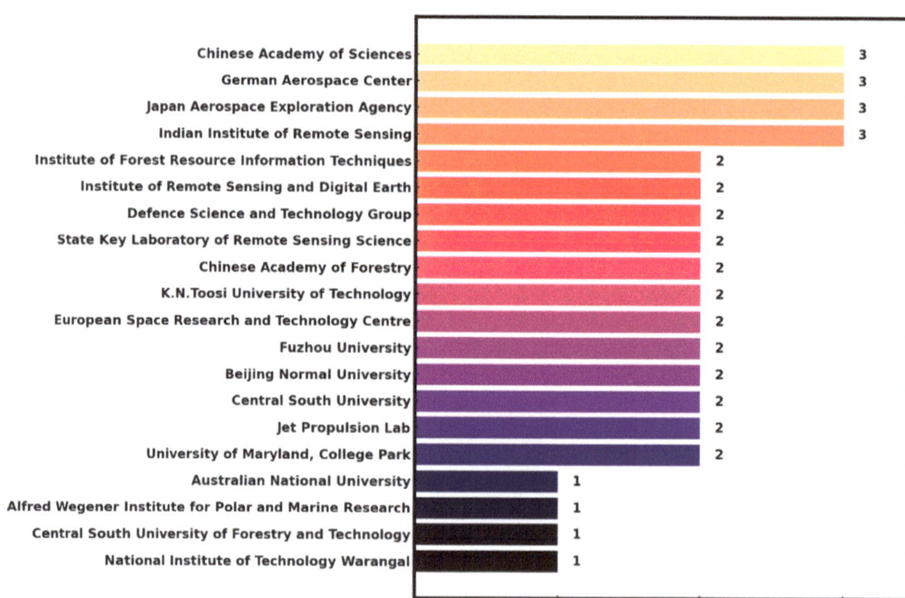

Figure 3. Top-20 research institutions that published SCI papers dealing with L-band SAR data and forest canopy penetration.

2.5. Publications by the First Author

From January 1978 to April 2024, the world's top-30 first authors who studied the use of L band SAR data and forest canopy penetration are shown in Figure 4. The author who produced the largest number of the papers in this field is Mark L. Williams, with three papers in total. Yasser Maghsoudi, Matteo Pardini, Mohammad Javad Valadan Zoej, Marco Lavalle, Tayebe Managhebi and Masanobu Shimada tied for second place, with two publications. The authors with one publication include Liming Jiang, Om Prakash Tripathi, A.K. Milne, Junli Chen, Chadi Abdallah, J. Jomaah, Masato Hayashi, Jyotishman Deka, Michael F. Toups, Kiran Dasari, P. S. Roy, Fulong Chen, Qingwei Tong, A.C. Lee, Wei Li, Ruixia Yang, Takeo Tadono, Shashi Kumar, Lal Bihari Singha, Jean Luc Betoulle, Nicolas Baghdadi, E. Mougin, and Van Nhu Le.

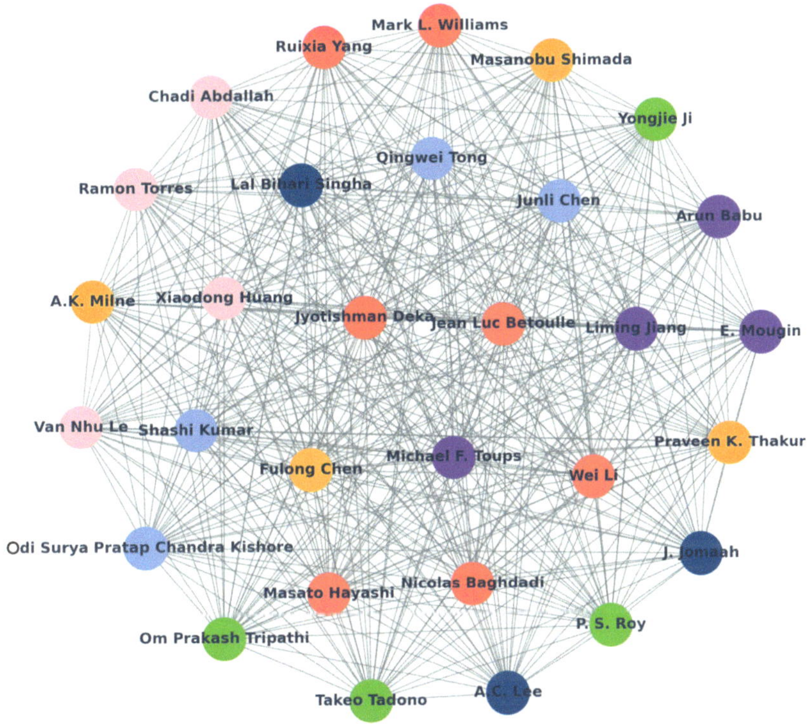

Figure 4. The-top 30 first authors who published SCI articles in the field of using L-band SAR data and dealing with forest canopy penetration.

2.6. Publication by Journal

From January 1978 to April 2024, the top 22 journals in terms of the publications dealing with L-band SAR data and forest canopy penetration are shown in Figure 5. *Remote Sensing* is the journal with the largest number of publications (10). *IEEE Transactions on Geoscience and Remote Sensing* ranked second, with three articles, and *Remote Sensing of Environment* ranked third, with two papers. There are another 19 journals with one publication.

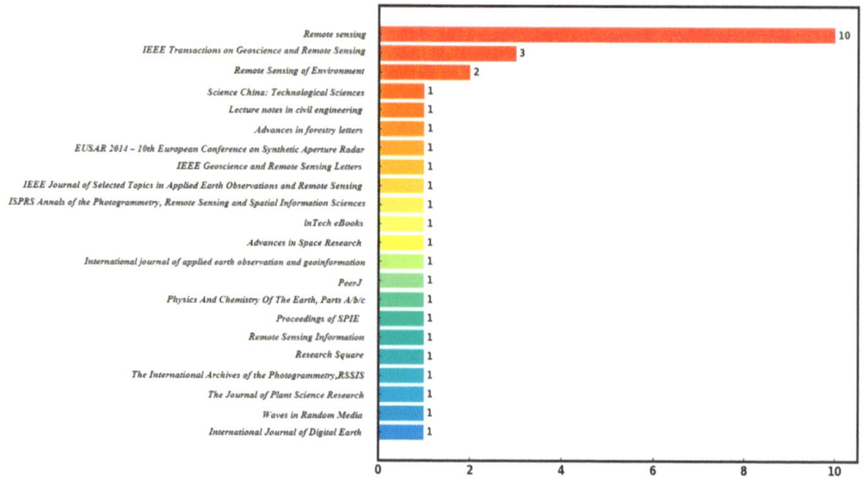

Figure 5. Publications by journal in the field of using L-band SAR data and dealing with forest canopy penetration.

3. Applications of L-Band SAR Data in Forestry

L-band SAR data have been used in forestry. Most of the data come from ALOS satellites with different polarization modes. The main applications deal with estimations of forest canopy height, moisture and forest stocks.

3.1. Forest Height

Kugler successfully estimated and verified the potential of L-band, P-band and X-band to estimate tree height in temperate forests by using airborne SAR sensors combined with constraints [6,7]. Zhang et al. used the improved RMoG model and ALOS-1 data to invert forest height with the estimation error reduced by 27.73% and 8.57% Asopa et al. used UAV SAR technology to estimate the tree height of tropical forests with a root mean square error (RMSE) of 4.21 m [8]. Huang et al. estimated tree height by using L-band SAR data and generated a digital terrain model (DTM) and digital surface model (DSM) validated by using UAV LiDAR data [9]. Thieu et al. proposed a new algorithm based on a mean set to increase phase, and combined it with the polarization characteristics of the VE-RVoG optimized set, developed to improve the estimation of forest height, and obtained an RMSE of 2.91 m and a correlation coefficient of 0.909 [10]. Xie et al. improved the accuracy of forest height estimation by using the new airborne PolInSAR data-processing strategy; the RMSE was significantly reduced, to 1.02 m, with a decrease of 12.86%, providing a feasible solution for forest height estimation with X-band waves [11]. Based on L-band single-baseline pooled SAR interferometric simulation data, Sui et al. proposed a standard scale optimization model suitable for various densities, successfully overcoming the failure of traditional methods in low-density regions, and effectively realized the estimation of forest height [12].

Moreover, Luo et al. used UAV SAR multi-baseline L-band data from the AfriSAR project to show better accuracy in forest height estimation, providing an improved method for estimating structural parameters of tropical rain forests [13]. Luo et al. conducted an estimation experiment using L-band multi-baseline fully polarized data from the AfriSAR project in the Lope Pongara pilot area and proposed a depth-based error-correction method that improved the accuracy of forest height estimation and demonstrated potential applications [14] of machine learning-interference feature prediction. Zhang et al. used the simulated L-band SAR data and combined it with the improved three-stage method to derive forest height with a significantly improved accuracy [15]. Integrating the TF-RVoG method based on time–frequency analysis and the improved single-baseline data decomposition method significantly improved the estimation accuracy of the forest canopy height model, with the RMSE dropping to 2.54 m [16]. Zang et al. combined ICESat-2 data and tree age information to propose a method to estimate the change in palm tree height in Peninsular Malaysia, and successfully produced a comprehensive map of tree height change from 2001 to 2020 [17]. The verification results showed that the estimated height was highly consistent with the actual height. Providing a spatially explicit tool with great potential for quantifying plantation stocks, Sa and Nei et al. used ALOS-2 L-band data to retrieve conifer tree height in Saihanba, Hebei Province and obtained an R^2 of 0.67 between the SAR-based estimated and the LiDAR-based conifer tree height [18]. These studies show that L-band SAR data have great potential in forest height estimation, but the estimation accuracy cannot meet the needs of forestry production.

3.2. Moisture

Grant et al. used an airborne L-band microwave radiometer to study the effect of forest cover on soil water estimation in Australia and utilized the L-MEB zero-order radiative transfer model to simultaneously estimate soil water and vegetation optical depth [19]. Richaume et al. found that SAR signal was highly correlated with the optical depth, roughness and canopy density of vegetation by using SMOS for large-scale moisture estimation, and the Hr value of the spatial pattern of soil moisture was correlated with land-cover types. Their results demonstrated that the evergreen broad-leaved mixed forests and the deciduous coniferous mixed forests had higher values of Hr, ranging from 0.32 to 0.39, desert, shrub and bare soil had lower values of Hr, ranging from 0.14 to 0.16, and the Hr values of grassland and tundra cultivated land changed to between 0.20 and 0.23 [20]. Konings et al. used SMAP data and a multi-temporal dual-channel retrieval algorithm (MT-DCA) to estimate the optical depth, moisture and reflectance of large-scale vegetation [21]. Lv et al. studied the relationship between optical depth, penetration and temperature of vegetation [22]. Since the L-band is sensitive to moisture, it is often used to estimate forest moisture. Holtzman et al. used L-band radiometer towers in Red Oak forests in Massachusetts, USA, to prove that the optical depth of vegetation measured by microwave radiometers is correlated with the amount of water in vegetation [23]. These studies indicate that the optical depth of vegetation is an important index to evaluate the microwave signal transmission process in forest canopy.

3.3. Forest Stocks

Forest stocks include stand volume, above-ground biomass (AGB) and carbon stocks. Balzter et al. estimated changes [24] in stand volume in Thetford, United Kingdom from 1910 to 1997 using Seasat and JELLS-1 satellite data. Santoro et al. used JELLS-1 L-band SAR data to study volume of forests in Sweden, Finland and Siberia, and achieved an estimation accuracy of greater than 75% [25]. Chowdhury et al. used ALOS L-band data to estimate forest volume in Siberia and obtained an R^2 of about 0.60 between the estimated and observed values, with an accuracy greater than 70% [26]. Santoro et al. conducted a comprehensive assessment of forest volume by using L-band ALOS data from 2006 to 2011; they found that HH-polarized SAR data had a good estimation effect, and obtained an error of less than 30% when the area was larger than 20 hectares [27,28]. Thiel et al. used ALOS data to estimate forest volume in central Siberia; the R^2 between the estimated and measured values reached 0.58, and the estimation accuracy was greater than 70% [29]. Christian Thiel et al. employed ALOS PALSAR L-band to estimate forest volume in central Siberia, and demonstrated that HV backscattering achieved a slightly higher accuracy than HH. They also found that the simple inversion method, coupled with multi-temporal SAR images, performed well in feature correction, providing a feasibility study for forest resource estimation in this region. Santoro et al. compared the forest volume-estimation potential of SAR data in X-, C- and L-bands, and demonstrated that L-band led to the highest accuracy, with a relative error of 31.3% [30]. Zhang et al. used ALOS-2 (PALSAR-

2) data to estimate forest volume in Huangfengqiao Forest Farm, Hunan Province, and obtained an R^2 of 0.61 [31].

Sandra Englhart et al. used X-L band SAR data to estimate forest AGB in Indonesian Borneo, and the results showed that the X-L band was suitable for the estimation when AGB was low, with an R^2 of 0.53 [32]. Oliver Cartus et al. used ALOS PALSAR data to carry out regional scale mapping of forest biomass in northeastern United States. They combined SAR data and optical remote-sensing calibration models, and the results showed that the accuracy and performance of the method was superior to the results from using SAR data alone, which are dependent on imaging conditions [33]. Peregon and Yamagata used L-band SAR data to carry out AGB estimation on deciduous forests in Western Siberia, and obtained an R^2 of 0.72 with an estimation accuracy of 85% [34]. Rahman et al. analyzed different observation models of ALOS PALSAR data, and conducted a regression analysis and estimation of natural forests in southeast Bangladesh [35]. Chaparro et al. studied the use of C- and X-band vegetation optical depth to estimate forest biomass and carbon balance, and concluded that vegetation optical depth from the L-band provided more accurate information because the penetration of microwaves through the canopy is higher at longer wavelengths and lower frequencies [36]. Berninger et al. used L-band and C-band SAR data for large-scale AGB monitoring, providing important information for accurately portraying forest loss [37,38]. Liu et al. compared airborne P-band and L-band TomoSAR measurements of the canopy-height model (CHM) and AGB over a tropical forest in Lope, Gabon, and found that the results of the CHM did not significantly differ, while the P-band was more sensitive than the L-band in the estimation. Maciej J. Soja et al. used P-band SAR data to estimate AGB in tropical forests and obtained an accuracy of 80%, based on the field measurements from 141 plots [39,40].

Ji et al. studied the sensitivity of L-band SAR backscattering with respect to forests with conditions of different mean canopy densities, different mean tree height, and different mean diameter at breast height (DBH), and found that the way of backscattering affected the improvement in biomass estimation accuracy. Hernandez-Stefanoni et al. improved the AGB map of tropical arid forests by integrating LiDAR, ALOS PALSAR, and climate data, and reduced the relative error of biomass estimation by 12.2% [41]. Zhang et al. used L-band ALOS data to estimate the AGB of Chinese fir forests in Huangfengqiao Forest Farm, Hunan Province, by extracting multiple rotation thresholds, and realized an estimation accuracy of 77.5% [42]. Ni et al. estimated the biomass of deciduous forests in mountainous areas with three-dimensional (3D) data and found that the season had a significant impact on the estimation results [43]. These studies show that the L-band has the potential to estimate forest stock and biomass, but the accuracy of estimation results is generally lower than 80%, due to inaccurate information of forest vertical structures.

4. Penetration of L-Band Signal and Its Influencing Factors

4.1. Penetration of L-Band Signal

Dal pointed out that microwave signals can penetrate vegetation and that the elevation bias caused by penetration is not exactly equal to the penetration depth, then proposing a penetration model [44]. Pardini and Papathanassiou carried out a forest canopy-penetration experiment using L- and P-band data, and their results showed that SAR penetration ability was closely related to band and canopy structure [45]. In Brazivella, Congo, Toochi et al. investigated the penetration capacity of six bands, including the K-band (1 cm), X-band (3 cm), C-band (5.6 cm), S-band (10 cm), L-band (23 cm), and P-band (75 cm), and further confirmed that the penetration of short wavelengths (X, K) was low and that the penetration depth was also dependent on the vertical structures of the forests [46]. Reginald R. Muskett progressively buried mesh reflectors underground in the Alaskan tundra, USA, and quantified the depth of L-band penetration into the soil [47]. Schlund et al. conducted penetration-depth and compensation experiments for temperate forests using an X-band based on LiDAR data. The estimated RMSE was less than 1 m, indicating great potential [48]. Teubner et al. explored the relationship between vegetation optical depth and gross primary production (GPP), and found that, overall, GPP was negatively correlated with vegetation optical depth in predominantly occurring both wet and dry areas, and that the correlation was similar to higher SIF [49].

Tanase et al. evaluated the effectiveness of C-band, L-band and P-band SAR sensors in Romanian coniferous forests using simulation models, and their results showed that different bands had different sensitivity to vegetation characteristics and disturbances. The authors emphasized the need for the comprehensive use of multi-band, dense time series and different types of sensors to compensate for the limitations of a single frequency and acquisition time [50]. Chaparro et al. quantified the contribution of ACD proportional vegetation optical depth/enhanced vegetation index signals, and their results confirmed an enhancement compared to higher frequency bands, indicating that the penetration depths of all bands were reduced in the densest forests. The 34% and 30% of variance could be explained with the proportional decrease in C- and X-band vegetation optical depth [51], respectively. Colliander et al. used airborne L-band SAR data to carry out a two-year forest-soil-moisture experiment under forest canopies in the northeastern United States [52]. Singh et al. found that the penetration depth of microwave signals into the ground varied significantly with the available content of soil moisture, with longer wavelengths having the stronger ability to penetrate the soil; however, the penetration capacity would be reduced as soil moisture content increased [53]. Liu et al. used ALOS L-band data to conduct penetration studies in extremely arid desert areas, and their results showed that the penetration depth of the L-band reached 2.98 m [54]. Qi et al. conducted an additional reference-height error analysis for baseline calibration based on a distributed Target DEM in TwinSAR-L in the arid region of eastern Xinjiang. Their results showed errors of 1.295 m and 1.39 m, respectively, which seriously reduced the product quality [55].

Wang et al. selected AGB, the leaf area index (LAI), and the normalized difference vegetation index (NDVI) to optimize effective scattering albedo (ω), surface roughness, and for estimation (VODini). When LAI was greater than 20.76 and NDVI larger than 0.83, the results were significantly improved, especially for dense vegetation [56]. Zhu et al. used ALOS data and a deep-learning algorithm to study penetration depth in desert areas, and found that the maximum penetration depth of the L-band reached 2.84 m, and that the penetration was also related to scattering coefficient, dielectric constant, surface roughness and mineral composition [57]. The simulation of Bai et al. showed the vegetation optical depth increased linearly with the decrease in LAI, while the results were similar to those from the satellite based L band, C-band and X-band. Their sensitivity tests indicated that polarization dependence become more pronounced at higher frequencies [58]. Olivares-Cabello et al., through unsupervised classification analysis on a global scale, found that the L-band is suitable for monitoring dense canopies, while X-band and LCX vegetation optical depths are more suitable for sparse tree canopy, savannas and grasslands [59]. Schmidt et al. analyzed the relationship between live fuel coverage and vegetation optical depth through random forest regression, using multi-band datasets and soil-moisture—marine-salinity sensors, providing important guidance for selecting suitable wavelengths for specific applications and algorithm development [60]. Baur et al. studied the attenuation characteristics of L-, C- and X-bands with respect to the conditions of different land-cover types, and concluded that shrub had a high transient peak value, while forest canopy had a low value [61]. In the areas with strong seasonal rainfall, the seasonal amplitude was greater in the C-band than in the L-band. The penetration characteristics of the L-band in various conditions and its comparison with the C-band and X-band are summarized in Table 2.

Table 2. Penetration characteristics of L-band in various conditions and its comparison with C-band and X-band (Note: √√—very good; √—good; *—not good).

Observed Object	L-Band	C-Band	X-Band
Sea Ice	*	√	√√
Snow (type and thickened layer)	√√	√√	√√
Soil moisture	√√	√√	√√
Soil roughness	√	√√	√√
Soils	√√	√	*
Water-land boundaries	√√	√	√√
Vegetation	√√	√√	√√
Vegetation moisture	√√	√√	√√
Ocean	√√	√√	*
Geological structure, structure	*	√	√
Desert underground	√√	√	*

4.2. Influencing Factors in L-Band Penetration

Singh et al. used the functional relationship between incidence angle and ground penetration depth and found that penetration ability was related to incidence angle, wavelength, and soil characteristics, including water moisture content and structural composition [62]. The longer the wavelength, the stronger the penetration ability, but the penetration ability decreased with soil depth. Ji et al. studied the sensitivity of different canopy densities, mean stand height, and mean diameter at breast height to L-band backscattering [63]. They concluded that canopy density had a greater effect on L-band backscattering than average height and mean diameter at breast height. HV is more sensitive to forest structural parameters than HH, also depending on the tree species. Richaume et al. found that the L-band signal was not only related to the optical depth of vegetation, but also to the canopy roughness of tree species and canopy density. Zhu et al. found that the L-band was related to ground roughness and even mineral composition, such as hematite reducing the penetration depth. These studies indicate that there are many factors affecting the penetration of the L-band in forests, mainly including incidence angle, polarization mode, forest canopy density, tree species (canopy roughness), crown height (age structure) and moisture. Moreover, slope, season, meteorological factors, forest LAI, leaf direction and so on, also affect the penetration effect.

5. Future Development

5.1. Integration of L-Band SAR Data and the Tomography Algorithm

Cazcarra-Bes et al. used TomoSAR technology to process L-band forest data obtained by monitoring at different times, extracted the distribution and spatial patterns of forests by a compressed sensing approach, and further utilized two complementary search methods to find the local maximum value and reconstruct the spatial patterns, so as to realize the inference of forest structure information and the assessment of the effect of delineating forests [64]. Minh et al. found that in the forests with a height of above 30 m and biomass up to 500 t/hm^2, the strong ground return of P-band can be seen in the tomography images [65]. Tello et al. used L-band SAR data in Trauenstein, Germany, to provide forest vertical-structure information, coupled with high spatial- and temporal-resolution images; they developed TomoSAR technology and reconstructed 3D models, and the results were verified based on airborne LiDAR data [66]. This experiment opened the door for 3D-based forest monitoring. Moussawi et al. compared P-band and L-band TomoSAR profiles, using the Land Vegetation Ice Sensor (LVIS), and discrete-return LiDAR to monitor and estimate tropical forest-structure parameters [67]. The extracted radar reflectance yielded RMSE values of 3.02 m and 3.68 m for P-band and L-band, respectively, and the corresponding determination coefficients were 0.95 and 0.93. Pardini et al. reviewed the features of L-band TomoSAR reconstruction and discussed the unique ability of reconstructing radar reflectance using TomoSAR to reveal the 3D structure and temporal changes of forests [68]. The authors emphasized the importance of penetration sensitivity to vegetation elements.

5.2. Integration of L-Band and P-Band

Minh D et al. believed that the further development of integrating the L-band and P-band would make it possible to use TomoSAR technology to more accurately extract forest vertical structures, which would not only solve the problem of forest classification, but also provide strong support for the next generation of Earth Explorer BIOMASS mission [69]. Lope, Gabon, Liu et al. compared the CHM and AGB models of a tropical forest obtained by an airborne P-band and L-band synthetic (SAR) Tomosar tomography [70]. In the forests located in Paracou, French Guiana and South America, Ngo et al. analyzed the applications of airborne P-band TomoSAR and LiDAR, showing that both could directly lead to high-resolution surface, height, and profile models. The results demonstrated that airborne-based products had higher quality due to stronger penetration. For the forest of an average height of 30 m at Paracou, a RMSE of less than 5 m for tree height estimation was obtained [71]. Moreover, Chuang et al. proposed a robust TomoSAR imaging procedure to obtain local high-resolution L-band images of forests for the areas of interest [72].

The aforementioned studies demonstrate the strong penetration of L-band SAR signal in forests, and reveal the key factors that affect the penetration ability of the L-band SAR signal in forests. However, the studies did not explain the attenuation mechanism of the SAR signal in forest canopies and also did not account for the response process of microwave transmittance and penetration depth. In the estimation of forest height, the improvement in estimation accuracy was mainly achieved by developing algorithms and enhancing computation performance. Obtaining the results, to some extent, was fortuitous.

In addition, the existing experiments focus mainly on temperate and boreal forests, and rarely take place in subtropical forests. Compared with the temperate and boreal forests, the structures of subtropical forests are more complex. Generally, subtropical forests often consist of three layers including tree, shrub, and grass, and have higher canopy closure and moisture inside the forests, which will lead to stronger impacts on the penetration of signals. Therefore, the results from the temperate and boreal forests will be less effective, and their applicability is limited in subtropical forests.

L-band SAR data provide the potential for advancing forestry remote sensing technologies by exploring forest vertical structures from the canopy surface to the interior of forests and from 2D- to 3D-model reconstruction l [73], and are also critical in realizing the technologies of TomoSAR and penetrating remote sensing in the future [74].

6. Conclusions and Discussion

Through bibliometric analysis of the related literature, this paper reviews the application status of L-band SAR data and their penetration in forests. The main conclusions are drawn as follows:

(1) The number of the publications is related to the availability of L-band SAR data. The earliest launched L-band SAR satellite took place in 1978; thus, this paper deals with the activities from 1972 to 2024. The first publication reviewed appeared in 1999, which indicates that, compared with other remote sensing technologies, the development of L-band-related technology started late. From 1999 to 2011, there was only one publication coming out each year, and less attention was paid. From 2012 to 2024, the number of the relevant publications increased rapidly and L-band SAR data (including airborne data) began to appear widely.

(2) The lack of L-band SAR data impedes their application. Compared with a large number of other remote-sensing data, L-band SAR data are relatively less available (see Table 1). Some satellites have ceased operation and others have just appeared recently, with

data such as SAOCOM and LT. There has been no relevant literature found. The existing publications mainly deal with the use of Japan's JERS and ALOS data. With the emergence of SAOCOM, LT and UAV data, this situation should be changed in the near future.

(3) The existing reports concentrate mainly on the temperate and boreal forests, while there have rarely been relevant L-band studies conducted in the vast subtropical forests. This situation is mainly related to the countries/regions in which the relevant research was carried out. The countries/regions with most of the existing studies are mainly distributed in the temperate and boreal zones, including China (mainly in the Chinese Academy of Sciences), Germany and France.

(4) The state of the art in applications of L-band SAR data: the existing studies show that the L-band signal has strong penetration in snow, soil moisture, water–land boundaries, vegetation, vegetation moisture, and in the ocean and desert underground. The applications of L-band SA data in forestry focus mainly on the estimation of forest height, moisture and forest stocks. The factors affecting the penetration of the L-band in forests mainly include forest-canopy closure, tree species, crown height and moisture. Slope, seasonality, meteorological features, LAI and leaf direction also have influence on the penetration.

(5) Directions for future efforts: the existing literature deals mainly with applied studies of L-band data, while theoretical research studies hardly exit. There are a lack of studies related to the mechanism of L-band signal working in various forests and the interactions between the L-band signal and the forests, especially subtropical forests. For example, the penetration capacity and attenuation process and characteristics of the L-band signal in various forests are unknown. The penetration depth, the transmission rate and response process of the L-band in different forests are also unclear. Moreover, there have rarely been reports dealing with the compensation mechanism of using L-band SAR data to estimate forest height. The gaps that exist currently imply the importance of carrying out the research on the mechanism of forest penetration of the L-band SAR signal. In addition, the integration of L-band and new technologies, such as P-band and chromatographic SAR data, will provide a technical way to improve forest height estimation.

Funding: This research was funded by the National Natural Science Foundation of China (Project number: 32171784).

Conflicts of Interest: The authors declare no conflicts of interest.

References

1. Fassnacht, F.E.; White, J.C.; A Wulder, M.; Næsset, E. Remote sensing in forestry: Current challenges, considerations and directions. *Int. J. For. Res.* **2024**, *97*, 11–37. [CrossRef]
2. Xiang, T.-Z.; Xia, G.-S.; Zhang, L. Mini-Unmanned Aerial Vehicle-Based Remote Sensing: Techniques, applications, and prospects. *IEEE Geosci. Remote Sens. Mag.* **2019**, *7*, 29–63. [CrossRef]
3. Guo, Q.; Liang, X.; Li, W.; Jin, S.; Guan, H.; Cheng, K.; Su, Y.; Tao, S. LiDAR Remote Sensing of Forest Ecosystems: Applications and Prospects. In *New Thinking in GIScience*; Springer Nature Singapore: Singapore, 2022; pp. 221–231.
4. Li, Z.; Zhao, L.; Li, K.; Chen, E.; Wan, X.; Xu, K. A review of forest resource monitoring technology by synthetic aperture radar. *J. Nanjing Univ. Inf. Sci. Technol. (Nat. Sci. Ed.)* **2019**, *12*, 150–158.
5. Li, X.; Ye, Z.; Long, J.; Zheng, H.; Lin, H. Inversion of Coniferous Forest Stock Volume Based on Backscatter and InSAR Coherence Factors of Sentinel-1 Hyper-Temporal Images and Spectral Variables of Landsat 8 OLI. *Remote Sens.* **2022**, *14*, 2754. [CrossRef]
6. Kugler, F. Pol-InSAR Forest Height Estimation at Different Frequencies: Opportunities and Limitations. Ph.D. Thesis, Technischen Universitat Munchen, Munich, Germany, 2015.
7. Zhang, Q.; Ge, L.; Du, Z. A Modified RMoG Model for Forest Height Inversion Using L-Band Repeat-Pass Pol-InSAR Data. In Proceedings of the IGARSS 2019—2019 IEEE International Geoscience and Remote Sensing Symposium, Yokohama, Japan, 28 July–2 August 2019; pp. 3217–3220.
8. Asopa, U.; Kumar, S. UAVSAR Tomography for Vertical Profile Generation of Tropical Forest of Mondah National Park, Gabon. *Earth Space Sci.* **2020**, *7*, e2020EA001230. [CrossRef]
9. Huang, Y.; Zhang, Q.; Ferro-Famil, L. Forest Height Estimation Using a Single-Pass Airborne L-Band Polarimetric and Interferometric SAR System and Tomographic Techniques. *Remote Sens.* **2021**, *13*, 487. [CrossRef]
10. Thieu, H.C.; Pham, M.N.; Le, V.N. Forest parameters inversion by mean coherence set from single-baseline PolInSAR data. *Adv. Space Res.* **2021**, *68*, 2804–2818. [CrossRef]

11. Xie, J.; Li, L.; Zhuang, L.; Zheng, Y. A New Strategy for Forest Height Estimation Using Airborne X-Band PolInSAR Data. *Remote Sens.* **2022**, *14*, 4743. [CrossRef]
12. Sui, A.; Michel, O.O.; Mao, Y.; Fan, W. An Improved Forest Height Model Using L-Band Single-Baseline Polarimetric InSAR Data for Various Forest Densities. *Remote Sens.* **2022**, *15*, 81. [CrossRef]
13. Luo, H.; Yue, C.; Xie, F.; Zhu, B.; Chen, S. A Method for Forest Canopy Height Inversion Based on Machine Learning and Feature Mining Using UAVSAR. *Remote Sens.* **2022**, *14*, 5849. [CrossRef]
14. Luo, H.; Yue, C.; Wang, N.; Luo, G.; Chen, S. Correcting Underestimation and Overestimation in PolInSAR Forest Canopy Height Estimation Using Microwave Penetration Depth. *Remote Sens.* **2022**, *14*, 6145. [CrossRef]
15. Zhang, J.; Zhang, Y.; Fan, W.; He, L.; Yu, Y.; Mao, X. A Modified Two-Steps Three-Stage Inversion Algorithm for Forest Height Inversion Using Single-Baseline L-Band PolInSAR Data. *Remote Sens* **2022**, *14*, 1986. [CrossRef]
16. Wang, L.; Zhou, Y.; Shen, G.; Xiong, J.; Shi, H. Forest Height Inversion Based on Time–Frequency RVoG Model Using Single-Baseline L-Band Sublook-InSAR Data. *Remote Sens.* **2022**, *15*, 166. [CrossRef]
17. Zang, J.; Ni, W.; Zhang, Y. Spatially explicit mapping annual oil palm heights in peninsular Malaysia combining ICESat-2 and stand age data. *Remote Sens. Environ.* **2023**, *295*, 113693. [CrossRef]
18. Sa, R.; Nei, Y.; Fan, W. Combining Multi-Dimensional SAR Parameters to Improve RVoG Model for Coniferous Forest Height Inversion Using ALOS-2 Data. *Remote Sens.* **2023**, *15*, 1272. [CrossRef]
19. Grant, J.P.; Van de Griend, A.A.; Wigneron, J.P.; Saleh, K.; Panciera, R.; Walker, J.P. Influence of forest cover fraction on L-band soil moisture retrievals from heterogeneous pixels using multi-angular observations. *Remote Sens. Environ.* **2010**, *114*, 1026–1037. [CrossRef]
20. Parrens, M.; Wigneron, J.-P.; Richaume, P.; Mialon, A.; Al Bitar, A.; Fernandez-Moran, R.; Al-Yaari, A.; Kerr, Y.H. Global-scale surface roughness effects at L-band as estimated from SMOS observations. *Remote Sens. Environ. Interdiscip. J.* **2016**, *181*, 122–136. [CrossRef]
21. Konings, A.G.; Piles, M.; Das, N.; Entekhabi, D. L-band vegetation optical depth and effective scattering albedo estimation from SMAP. *Remote Sens. Environ. Interdiscip. J.* **2017**, *198*, 460–470. [CrossRef]
22. Lv, S.; Zeng, Y.; Su, Z.; Wen, J. A Closed-Form Expression of Soil Temperature Sensing Depth at L-Band. *IEEE Trans. Geosci. Remote Sens.* **2019**, *57*, 4889–4897. [CrossRef]
23. Holtzman, N.M.; Anderegg, L.D.; Kraatz, S.; Mavrovic, A.; Sonnentag, O.; Pappas, C.; Cosh, M.H.; Langlois, A.; Lakhankar, T.; Tesser, D.; et al. L-band vegetation optical depth as an indicator of plant water potential in a temperate deciduous forest stand. *Biogeosciences* **2021**, *18*, 739–753. [CrossRef]
24. Balzter, H.; Skinner, L.; Luckman, A.; Brooke, R. Estimation of tree growth in a conifer plantation over 19 years from multi-satellite L-band SAR. *Remote Sens. Environ.* **2003**, *84*, 184–191. [CrossRef]
25. Santoro, M.; Schmullius, C.; Askne, J.; Eriksson, L. Evaluation of JERS-1 L-band SAR backscatter for stem volume retrieval in boreal forest. In Proceedings of the IGARSS 2004. 2004 IEEE International Geoscience and Remote Sensing Symposium, Anchorage, AK, USA, 20–24 September 2004.
26. Chowdhury, T.A.; Thiel, C.; Schmullius, C. Growing stock volume estimation from L-band ALOS PALSAR polarimetric coherence in Siberian forest. *Remote Sens. Environ.* **2014**, *155*, 129–144. [CrossRef]
27. Santoro, M.; Eriksson, L.E.B.; Fransson, J.E.S. Reviewing ALOS PALSAR backscatter observations for stem volume retrieval in Swedish forest. *Remote Sens.* **2015**, *7*, 4290–4317. [CrossRef]
28. Khati, U.; Singh, G.; Tebaldini, S. Stock Volume Loss Estimation in Poplars using Regression Models and ALOS-2/PALSAR-2 backscatter. In Proceedings of the IGARSS 2019—2019 IEEE International Geoscience and Remote Sensing Symposium, Yokohama, Japan, 28 July–2 August 2019.
29. Thiel, C.; Schmullius, C. The potential of ALOS PALSAR backscatter and InSAR coherence for forest growing stock volume estimation in Central Siberia. *Remote Sens. Environ.* **2016**, *173*, 258–273. [CrossRef]
30. Santoro, M.; Cartus, O.; Fransson, J.E.; Wegmüller, U. Complementarity of X-, C-, and L-band SAR Backscatter Observations to Retrieve Forest Stem Volume in Boreal Forest. *Remote Sens.* **2019**, *11*, 1563. [CrossRef]
31. Zhang, H.; Zhu, J.; Wang, C.; Lin, H.; Long, J.; Zhao, L.; Fu, H.; Liu, Z. Forest Growing Stock Volume Estimation in Subtropical Mountain Areas Using PALSAR-2 L-Band PolSAR Data. *Forests* **2019**, *10*, 276. [CrossRef]
32. Englhart, S.; Keuck, V.; Siegert, F. Aboveground biomass retrieval in tropical forests—The potential of combined X- and L-band SAR data use. *Remote Sens. Environ. Interdiscip. J.* **2011**, *115*, 1260–1271. [CrossRef]
33. Cartus, O.; Santoro, M.; Kellndorfer, J. Mapping forest aboveground biomass in the Northeastern United States with ALOS PALSAR dual-polarization L-band. *Remote Sens. Environ.* **2012**, *124*, 466–478. [CrossRef]
34. Peregon, A.; Yamagata, Y. The use of ALOS/PALSAR backscatter to estimate above-ground forest biomass: A case study in Western Siberia. *Remote Sens. Environ.* **2013**, *137*, 139–146. [CrossRef]
35. Rahman, M.M.; Sumantyo, J.T.S. Retrieval of tropical forest biomass information from ALOS PALSAR data. *Geocarto Int.* **2013**, *28*, 382–403. [CrossRef]
36. Chaparro, D.; Duveiller, G.; Piles, M.; Cescatti, A.; Vall-Llossera, M.; Camps, A.; Entekhabi, D. Sensitivity of L-band vegetation optical depth to carbon stocks in tropical forests: A comparison to higher frequencies and optical indices. *Remote Sens. Environ.* **2019**, *232*, 111303. [CrossRef]

37. Berninger, A.; Lohberger, S.; Stängel, M.; Siegert, F. SAR-Based Estimation of Above-Ground Biomass and Its Changes in Tropical Forests of Kalimantan Using L- and C-Band. *Remote Sens.* **2018**, *10*, 831. [CrossRef]
38. Ramachandran, N.; Saatchi, S.; Tebaldini, S.; d'Alessandro, M.M.; Dikshit, O. Mapping tropical forest aboveground biomass using airborne SAR tomography. *Sci. Rep.* **2023**, *13*, 6233. [CrossRef] [PubMed]
39. Soja, M.J.; Quegan, S.; D'alessandro, M.M.; Banda, F.; Scipal, K.; Tebaldini, S.; Ulander, L.M. Mapping above-ground biomass in tropical forests with ground-cancelled P-band SAR and limited reference data. *Remote Sens. Environ.* **2021**, *253*, 112153. [CrossRef]
40. Cazcarra-Bes, V.; Pardini, M.; Papathanassiou, K. Definition of Tomographic SAR Configurations for Forest Structure Applications at L-Band. *IEEE Geosci. Remote Sens. Lett.* **2022**, *19*, 1–5. [CrossRef]
41. Hernández-Stefanoni, J.L.; Castillo-Santiago, M.Á.; Mas, J.F.; Wheeler, C.E.; Andres-Mauricio, J.; Tun-Dzul, F.; George-Chacón, S.P.; Reyes-Palomeque, G.; Castellanos-Basto, B.; Vaca, R.; et al. Improving aboveground biomass maps of tropical dry forests by integrating LiDAR, ALOS PALSAR, climate and field data. *Carbon Balance Manag.* **2020**, *15*, 1–17. [CrossRef] [PubMed]
42. Zhang, T.; Lin, H.; Long, J.; Zheng, H.; Ye, Z.; Liu, Z. Evaluating the Sensitivity of Polarimetric Features Related to Rotation Domain and Mapping Chinese Fir AGB Using Quad-Polarimetric SAR Images. *Remote Sens.* **2023**, *15*, 1519. [CrossRef]
43. Ni, W.; Yu, T.; Pang, Y.; Zhang, Z.; He, Y.; Li, Z.; Sun, G. Seasonal effects on aboveground biomass estimation in mountainous deciduous forests using ZY-3 stereoscopic imagery. *Remote Sens. Environ.* **2023**, *289*, 113520. [CrossRef]
44. Dall, J. InSAR Elevation Bias Caused by Penetration into Uniform Volumes. *IEEE Trans. Geosci. Remote Sens.* **2007**, *45*, 2319–2324. [CrossRef]
45. Pardini, M.; Papathanassiou, K. Sub-canopy topography estimation: Experiments with multibaseline SAR data at L-band. In Proceedings of the 2012 IEEE International Geoscience and Remote Sensing Symposium, Munich, Germany, 22–27 July 2012.
46. Toochi, E.C.; John, O.; Nwaihu, E.C.; Oboho, E.G. Environmental Sensing and Forest Degradation Using Mono-Temporal Analyses of Logging Roads. *Adv. For. Lett.* **2015**, *4*, 13. [CrossRef]
47. Muskett, R.R. L-Band InSAR Penetration Depth Experiment, North Slope Alaska. *J. Geosci. Environ. Prot.* **2017**, *5*, 14. [CrossRef]
48. Schlund, M.; Baron, D.; Magdon, P.; Erasmi, S. Canopy penetration depth estimation with TanDEM-X and its compensation in temperate forests. *ISPRS J. Photogramm. Remote Sens.* **2019**, *147*, 232–241. [CrossRef]
49. Teubner, I.E.; Forkel, M.; Jung, M.; Liu, Y.Y.; Miralles, D.G.; Parinussa, R.M.; van der Schalie, R.; Vreugdenhil, M.; Schwalm, C.R.; Tramontana, G.; et al. Identifying the link between microwave vegetation optical depth and gross primary production. In Proceedings of the 20th EGU General Assembly, EGU2018, Vienna, Austria, 4–13 April 2018; p. 4622.
50. Tanase, M.A.; Villard, L.; Pitar, D.; Apostol, B.; Petrila, M.; Chivulescu, S.; Leca, S.; Borlaf-Mena, I.; Pascu, I.S.; Dobre, A.C.; et al. Synthetic aperture radar sensitivity to forest changes: A simulations-based study for the Romanian forests. *Sci. Total Environ.* **2019**, *689*, 1104–1114. [CrossRef] [PubMed]
51. Chaparro, D.; Feldman, A.F.; Chaubell, M.J.; Yueh, S.H.; Entekhabi, D. Robustness of Vegetation Optical Depth Retrievals Based on L-Band Global Radiometry. *IEEE Trans. Geosci. Remote Sens.* **2022**, *60*, 4413417. [CrossRef]
52. Colliander, A.; Cosh, M.H.; Kelly, V.; Kraatz, S.; Bourgeau-Chavez, L.L.; Siqueira, P.; Roy, A.; Lakhankar, T.; Konings, A.G.; Holtzman, N.; et al. Detecting Soil Moisture under Temperate Forest Canopies: SMAP Validation Experiment 2019–2021 (SMAPVEX19-21). In *AGU Fall Meeting Abstracts*; American Geophysical Union: Washington, DC, USA, 2020.
53. Singh, A.; Meena, G.K.; Kumar, S.; Gaurav, K. Evaluation of the Penetration Depth of L- and S-Band (NISAR mission) Microwave SAR Signals into Ground. In Proceedings of the 2019 URSI Asia-Pacific Radio Science Conference (AP-RASC), New Delhi, India, 9–15 March 2019.
54. Liu, G.; Fu, H.; Zhu, J.; Wang, C.; Xie, Q. Penetration Depth Inversion in Hyperarid Desert From L-Band InSAR Data Based on a Coherence Scattering Model. *IEEE Geosci. Remote Sens. Lett.* **2021**, *18*, 1981–1985. [CrossRef]
55. Qi, Y.; Wang, Y.; Hong, J.; Du, S. Additional Reference Height Error Analysis for Baseline Calibration Based on a Distributed Target DEM in TwinSAR-L. *Remote Sens.* **2021**, *13*, 2750. [CrossRef]
56. Wang, M.; Fan, L.; Frappart, F.; Ciais, P.; Sun, R.; Liu, Y.; Li, X.; Liu, X.; Moisy, C.; Wigneron, J.P. An alternative AMSR2 vegetation optical depth for monitoring vegetation at large scales. *Remote Sens. Environ.* **2021**, *263*, 112556. [CrossRef]
57. Zhu, J.; Liu, G.; Zhao, R.; Ding, X.; Fu, H. ML based approach for inverting penetration depth of SAR signals over large desert areas. *Remote Sens. J. Environ.* **2023**, *295*, 113643. [CrossRef]
58. Bai, X.; Zheng, D.; Li, X.; Wigneron, J.P.; van der Velde, R.; Zhang, P.; Su, Z. Simulation of SMAP and AMSR2 observations and estimation of multi-frequency vegetation optical depth using a discrete scattering model in the Tibetan grassland. *Remote Sens. Environ.* **2023**, *292*, 113592. [CrossRef]
59. Olivares-Cabello, C.; Chaparro, D.; Vall-llossera, M.; Camps, A.; López-Martínez, C. Global Unsupervised Assessment of Multifrequency Vegetation Optical Depth Sensitivity to Vegetation Cover. *IEEE J. Sel. Top. Appl. Earth Obs. Remote Sens.* **2023**, *16*, 538–552. [CrossRef]
60. Schmidt, L.; Forkel, M.; Zotta, R.M.; Scherrer, S.; Dorigo, W.A.; Kuhn-Régnier, A.; van der Schalie, R.; Yebra, M. Assessing the sensitivity of multi-frequency passive microwave vegetation optical depth to vegetation properties. *Biogeosciences* **2023**, *20*, 1027–1046. [CrossRef]
61. Baur, M.J.; Jagdhuber, T.; Feldman, A.F.; Akbar, R.; Entekhabi, D. Estimation of relative canopy absorption and scattering at L-, C- and X-bands. *Remote Sens. Environ.* **2019**, *233*, 111384. [CrossRef]

62. Singh, A.; Meena, G.K.; Kumar, S.; Gaurav, K. Analysis of the Effect of Incidence Angle and Moisture Content on the Penetration Depth of L- and S-Band Sar Signals into the Ground Surface. *SPRS Ann. Photogramm. Remote Sens. Spat. Inf. Sci.* **2018**, *IV-5*, 197–202. [CrossRef]
63. Ji, Y.; Huang, J.; Ju, Y.; Guo, S.; Yue, C. Forest structure dependency analysis of L-band SAR backscatter. *PeerJ* **2020**, *8*, e10055. [CrossRef] [PubMed]
64. Cazcarra-Bes, V.; Tello-Alonso, M.; Fischer, R.; Heym, M.; Papathanassiou, K. Monitoring of Forest Structure Dynamics by Means of L-Band SAR Tomography. *Remote Sens.* **2017**, *9*, 1229. [CrossRef]
65. Minh, D.H.; Le Toan, T.; Tebaldini, S.; Rocca, F.; Iannini, L. Assessment of the P- and L-band SAR tomography for the characterization of tropical forests. In Proceedings of the 2015 IEEE International Geoscience and Remote Sensing Symposium (IGARSS), Milan, Italy, 26–31 July 2015.
66. Tello, M.; Cazcarra-Bes, V.; Pardini, M.; Papathanassiou, K. Forest Structure Characterization from SAR Tomography at L-Band. *IEEE J. Sel. Top. Appl. Earth Obs. Remote Sens.* **2018**, *11*, 3402–3414. [CrossRef]
67. El Moussawi, I.; Ho Tong Minh, D.; Baghdadi, N.; Abdallah, C.; Jomaah, J.; Strauss, O.; Lavalle, M.; Ngo, Y.N. Monitoring Tropical Forest Structure Using SAR Tomography at L- and P-Band. *Remote Sens.* **2019**, *11*, 1934. [CrossRef]
68. Pardini, M.; Cazcarra-Bes, V.; Papathanassiou, K. TomoSAR Mapping of 3D Forest Structure: Contributions of L-Band Configurations. *Remote Sens.* **2021**, *13*, 2255. [CrossRef]
69. Ho Tong Minh, D.; Ngo, Y.N.; Le, T.T. Potential of p-band SAR tomography in forest type classification. *Remote Sens.* **2021**, *13*, 696. [CrossRef]
70. Liu, X.; Zhang, L.; Yang, X.; Liao, M.; Li, W. Retrieval of Tropical Forest Height and Above-Ground Biomass Using Airborne P- and L-Band SAR Tomography. *IEEE Geosci. Remote Sens. Lett.* **2022**, *19*, 1–5. [CrossRef]
71. Ngo, Y.N.; Huang, Y.; Minh, D.H.; Ferro-Famil, L.; Fayad, I.; Baghdadi, N. Tropical Forest Vertical Structure Characterization: From GEDI to P-Band SAR Tomography. *IEEE Geosci. Remote Sens. Lett.* **2022**, *19*, 1–5. [CrossRef]
72. Chuang, H.Y.; Kiang, J.F. High-Resolution L-Band TomoSAR Imaging on Forest Canopies with UAV Swarm to Detect Dielectric Constant Anomaly. *Sensors* **2023**, *23*, 8335. [CrossRef] [PubMed]
73. Huang, H. Interpretation of Nature 2020 Top Ten scientific Discoveries. *Sci. Found. China* **2019**, *35*, 239–240. (In Chinese)
74. Wang, L.; Zuo, B.; Le, Y.; Chen, Y.; Li, J. Penetrating remote sensing: Next-generation remote sensing for transparent earth. *Innovation* **2023**, *4*, 100519. [CrossRef]

plants

MDPI

Article

Influence of Time-Lag Effects between Winter-Wheat Canopy Temperature and Atmospheric Temperature on the Accuracy of CWSI Inversion of Photosynthetic Parameters

Yujin Wang [1,2], Yule Lu [1,2], Ning Yang [1,2], Jiankun Wang [1,2], Zugui Huang [1,2], Junying Chen [1,2] and Zhitao Zhang [1,2,*]

[1] College of Water Resources and Architectural Engineering, Northwest A&F University, Xianyang 712100, China; wangyujin@nwafu.edu.cn (Y.W.); luyule@nwafu.edu.cn (Y.L.); yangning1228@126.com (N.Y.); wangjiankun@nwafu.edu.cn (J.W.); hzgui@nwafu.edu.cn (Z.H.); junyingchen@nwafu.edu.cn (J.C.)

[2] Key Laboratory of Agricultural Soil and Water Engineering in Arid and Semiarid Areas, Ministry of Education, Northwest A&F University, Xianyang 712100, China

* Correspondence: zhitaozhang@126.com

Abstract: When calculating the CWSI, previous researchers usually used canopy temperature and atmospheric temperature at the same time. However, it takes some time for the canopy temperature (Tc) to respond to atmospheric temperature (Ta), suggesting the time-lag effects between Ta and Tc. In order to investigate time-lag effects between Ta and Tc on the accuracy of the CWSI inversion of photosynthetic parameters in winter wheat, we conducted an experiment. In this study, four moisture treatments were set up: T1 (95% of field water holding capacity), T2 (80% of field water holding capacity), T3 (65% of field water holding capacity), and T4 (50% of field water holding capacity). We quantified the time-lag parameter in winter wheat using time-lag peak-seeking, time-lag cross-correlation, time-lag mutual information, and gray time-lag correlation analysis. Based on the time-lag parameter, we modified the CWSI theoretical and empirical models and assessed the impact of time-lag effects on the accuracy of the CWSI inversion of photosynthesis parameters. Finally, we applied several machine learning algorithms to predict the daily variation in the CWSI after time-lag correction. The results show that: (1) The time-lag parameter calculated using time-lag peak-seeking, time-lag cross-correlation, time-lag mutual information, and gray time-lag correlation analysis are 44–70, 32–44, 42–58, and 76–97 min, respectively. (2) The CWSI empirical model corrected by the time-lag mutual information method has the highest correlation with photosynthetic parameters. (3) GA-SVM has the highest prediction accuracy for the CWSI empirical model corrected by the time-lag mutual information method. Considering time lag effects between Ta and Tc effectively enhanced the correlation between CWSI and photosynthetic parameters, which can provide theoretical support for thermal infrared remote sensing to diagnose crop water stress conditions.

Keywords: time-lag effects; winter wheat; CWSI; photosynthetic rate; transpiration rate; stomatal conductance

Citation: Wang, Y.; Lu, Y.; Yang, N.; Wang, J.; Huang, Z.; Chen, J.; Zhang, Z. Influence of Time-Lag Effects between Winter-Wheat Canopy Temperature and Atmospheric Temperature on the Accuracy of CWSI Inversion of Photosynthetic Parameters. *Plants* **2024**, *13*, 1702. https://doi.org/10.3390/plants13121702

Academic Editor: Stefano Martellos

Received: 7 May 2024
Revised: 12 June 2024
Accepted: 16 June 2024
Published: 19 June 2024

1. Introduction

The timely and accurate diagnosis of crop water stress conditions effectively determines the timing of irrigation and facilitates precision irrigation, crucial for enhancing water use efficiency (WUE) and increasing yield [1,2]. When crops experience water stress, their physiological indicators, such as photosynthetic parameters, leaf water potential, and external morphology, undergo changes. These changes include a decrease in the leaf area index, a reduction in chlorophyll concentration, and a diminution in leaf length and width [3,4]. Physiological indicators, including photosynthetic parameters, leaf water potential, and stem water potential, focus on the crop itself for research and offer a straightforward, scientific method to diagnose the status of crop water deficit. These indicators have proven effective in monitoring the water status of crops [5]. Moisture stress impacts photosynthetic parameters across the reproductive period of crops, displaying consistent trends in the net photosynthesis rate (Pn), transpiration rate (Tr), and stomatal conductance (gs), which all decrease with increasing moisture stress [6]. The extent of changes in photosynthetic parameters varies with different levels of water stress, with minor reductions under mild water stress and significant declines under moderate to severe water stress [2].

Beyond the physiological indicators of crop water deficit, the crop water stress index (CWSI), sensitive to soil moisture, stands as a reliable indirect metric for monitoring crop water status. Under water stress conditions, crops exhibit reduced stomatal conductance and diminished transpiration cooling, leading to an increase in canopy temperature. Idso et al. [7] found that the differential in canopy temperature following noon effectively measures the crop water deficit, revealing a consistent linear relationship between the canopy air temperature differential (CTD) and vapor pressure deficit (VPD) under clear sky conditions, which is not affected by environmental factors, such as wind speed. In response to the above phenomenon, Idso, Jackson, Pinter, Reginato and Hatfield [7] propose the CWSI empirical model, which has the advantages of fewer computational parameters, ease of measurement, and sensitivity to crop varieties. Jackson et al. [8] proposed the CWSI theoretical model based on the principle of canopy energy balance, taking into account environmental factors such as aerodynamic resistance, crop minimum canopy resistance, and net radiation, making the CWSI model more theoretical.

The CWSI is a sensitive indicator used to reflect water stress caused by the stomatal function of the crop, and continuous water stress results in an increasing trend of the CWSI and a decreasing trend of Pn, Tr, and gs [9]. There is a good negative correlation between the CWSI and photosynthetic parameters [6,10]. The results of Ramos-Fernández et al. [11] showed a strong correlation between the CWSI and gs ($R^2 = 0.91$). When the crop is subjected to water stress, the soil–root hydraulic resistance increases [12], which reduces root water transport and eventually leads to the reduction in or closure of plant stomata and a decrease in photosynthetic parameters [13]. Different physiological characteristics of wheat have different sensitivities to soil moisture [14]; therefore, the correlation between Pn, gs, Tr, and CWSI varies.

In calculating the CWSI, previous researchers always used atmospheric temperature (Ta) and canopy temperature (Tc) at the same moment [15]. However, there is a time-lag effect in the response of Tc to Ta [16]. Therefore, it is more accurate to use atmospheric temperature that actually influences the canopy temperature. Zhang et al. [17] discovered that accounting for the time-lag effect significantly enhances the accuracy of the CWSI in estimating soil water content. Currently, research on the impact of this time lag on the accuracy of the CWSI in the inversion of photosynthetic parameters remains unexplored.

We hypothesized that the time-lag effects between the canopy temperature and atmospheric temperature have a significant impact on the model accuracy of the CWSI inversion of photosynthetic parameters. Therefore, we conducted an experiment with winter wheat, where we continuously monitored the canopy temperature and environmental factors of winter wheat. We quantified the time-lag parameters between Ta and Tc using time-lag peak-finding, time-lag cross-correlation, time-lag mutual information, and time-lag gray correlation analysis. We then modified the theoretical and empirical CWSI models based on these time-lag parameters. Finally, we investigated the implication and mechanisms of Ta and Tc time-lag effects on the accuracy of the CWSI inversion of photosynthesis parameters.

2. Results

2.1. Time-Lag Parameters of Winter Wheat under Different Water Stresses

As depicted in Figure 1, the CCE equation fitted the daily variation process of winter-wheat canopy temperature smoothed by S-G filtering with excellent accuracy ($R^2 = 0.98$), and the ECS equation fitted the daily variation process of atmospheric temperature smoothed by S-G filtering with equal precision ($R^2 = 0.98$). As seen in Figures 2–6, among the time-lagged parameters of Tc and Ta obtained by different methods, the gray time-lag correlation analysis was the largest. The time-lag peak-seeking method and the time-lag mutual information method were the second largest, followed by the time-lag cross-correlation method.

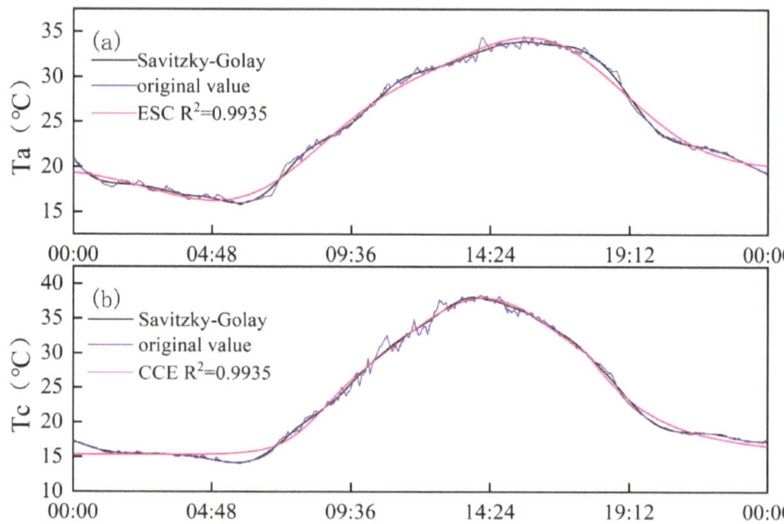

Figure 1. (**a**) ESC equations fitted to S-G filter-smoothed Ta; (**b**) CCE equation fitted to S-G filter-smoothed Tc.

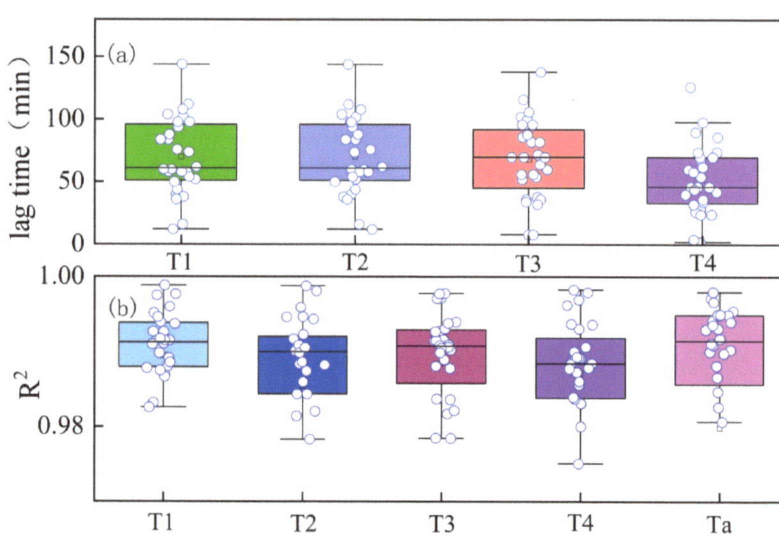

Figure 2. (**a**) Time-lag parameters of T1 (fully irrigated), T2 (mild water stress), T3 (moderate water stress), and T4 (severe water stress) calculated by time-lag peak-finding method. (**b**) Coefficient of determination (R^2) for T1, T2, T3, and T4 fitted by the time-lag peak-finding method.

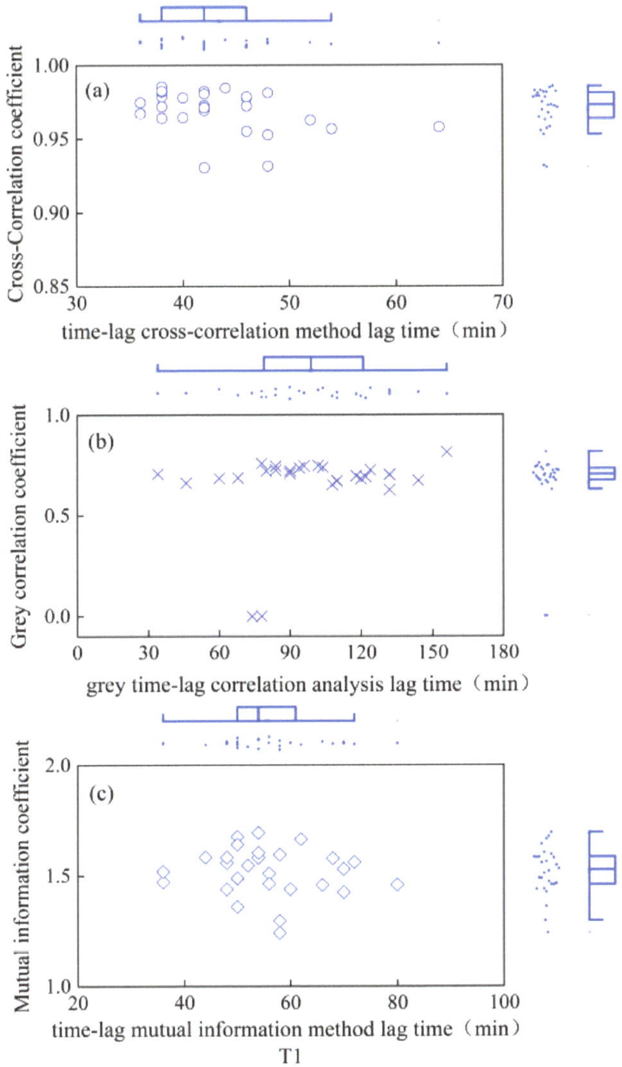

Figure 3. Time-lag parameters and corresponding coefficients for fully irrigated treatment. (**a**) is the time-lag parameter calculated by the time-lag cross-correlation method and the cross-correlation coefficient between Ta and Tc after the corrected time-lag; (**b**) is the time-lag parameter calculated by the time-lag grey correlation analysis and the time-lag grey correlation coefficient between Ta and Tc after the corrected time-lag; (**c**) is the time-lag parameter calculated by the time-lag mutual information method and the mutual information coefficient between Ta and Tc after the corrected time-lag. Circles indicate the results of the time-lag cross-correlation method under the four moisture treatments; cross sign indicates the results of the time-lag grey correlation analysis under the four moisture treatments; squares indicate the results of the time- lag mutual information method.

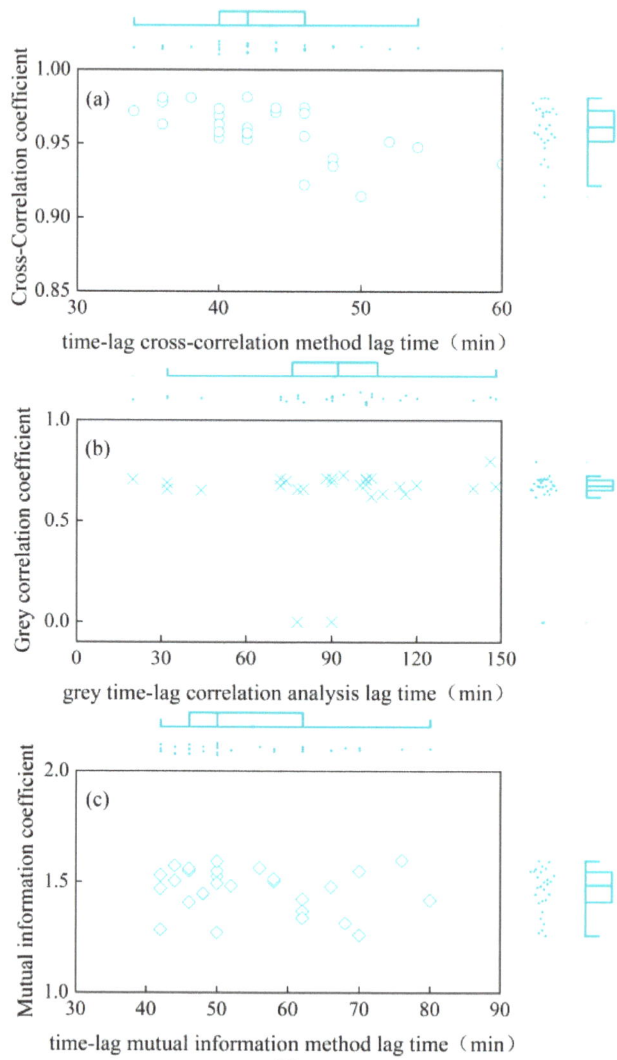

Figure 4. Time-lag parameters and corresponding coefficients for mild water stress treatment. (**a**) is the time-lag parameter calculated by the time-lag cross-correlation method and the cross-correlation coefficient between Ta and Tc after the corrected time-lag; (**b**) is the time-lag parameter calculated by the time-lag grey correlation analysis and the time-lag grey correlation coefficient between Ta and Tc after the corrected time-lag; (**c**) is the time-lag parameter calculated by the time-lag mutual information method and the mutual information coefficient between Ta and Tc after the corrected time-lag. Circles indicate the results of the time-lag cross-correlation method under the four moisture treatments; cross sign indicates the results of the time-lag grey correlation analysis under the four moisture treatments; squares indicate the results of the time- lag mutual information method.

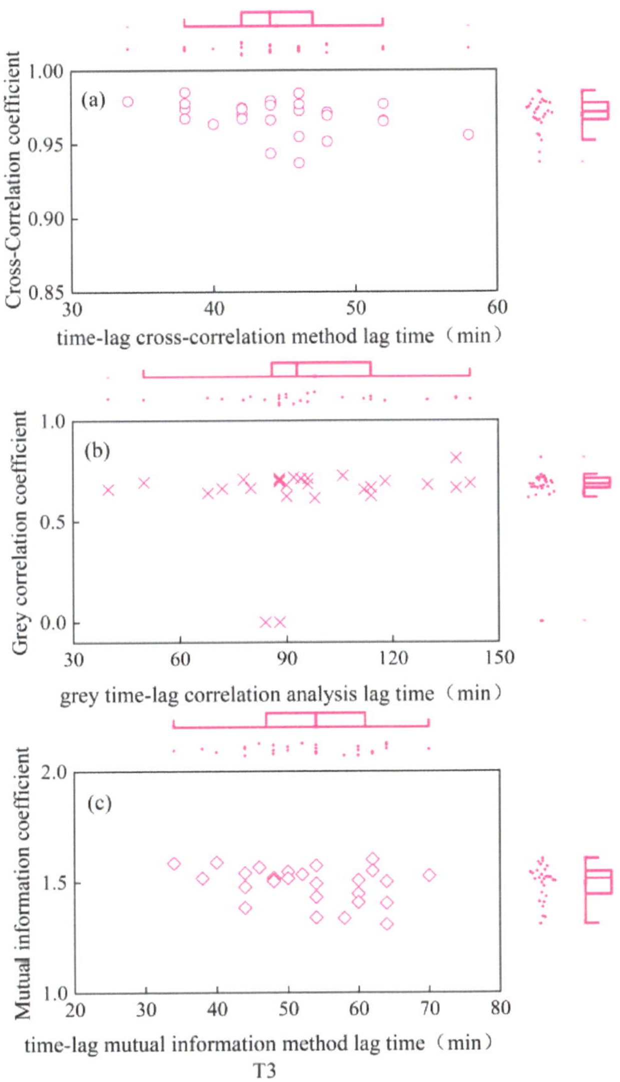

Figure 5. Time-lag parameters and corresponding coefficients for moderate water stress treatment. (**a**) is the time-lag parameter calculated by the time-lag cross-correlation method and the cross-correlation coefficient between Ta and Tc after the corrected time-lag; (**b**) is the time-lag parameter calculated by the time-lag grey correlation analysis and the time-lag grey correlation coefficient between Ta and Tc after the corrected time-lag; (**c**) is the time-lag parameter calculated by the time-lag mutual information method and the mutual information coefficient between Ta and Tc after the corrected time-lag. Circles indicate the results of the time-lag cross-correlation method under the four moisture treatments; cross sign indicates the results of the time-lag grey correlation analysis under the four moisture treatments; squares indicate the results of the time- lag mutual information method.

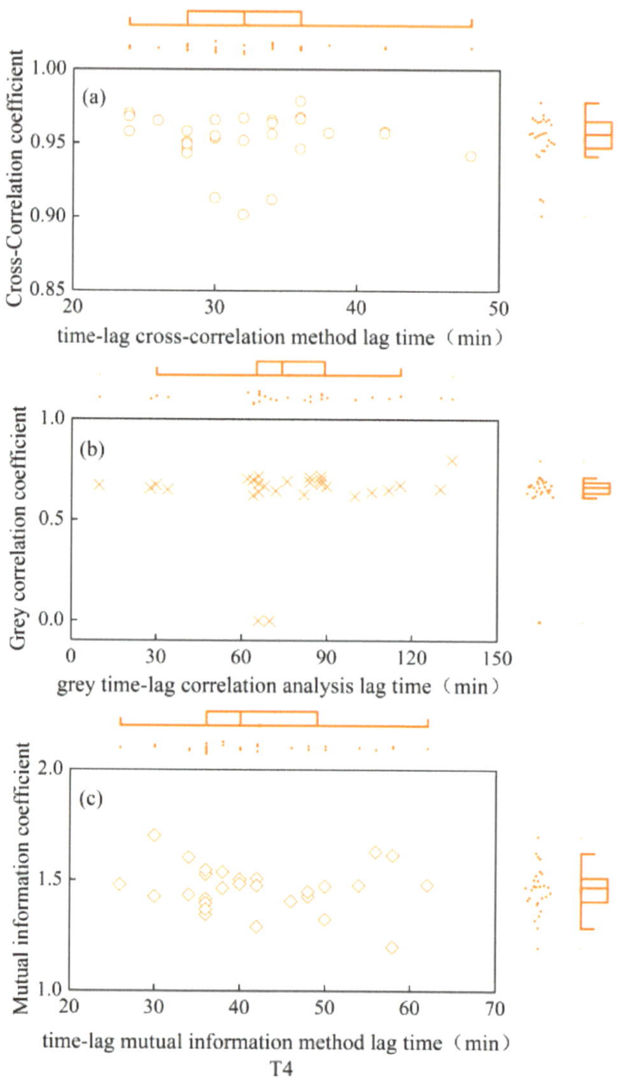

Figure 6. Time-lag parameters and corresponding coefficients for severe water stress treatment. (**a**) is the time-lag parameter calculated by the time-lag cross-correlation method and the cross-correlation coefficient between Ta and Tc after the corrected time-lag; (**b**) is the time-lag parameter calculated by the time-lag grey correlation analysis and the time-lag grey correlation coefficient between Ta and Tc after the corrected time-lag; (**c**) is the time-lag parameter calculated by the time-lag mutual information method and the mutual information coefficient between Ta and Tc after the corrected time-lag. Circles indicate the results of the time-lag cross-correlation method under the four moisture treatments; cross sign indicates the results of the time-lag grey correlation analysis under the four moisture treatments; squares indicate the results of the time- lag mutual information method.

The time-lag parameters between the canopy temperature (Tc) and atmospheric temperature (Ta), calculated using four different methods, exhibited distinct values across varying irrigation treatments. For the fully irrigated treatment, the time-lag parameters were approximately 53 min, 44 min, 58 min, and 97 min when calculated using the time-lag peak-finding method, time-lag cross-correlation method, time-lag mutual information method, and gray time-lag correlation analysis, respectively; for the mild water stress treatment, these time-lag parameters were about 52 min, 43 min, 55 min, and 92 min, respectively. For the moderate water stress treatment, the parameters were approximately 55 min, 44 min, 54 min, and 98 min. Lastly, for the severe water stress treatment, the parameters were around 44 min, 32 min, 42 min, and 76 min. These results highlight the variability in time-lag parameters across different irrigation treatments, as well as the influence of the chosen calculation method.

This indicates that the time lag between the Tc and Ta obtained from different calculation methods for the fully irrigated, mild water stress, and moderate water stress treatments did not differ significantly. However, for the severe water stress treatment, the Tc reached its peak time later, resulting in a decrease in the time-lag parameter between the Ta and Tc by approximately 10 to 22 min. This phenomenon might be associated with the soil moisture threshold [18]. When the soil moisture threshold was reached, the water lost through transpiration in winter wheat could not be replenished promptly. To ensure the normal life activities of the crop, the expansion rate of the crop leaves was reduced, stomatal conductance decreased significantly, transpiration rate declined, and canopy temperature continued to increase, reaching the peak time later. This led to a shorter time lag between the atmospheric temperature and canopy temperature [19].

In addition, the cross-correlation coefficient, mutual information coefficient, and gray correlation coefficient values corresponding to the peak moments for the fully irrigated, mild water stress, moderate water stress, and severe water stress treatments did not differ significantly. This indicated that the linear correlation [20], nonlinear correlation [21], and curve similarity [22] of Tc and Ta under the four moisture treatments after a time-lag correction did not differ much.

The accuracy of the CWSI inversion of photosynthetic parameters before and after time-lag effects was considered.

2.2. Time-Lag Peak-Seeking Method, Time-Lag Cross-Correlation Method, Time-Lag Mutual Information Method, and Gray Time-Lag Correlation Analysis

As shown in Figures 7 and 8, correcting the time lag between the Ta and Tc improved the accuracy of the CWSI inversion for Pn. After CWSI empirical and theoretical models were corrected using the time-lag peak-seeking method, time-lag cross-correlation method, time-lag mutual information method, and gray time-lag correlation analysis, the correlation between the CWSI and Pn improved for all methods, with the empirical model showing a more significant improvement. This indicated that the time-lag effect had a substantial impact on the accuracy of the CWSI empirical model in inverting Pn, while its impact on the accuracy of the CWSI theoretical model in inverting Pn was small and negligible.

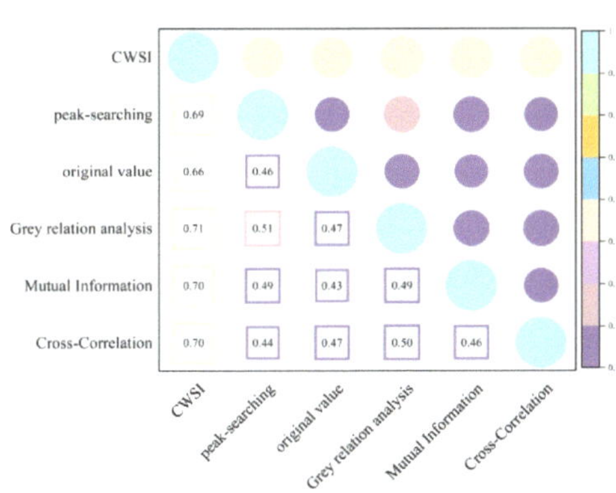

Figure 7. Heat map of the CWSI theoretical model and Pn before and after considering time-lag effects.

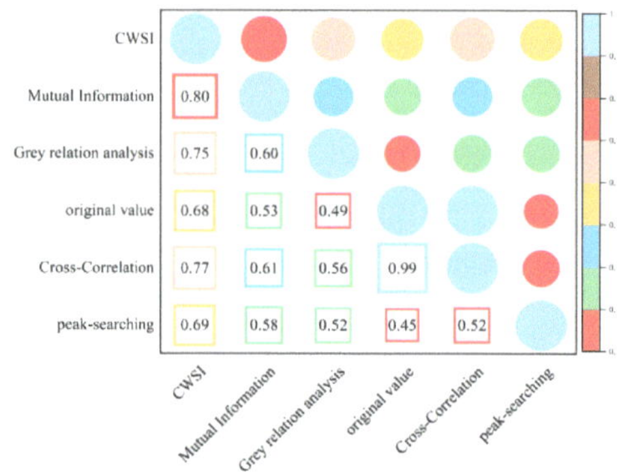

Figure 8. Heat map of the CWSI empirical model and Pn before and after considering time-lag effects.

As shown in Figures 9 and 10, the time-lag effect has a small impact on the accuracy of the CWSI theoretical model inverting Tr and a large impact on the accuracy of the CWSI empirical model inverting Tr. The correlation between the CWSI and Tr does not change after correcting the CWSI theoretical model by using the time-lag peak-seeking method. The correlation between the CWSI and Tr is improved by correcting the CWSI theoretical model by using time-lag cross-correlation method, time-lag mutual information method, and gray time-lag correlation analysis. The correlation between Tr and the CWSI theoretical model corrected based on gray time-lag correlation analysis and time-lag mutual information is the best (R^2 = 0.90). The correlation between the CWSI empirical model and Tr decreases after correcting the CWSI empirical model using the time-lag mutual information method and gray time-lag correlation analysis. The accuracy of the CWSI inversion of Tr improves after correcting the CWSI empirical model using the time-lag peak-seeking method and time-lag mutual correlation method. The correlation between Tr and the CWSI empirical model corrected based on the time-lag mutual correlation method is the highest (R^2 = 0.94).

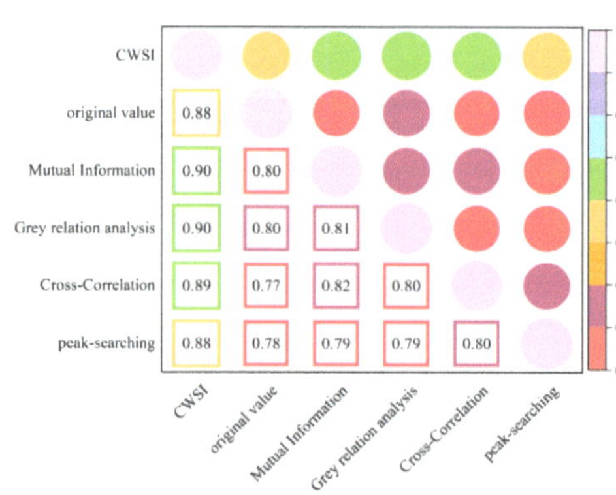

Figure 9. Heat map of the CWSI theoretical model and Tr before and after considering time-lag effects.

As demonstrated in Figures 11 and 12, the correlation between the CWSI and gs remains unchanged after the CWSI theoretical model was corrected by applying the time-lag peak-seeking method. However, the correlation between the CWSI and gs improved after the CWSI theoretical model was corrected using the time-lag cross-correlation method, the time-lag mutual information method, and the gray time-lag correlation analysis. Notably, the time-lag mutual information method enhanced the accuracy of the CWSI theoretical model inversion of gs the most (R^2 = 0.96). The time-lag effect significantly impacted the accuracy of the CWSI empirical model inversion of gs. The correlation between the CWSI and gs increased after correcting the CWSI empirical model using the time-lag peak-seeking method, the time-lag mutual correlation method, the time-lag mutual information method, and the gray time-lag correlation analysis. gs showed the highest correlation with the CWSI empirical model based on the time-lag mutual information method (R^2 = 0.96).

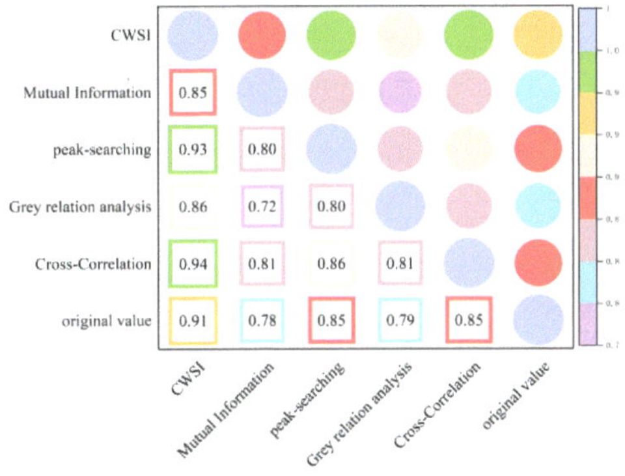

Figure 10. Heat map of the CWSI empirical model and Tr before and after considering time-lag effects.

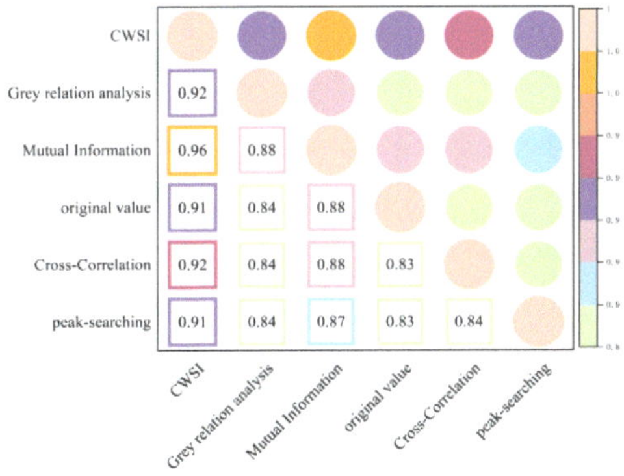

Figure 11. Heat map of the CWSI theoretical model and gs before and after considering time-lag effects.

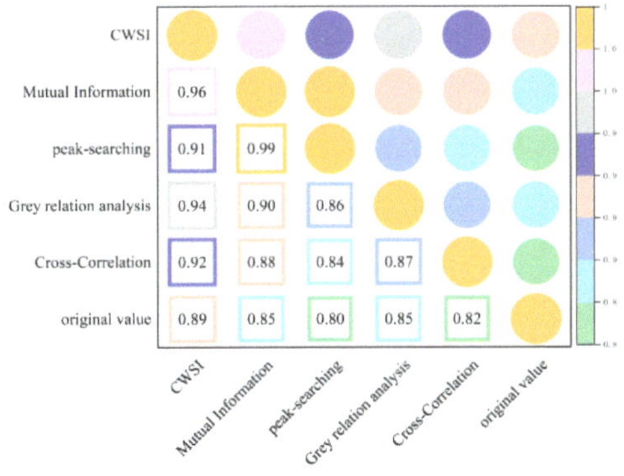

Figure 12. Heat map of the CWSI empirical model and gs before and after considering time-lag effects.

In summary, it was observed that time-lag effect between the Ta and Tc caused a significant impact on the accuracy of the CWSI inversion of photosynthetic parameters. The impact was more substantial on the accuracy of the CWSI empirical model for inverting photosynthetic parameters, and the time-lag-corrected CWSI empirical model demonstrated a higher correlation with the photosynthetic parameters. This indicated that the CWSI empirical model was more sensitive to the time-lag effect than the CWSI theoretical model. The reason for this phenomenon might be that the CWSI theoretical model required measurements of net radiation, soil heat flux, wind speed, and canopy resistance, making the theoretical models less volatile [23]. The correlation of the time-lag-corrected CWSI with Pn, Tr, and gs was in the order of gs > Tr > Pn. Pn showed the highest correlation with the empirical/theoretical CWSI models corrected by the time-lag mutual information method (R^2 = 0.8); Tr had the best correlation with the CWSI empirical model corrected by the time-lag mutual information method (R^2 = 0.93); and gs exhibited the best correlation with the CWSI empirical model corrected by the time-lag mutual information method (R^2 = 0.93). Occasionally, time-lag correction reduced the accuracy of the CWSI inversion of photosynthetic parameters. This reduction could be attributed to the fact that the time-lag effect between the Tc and environmental factors, such as relative humidity and solar radiation, was not accounted for [24].

Meanwhile, the time-lag effect was the result of the continuous direct or indirect influence of previous environmental factors on crops, representing an accumulative process [25–27]. The time-lag peak-finding method, which utilized a function to fit the daily change curves of the Tc and Ta and defined the time-lag parameter solely by the time difference between its peak points, exhibited certain limitations.

2.3. *Machine Learning Algorithms for Predicting CWSI Empirical Models Based on Time-Lag Mutual Information Correction*

The accuracy of the CWSI empirical model corrected based on the time-lag mutual information method for the inversion of photosynthetic parameters was overall high. It was investigated by using Genetic Algorithm Optimized Support Vector Machines (GA-SVMs) based on genetic algorithms, Bayesian Optimized Long and Short-Term Memory Neural Networks (Bayes-LSTMs), Particle Swarm Algorithm Optimized Long and Short-Term Memory (PSO-LSTM) based on particle swarm algorithms, Convolutional Bi-directional Long and Short-Term Memory Neural Networks (CNN-BILSTMs), Attention Mechanism Long Short-Term Memory Neural Networks (attention-LSTMs), and Attention Mechanism Gated Recurrent Unit (attention-GRU) machine learning algorithms for predictions. The prediction accuracies are shown in Table 1.

Table 1. Prediction accuracy of machine learning algorithms for the CWSI empirical model based on time-lag mutual information correction.

Machine Learning Algorithm	R^2	RMSE
attention-LSTM	0.88928	0.035052
GRU-attention	0.80197	0.037266
CNN-BILSTM	0.9148	0.031886
GA-SVM	0.98237	0.016596
PSO-LSTM	0.9466	0.028885
Bayes-LSTM	0.97865	0.018266

With GA-SVM > Bayes-LSTM > PSO-LSTM > CNN-BILSTM > attention-LSTM > GRU-attention. The prediction accuracy of the above models for the CWSI empirical model corrected by the time-lag mutual information method was higher overall. Predicted effect diagrams are shown in Figures 13–24. The GA-SVM model had the highest prediction accuracy (R^2 = 0.982, RMSE = 0.017).

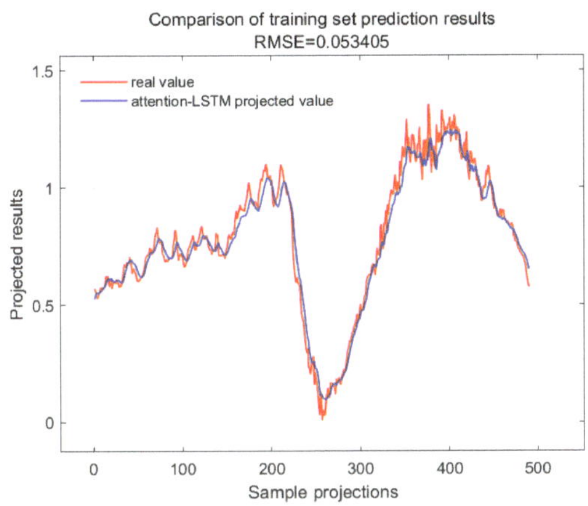

Figure 13. Training set for attention-LSTM.

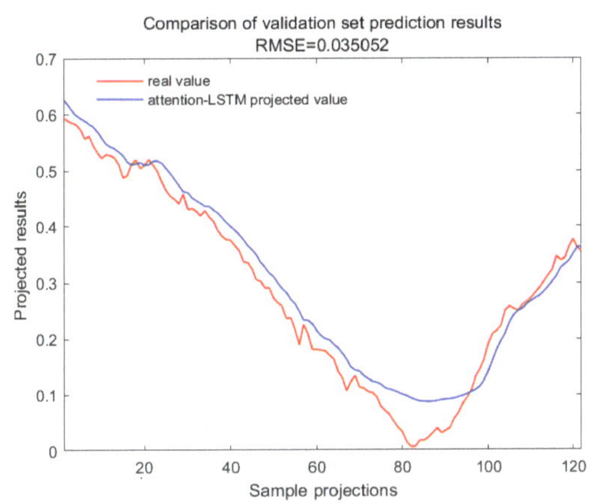

Figure 14. Validation set for attention-LSTM.

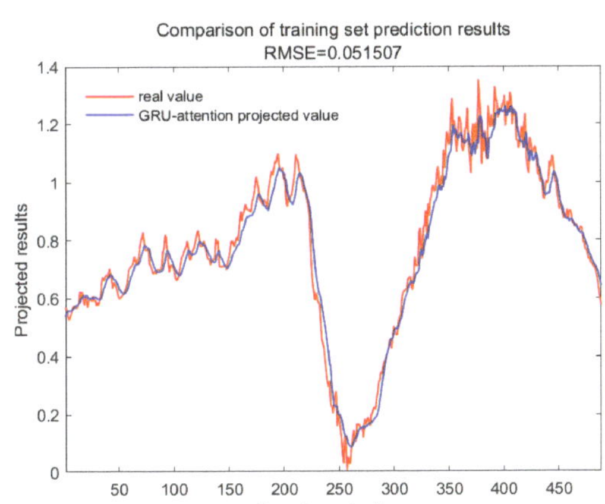

Figure 15. Training set for GRU-attention.

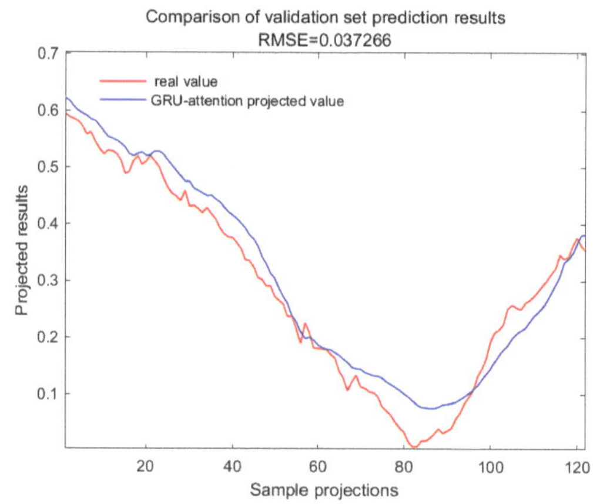

Figure 16. Validation set for GRU-attention.

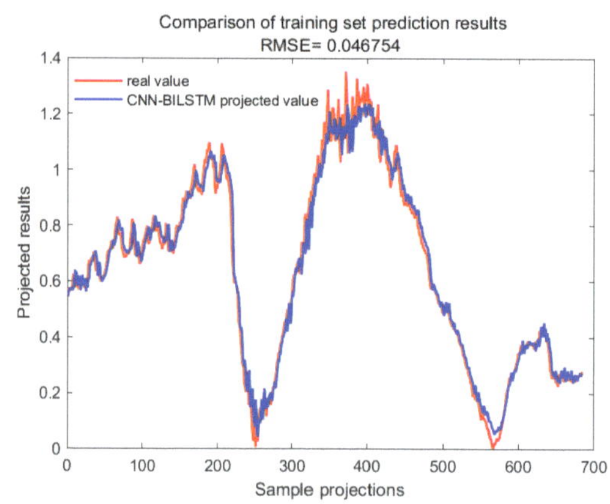

Figure 17. Training set for CNN-BILSTM.

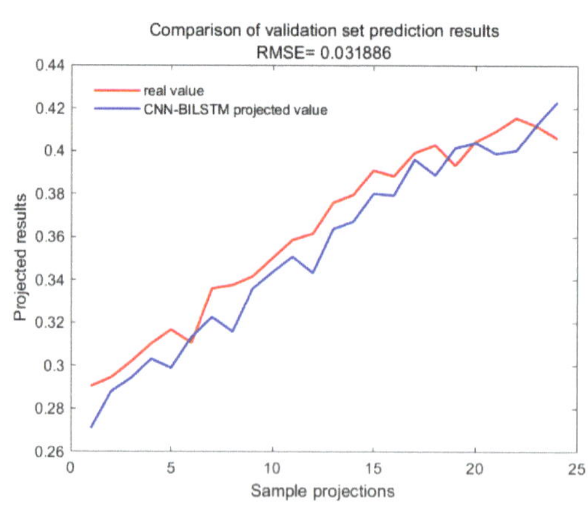

Figure 18. Validation set for CNN-BILSTM.

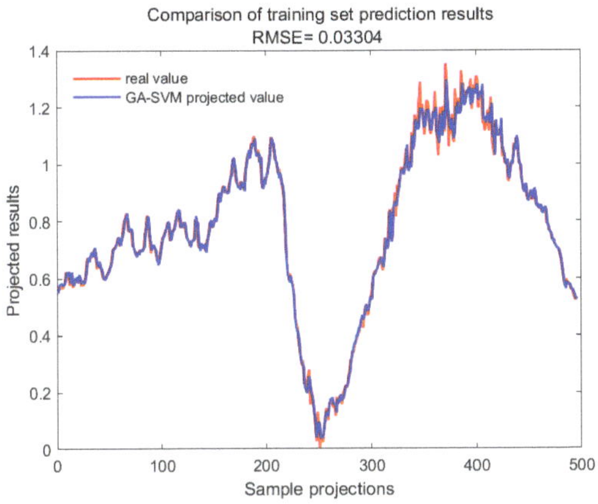

Figure 19. Training set for GA-SVM.

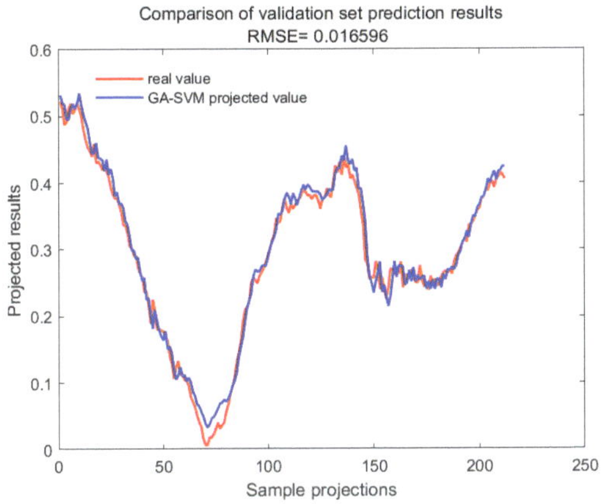

Figure 20. Validation set for GA-SVM.

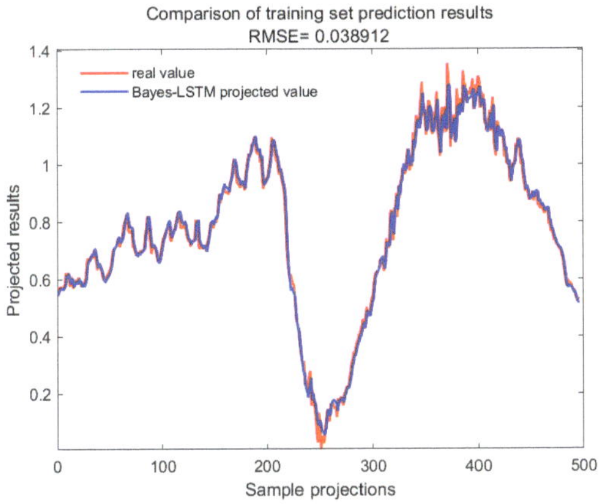

Figure 21. Training set for Bayes-LSTM.

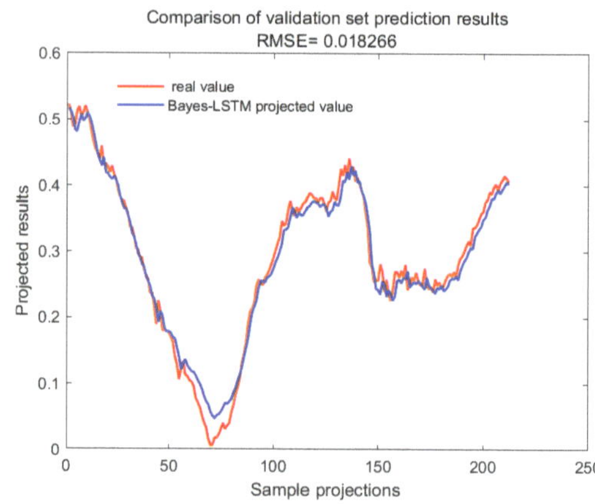

Figure 22. Validation set for Bayes-LSTM.

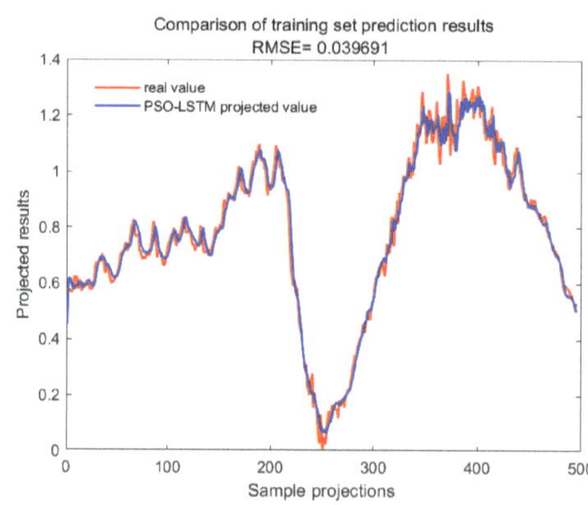

Figure 23. Training set for PSO-LSTM.

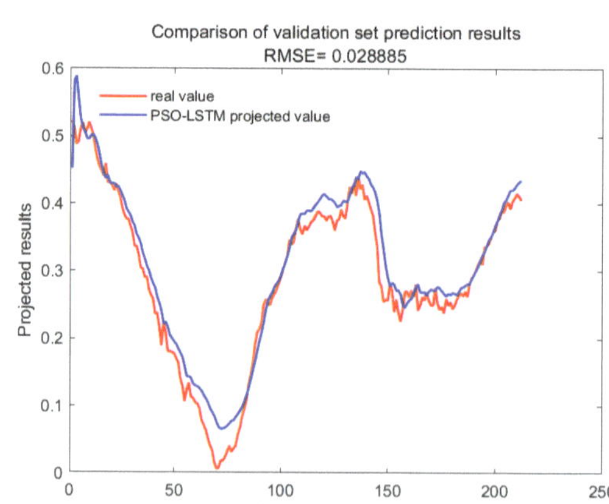

Figure 24. Validation set for PSO-LSTM.

3. Materials and Methods

3.1. Study Site Description

The experimental site is located at the Institute of Water Saving Agriculture in Arid Regions, Northwest Agriculture and Forestry University, Yangling District, Shaanxi Province,

China (108°24′ E, 34°20′ N). This region is characterized by a warm, temperate, semi-humid monsoon climate, distinguished by four distinct seasons and moderate rainfall. The average annual temperature ranges from approximately 13 °C to 15 °C. Rainfall predominantly occurs in July and August, driven by the southeast monsoon, with an annual average between 600 mm and 800 mm. The effect of groundwater recharge is not considered in this experiment.

3.2. Experiment Design

The experimental site was 32.5 m × 10.5 m and divided into 12 plots, each measuring 4 m × 4 m. Protected row treatments were used to mitigate the effects of water infiltration (Figure 25). Four moisture treatments were used in the experiment to obtain generalizable results: T1 (fully irrigated), T2 (mild water stress), T3 (moderate water stress), and T4 (severe water stress). The upper irrigation limits were set at 95%, 80%, 65%, and 50% of the field water holding capacity for T1, T2, T3, and T4, respectively. Each moisture treatment was replicated three times. Instruments for the continuous monitoring of canopy temperature and environmental factors were positioned above experimental plots 2, 5, 8, and 11. The cultivar used was "Genmai 68" winter wheat sown at 25 cm spacing with 30 g of seed per row. The sowing date was 19 October 2022 and the harvest date was 1 June 2023. Irrigation was carried out by drip irrigation system, and the irrigation quota is detailed in Table 2. The measured volumetric soil water content was calculated by oven-drying method and the irrigation quota is calculated as follows:

$$m = H \cdot (\theta_s - \theta_o) \cdot p \cdot s \tag{1}$$

where m is the irrigation quota (mm); H is the planned wetted layer depth (m): 0.4–0.5 m (green-up stage), 0.5–0.6 m (jointing stage), 0.6–0.8 m (tasseling stage), and 0.8–1.0 m (grouting period); θ_s is the field capacity (%), which is the upper limit of soil moisture content; θ_o is the measured volumetric soil moisture content (%); s is the trial plot size (m); and p is the drip irrigation wetting ratio, 0.6.

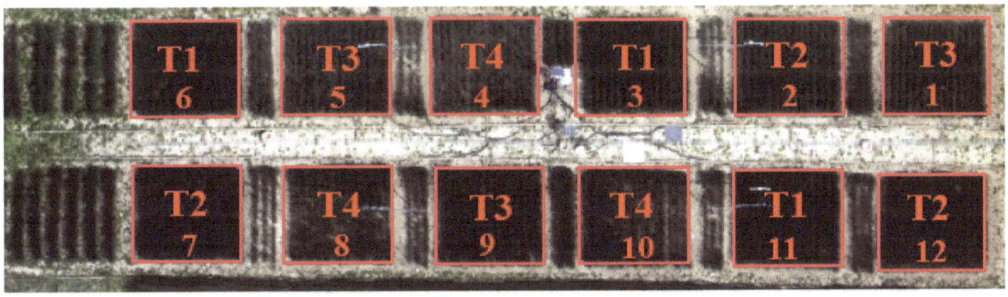

Figure 25. Overview of the experimental site.

Table 2. Irrigation quota of winter wheat.

Irrigation Date	Irrigation Quota (mm)			
	T1	T2	T3	T4
17 February 2023	46.7	43.9	19.2	18.2
25 February 2023	50.3	23.8	23.0	16.0
5 March 2023	63.0	28.5	13.4	31.2
12 March 2023	58.1	19.4	17.7	51.0
19 March 2023	64.0	54.6	40.0	16.3
29 March 2023	64.0	44.8	33.0	16.3
7 April 2023	90.3	36.0	15.8	38.0
18 April 2023	91.9	44.9	40.6	42.0
28 April 2023	29.5	25.2	19.4	17.0
12 May 2023	53.8	69.0	88.4	14.0
23 May 2023	53.9	65.0	30.6	17.9

3.3. *Data Acquisition*

3.3.1. Tc Measurements

In this study, the canopy temperature of winter wheat was continuously monitored using an SI-411 infrared thermometer. The monitoring interval was set at 2 min. Considering the effect of crop cover on the instrumental monitoring of canopy temperature, the time-lag parameter was calculated in this experiment starting from 16 February 2023. The canopy temperature of winter wheat for the four moisture treatments is shown in Figure 26.

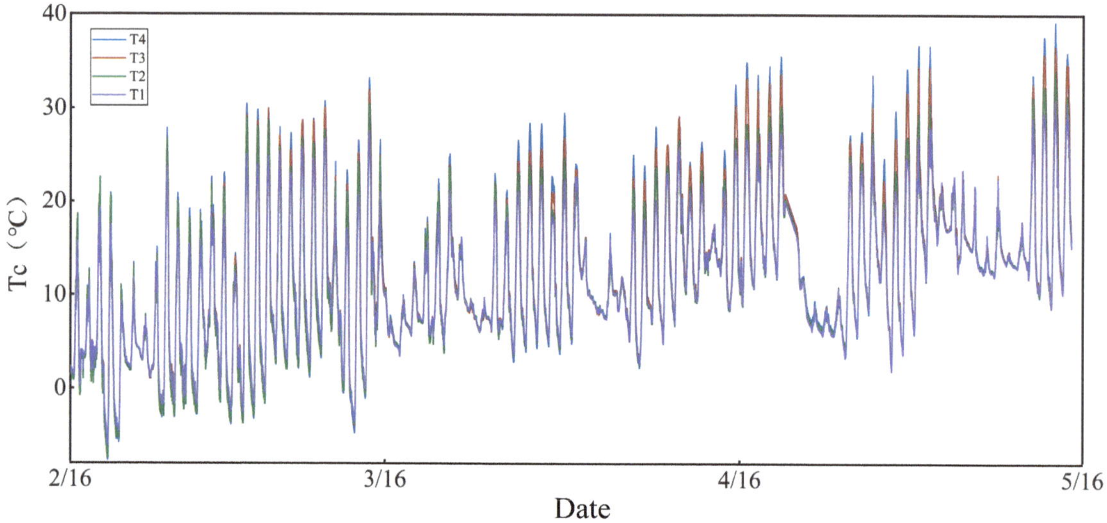

Figure 26. Canopy temperature of winter wheat under four moisture treatments.

3.3.2. Environmental

In this experiment, meteorological factors were continuously monitored using the AWS-CR1000 scientific-grade automatic meteorological monitoring system, as detailed in Table 3. Data collection intervals were set at 2 min. Meteorological factors are shown in Figure 27.

Table 3. Summary of environmental factors observed by the weather station.

Variables	Sensor Number	Instrument Height (m)	Abbreviation	Unit
Solar radiation	SN-500	3.5	Rs	$W \cdot m^{-2}$
Soil heat fux	HFP01	−0.10	G	$W \cdot m^{-2}$
Atmospheric temperature	HC2AS3	2.5	Ta	°C
Relative humidity	HC2AS3	2.5	RH	%
Wind speed	HC2AS3	2	u	$m \cdot s^{-1}$

3.3.3. Photosynthetic Parameters Measurements

The differences in crop physiological indicators at different irrigation levels were small in the morning and evening, and the differences were largest around midday, which could accurately reflect the crop water status [28,29]. Therefore, we chose sunny and windless days to collect the photosynthetic parameters of winter wheat: stomatal conductance gs (mol/($m^2 \cdot s$)), net photosynthesis rate Pn (μmol/($m^2 \cdot s$)), and transpiration rate Tr (mmol/($m^2 \cdot s$)) at 14:00 using a portable photosynthesizer model Li-6800 from LICOR, Lincoln, NE, USA. Three wheat plants were randomly selected from each plot, and the measurements were repeated three times for each wheat flag leaf, and the average value was taken as the photosynthetic parameters of the crop under the moisture treatment; to ensure the accuracy of the acquired data, the CO_2 concentration of the Li-6800 portable photosynthesizer reached 400 μmol/mol, and the intensity of the light reached 1000 μmol/(m^2·s) during the measurement. The data of photosynthetic parameters were collected 12 times in this experiment (Figure 28).

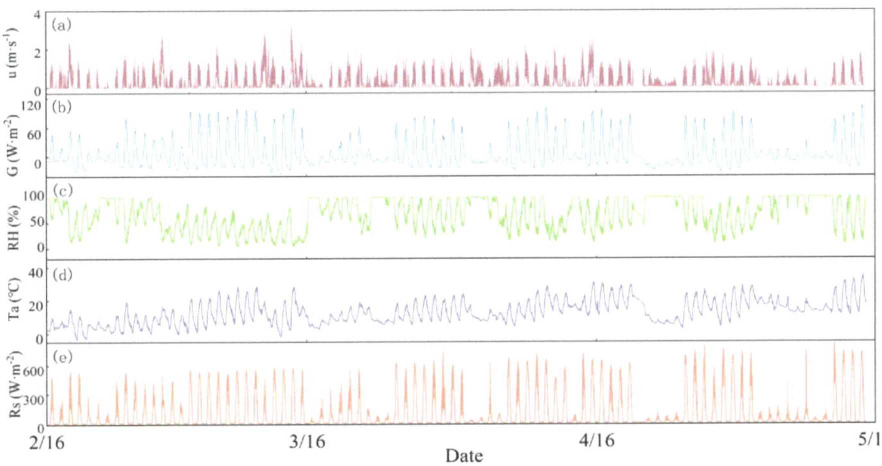

Figure 27. Dynamic variations in (**a**) u (m·s^{-1}); (**b**) G (W·m^{-2}); (**c**) RH (%); (**d**) Ta (°C); (**e**) Rs (W·m^{-2}).

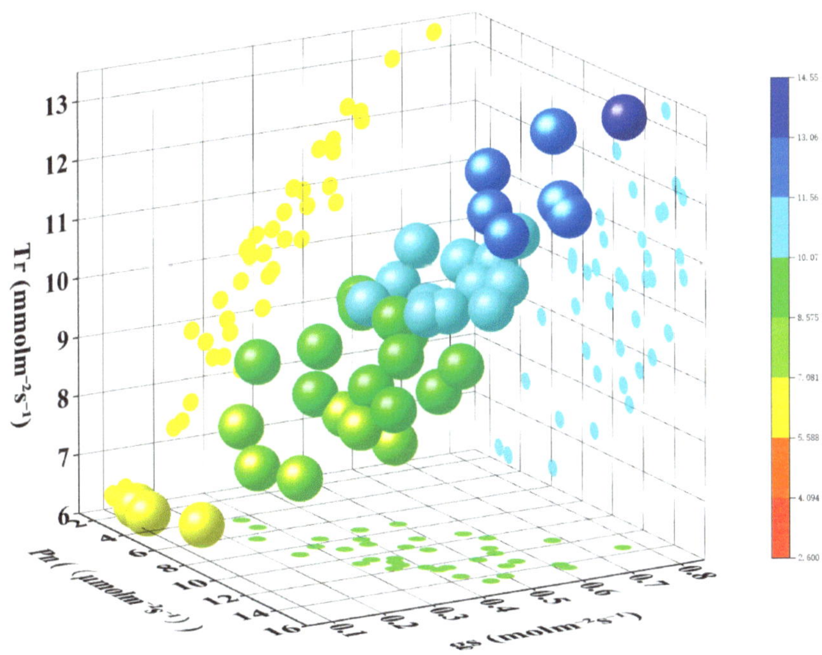

Figure 28. Pn (μmol/(m^2·s)), gs (mol/(m^2·s)), and Tr (mmol/(m^2·s)) in winter wheat.

3.4. Data Processing

3.4.1. Savitzky–Golay (S-G) Filter

The Savitzky–Golay (S-G) filter [30] is a smoothing filtering technique that employs local least squares to eliminate noise from time-series data. This method achieves its smoothing effect by fitting a polynomial to the data, which effectively removes noise while preserving the signal's original shape as closely as possible. Consequently, the S-G filter maintains the integrity of the signal, ensuring effective smoothing.

3.4.2. Z-Score Normalization

Employing the Z-Score standardization method [31], the dimensionless standardization of raw indicator data effectively mitigates the impact of discrepancies in data size, characteristics, and distribution. This approach eliminates unit differences across the data, enabling comparability among variables with diverse characteristics, while preserving the original distribution pattern of the data.

3.4.3. Time-Lag Peak-Seeking Method

The time-lag peak-seeking method [32,33] selects the appropriate function to fit the Ta and Tc and determine the peak position of the fitted curve. The time difference corresponding to this peak point represents the time-lag parameter between the Ta and Tc. Zhang and Wu [34] used the Gaussian function to fit the canopy temperature and atmospheric temperature of summer maize and achieved good accuracy. However, the Gaussian function fits the canopy temperature and atmospheric temperature of winter wheat with lower accuracy. The CCE equation has a higher fitting accuracy for the canopy temperature after smoothing by S-G filtering, and the ECS equation has a higher fitting accuracy for the atmospheric temperature after smoothing by S-G filtering.

The CCE equation is expressed as follows:

$$double1 \quad z = x - xc_1 \tag{2}$$

$$\begin{aligned} y = y_0 + A \times (\exp(-z \times z/(2 \times w)) + (1 - 0.5 \times (1 - \tan(k_2 \times (x - xc_2)))) \times \\ B \times \exp(-0.5 \times k_3 \times (abs(x - xc_3) + (x - xc_3)))) \end{aligned} \tag{3}$$

where xc_1 is the peak moment of winter-wheat canopy temperature. The fitting accuracy was judged by the coefficient of determination (R^2).

The ECS equation is expressed as follows:

$$\begin{aligned} y = y_0 + A/(w \times sqrt(2 \times pi)) \times (exp(-0.5 \times ((x - xc)/w)^2) \times (1 + (a_3/(3 \times 2 \times 1)) \times ((x - xc)/w) \times \\ (((x - xc)/w)^2 - 3) + (a_4/(4 \times 3 \times 2 \times 1)) \times (((x - xc)/w)^4 - 6 \times ((x - xc)/w)^3 + 3) + \\ ((10 \times a_3^2)/(6 \times 5 \times 4 \times 3 \times 2 \times 1)) \times (((x - xc)/w)^6 - 15 \times ((x - xc)/w)^4 + \\ 45 \times ((x - xc)/w)^2 - 15))) \end{aligned} \tag{4}$$

where xc is the moment of peak atmospheric temperature. The accuracy of the fit is judged by the coefficient of determination (R^2), and the peak time difference between Tc and Ta is the time-lag parameter between Tc and Ta.

3.4.4. Time-Lag Cross-Correlation Method

Zhang et al. [35] used the time-lag cross-correlation method to calculate the time lag between the canopy temperature and atmospheric temperature in winter wheat. They then found that correcting the time-lag effect between Tc and Ta by the time-lag cross-correlation method can improve the accuracy of the CWSI inversion of SWC. X (Tc) is first mapped to Y (Ta) in the chronological order of observations. Then, Tc is shifted in steps of 2 min and the Pearson correlation coefficients of the two series are calculated. When the Pearson correlation coefficient attains its maximum value, the corresponding shift duration is designated as the time-lag parameter for the two series [16,36], where the correlation coefficient is calculated as:

$$R_k = \frac{\sum_{i=1}^{n-k} (x_i - \overline{x_i})(y_{i+k} - \overline{y_{i+k}})}{\sqrt{\sum_{i=1}^{n-k} (x_i - \overline{x_i})^2}\sqrt{\sum_{i=1}^{n-k} (y_{i+k} - \overline{y_{i+k}})^2}} \tag{5}$$

$$R_m = \max(R_k) \tag{6}$$

$$T_L = 2m \tag{7}$$

where R_k is Pearson correlation coefficient for a sliding shift number of K; n is the sample size; x_i is the canopy temperature (°C); y_{i+k} is the atmospheric temperature (°C); $\overline{x_i}$ is the mean of canopy temperature series (°C); $\overline{y_{i+k}}$ is the mean of atmospheric temperature series (°C); R_m is the maximum correlation coefficient; m is the sliding shifts in the canopy temperature series that correspond to the maximum Pearson correlation coefficient; $k = 0, \pm 1, \pm 2, \ldots, \pm n$, $k > 0$ indicates the canopy temperature change ahead of atmospheric temperature; and $k < 0$ indicates that canopy temperature changes lag behind atmospheric temperature. T_L is the time-lag parameter (min).

3.4.5. Time-Lag Mutual Information Method

To date, no researcher has calculated the time-lag parameter between Tc and Ta using the time-lag mutual information method. Therefore, this study investigates it. Employing the time-lag mutual information method, the time-lag parameter between the canopy temperature, X, and atmospheric temperature, Y, is determined [37]. The formula is presented as follows:

$$I(X, Y, \tau) = \sum_x \sum_y p(x_t, y_{t+\tau}) \log \frac{p(x_t, y_{t+\tau})}{p(x_t,)p(y_{t+\tau})} \tag{8}$$

where $P(x_t, y_{t+\tau})$ is the $X = x_t, Y = y_{t+\tau}$ joint distribution probability. τ is the time-lag parameter. The time-lag parameter τ is determined when the mutual information coefficient reaches its peak. A positive τ means that x changes before y, while a negative τ indicates that x changes after y.

3.4.6. Gray Time-Lag Correlation Analysis

Currently, no researcher has employed the gray the time-lag correlation analysis to investigate the time-lag effect between the Ta and Tc. Therefore, this study pioneers the use of gray time-lag correlation analysis to calculate the time-lag parameter between Ta and Tc. The methodology is outlined as follows:

① The reference sequence canopy temperature (Tc) is

$$X = (x(1), x(2), \ldots, x(n)) \tag{9}$$

The comparison of the sequence group atmospheric temperature (Ta) is

$$Y_\tau = (y(1+\tau), y(2+\tau), \ldots, y(n+\tau)) \tag{10}$$

where τ is the time-lag parameter.

② Calculate the correlation coefficient $\zeta(x(k), y_\tau(k+\tau))$ between X and Y_τ with the following formula:

$$\zeta(x(k), y_\tau(k+\tau)) = \frac{\min\tau\min k|x(k) - y_\tau(k+\tau)| + \rho\max\tau\max k|x(k) - y_\tau(k+\tau)|}{|x(k) - y_\tau(k+\tau)| + \rho\max\tau\max k|x(k) - y_\tau(k+\tau)|} \tag{11}$$

$$k = 1, 2, 3, \ldots, n \tag{12}$$

$$\tau = 0, 1, \ldots, T - n \tag{13}$$

$$\gamma(\tau) = \frac{1}{n}\sum_{k=1}^{n} \zeta(x(k), y_\tau(k+\tau)) \tag{14}$$

$$\tau = 0, 1, \ldots, T - n \tag{15}$$

where ρ is the resolution factor, $\rho = 0.5$; T is the time span of the time series.

③The time lag parameter τ between Ta and Tc is identified as the time at which $\gamma(\tau)$ peaks.

$$\gamma(\tau^*) = \max_{0 \le \tau \le T-n} \gamma(\tau) \tag{16}$$

where $\gamma(\tau^*)$ is the gray correlation between X and Y, τ^* is the time-lag parameter of Y and X.

3.4.7. CWSI Theoretical Model

Based on the canopy energy balance theory, Jackson, Idso, Reginato and Pinter Jr [8] developed a theoretical model of the CWSI. The formula is as follows:

$$CWSI = \frac{\gamma(1 + \frac{r_c}{r_a}) - \gamma^*}{\Delta + \gamma(1 + \frac{r_c}{r_a})} \tag{17}$$

$$\gamma = 0.665 \times 101.3 \times (\frac{293 - 0.0065Z}{293})^{5.26} \tag{18}$$

$$\gamma^* = \gamma \times (1 + \frac{r_c}{r_a}) \tag{19}$$

$$\Delta = 45.03 + 3.014T + 0.05345T^2 + 0.00224T^3 \tag{20}$$

$$T = \frac{Tc + Ta}{2} \tag{21}$$

$$r_a = \frac{4.72[\ln(\frac{z-d}{z_0})]^2}{(1 + 0.54u)} \tag{22}$$

where the $CWSI$ is crop water stress index; γ is psychrometric coefficient (Pa·°C^{-1}); r_c is canopy resistance (s·m^{-1}); r_a is aerodynamics resistance (s·m^{-1}); Δ is slope of the water vapor pressure curve (Pa·°C^{-1}); Z is height above sea level (m); d is zero-plane displacement (m), d = 0.63 h; z_0 is roughness (m), z_0 = 0.13 h; h is crop height (m); u is reference height wind speed (m·s^{-1}); z is reference height (m), z = 2; and r_c is canopy resistance (s·m^{-1}), displayed in Table 4.

Table 4. r_c of winter wheat at different fertility stages [38].

Growth Period	r_c (s·m^{-1})
Regreening stage-jointing stage	13.01
Jointing stage-tasseling stage	18.03
Tasseling stage-filling stage	26.85

3.4.8. CWSI Empirical Model

The CWSI empirical model was first constructed by Idso et al. [7]. The formula is as follows:

$$CWSI = \frac{(Tc - Ta) - NWSB}{NTB - NWSB} \tag{23}$$

$$CTD = Tc - Ta \tag{24}$$

$$VPD = 0.6108 \times \exp(\frac{17.27 \times Ta}{Ta + 237.7}) \times (\frac{100 - RH}{100}) \tag{25}$$

$$VPG = 0.6108 \times \exp(\frac{17.27 \times Ta}{Ta \times 237.7}) - 0.6108 \times \exp(\frac{17.27 \times (Ta + b)}{(Ta + b) + 237.7}) \tag{26}$$

$$NWSB = a \times VPD + b \tag{27}$$

$$NTB = a \times VPG + b \tag{28}$$

where *Tc* is canopy temperature (°C); *Ta* is atmospheric temperature (°C); *NWSB* is lower bound (no water stress); *NTB* is upper bound (no transpiration); *CTD* is canopy air temperature differential (°C); and *a*, *b* are the slope and intercept of CTD and VPD linear fits, respectively.

Solar radiation intensifies during the period from 13:00 to 15:00, when the discrepancy between crop and soil water supply conditions becomes more pronounced, and the linear relationship between the canopy temperature difference (CTD) and vapor pressure deficit (VPD) is distinct [39]. Consequently, this study opts for a linear fitting of the CTD and VPD specifically for the 13:00–15:00 interval. The results are presented in Figure 29.

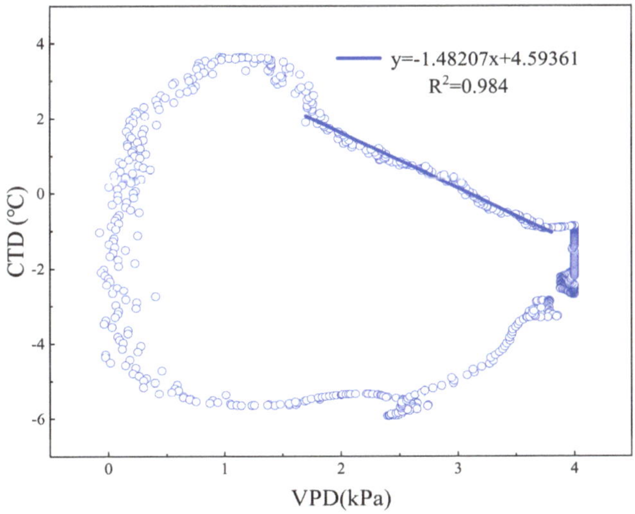

Figure 29. The blue circle represents the trajectory of the VPD (kPa) and the corresponding CTD (°C) over the course of a day; The blue line represents the outcome of a linear regression analysis of the relationship between VPD (kPa) and CTD (°C) between 13:00 and 15:00.

3.4.9. Evaluation Indicators

In this study, the accuracy of the CWSI inversion of photosynthetic parameters, both before and after time-lag corrections, is assessed using the coefficient of determination (R^2). An R^2 value closer to 1 indicates a higher inversion accuracy.

Similarly, the prediction accuracy of the machine learning algorithm is evaluated through the coefficient of determination (R^2) and the root-mean-square error (RMSE), with R^2 values nearing 1 and RMSE values approaching 0 denoting enhanced prediction accuracy.

3.4.10. Machine Learning Algorithms

In this study, various machine learning and deep learning methods were employed to process and predict the crop water stress index (CWSI), including Genetic Algorithm Optimized Support Vector Machines (GA-SVMs), Bayesian Optimized Long Short-Term Memory Neural Networks (Bayes-LSTMs), Particle Swarm Algorithm Optimized Long Short-Term Memory (PSO-LSTM), Convolutional Bi-directional Long Short-Term Memory Neural Networks (CNN-BILSTMs), Attention Mechanism Long Short-Term Memory Neural Networks (Attention-LSTMs), and Attention Mechanism Gated Recurrent Units (Attention-GRUs). The GA-SVM optimizes SVM parameters using a genetic algorithm, effectively enhancing the model's classification and prediction performance, making it suitable for small but complex datasets. PSO-LSTM employs particle swarm optimization to find the optimal parameters for LSTM, improving prediction performance and training efficiency, suitable for scenarios with a large parameter space. The CNN-BILSTM combines a CNN and bi-directional LSTM to simultaneously extract spatial and temporal features, enhancing the prediction capability for complex long time-series data with spatial dependencies. The Attention-LSTM incorporates an attention mechanism into LSTM, enhancing the model's focus on important time steps and improving prediction accuracy, particularly for long time-series data with significant features. The Attention-GRU introduces an attention mechanism into the GRU, simplifying the network structure while improving the focus on important time steps, making it suitable for the efficient prediction of long time-series data. Overall, the introduction of the attention mechanisms (Attention-LSTM and Attention-GRU) significantly enhances the model's ability to capture important information, thus improving prediction accuracy. Bayes-LSTM enhances model robustness by addressing parameter uncertainty. Both PSO-LSTM and GA-SVM improve model performance through optimization algorithms, but are sensitive to initial settings and optimization processes.

4. Discussions

4.1. Time-Lag Parameters between the Tc and Ta Calculated by Different Models

The essence of the peak-finding method is to find a suitable function for fitting [40], and the time difference of the peak of the curve is the time-lag parameter between the two series. In this study, the CCE equation was applied to fit the Tc of winter wheat after S-G filter smoothing, and the ECS equation was applied to fit the Ta after S-G filter smoothing. This is in general agreement with the time-lag parameter between the Tc and Ta for summer maize obtained by Zhang et al. [34]. Considering that the time-lag effect was the persistent influence of previous climatic conditions on current crop growth as a result of the cumulative effects of meteorological factors and soil moisture content on the crop [41,42], there were limitations in determining the time-lag parameter through isolated points. Therefore, the time-lag parameter between the Ta and Tc was calculated using the time-lag cross-correlation method [43]. The time-lag parameter calculated in this study was about 32–44 min, consistent with the findings of Zhang et al. [44].

Meanwhile, this study innovatively utilized the time-lag mutual information method [45] and gray time-lag correlation analysis [46] to calculate the time-lag parameter between the Ta and Tc. The time-lag parameter calculated by the time-lag mutual information method ranged from 42 to 58 min, while the gray time-lag correlation analysis-calculated time-lag parameter ranged from 76 to 98 min. Additionally, the time-lag parameter between the Tc and Ta in winter wheat calculated by the four methods all experienced a significant sudden drop under the heavy water stress treatment. Pn, Tr, and gs all exhibited a decreasing trend with diminishing soil moisture [47], and a sudden drop occurred during the severe water stress treatment (T4) [2]. This phenomenon might be related to the soil moisture quench value [5].

Mild water stress does not affect the normal life activities of the crop, and the physiological activities of the plant are limited only when the degree of drought stress exceeds the drought threshold. When the soil moisture threshold is reached, stomata are reduced or closed, and water lost through stomatal transpiration and CO_2 entering the chloroplasts is reduced [48]. As a result, Pn, Tr, and gs undergo varying degrees of reduction [49]. Wu et al. [50] found that, when the soil volumetric water content was lower than 60% of the field holding capacity for a long period of time, leaf enlargement was restricted, the total leaf area for light energy interception was reduced, and the gas exchange process of winter wheat was limited, which was the reason for the sudden decrease in photosynthetic parameters under severe water stress. At the same time, the decrease in stomatal conductance reduces crop transpiration, evaporative cooling was reduced, and canopy temperatures continue to rise [51], reaching their peaks later, resulting in a decrease in the time-lag parameter between the Tc and Ta in the heavy water stress treatment. Liu et al. [52] found that the soil moisture quench value of winter wheat was about 43.5–52.2% of the soil water content, which was consistent with the soil moisture treatments in this experiment, where there were abrupt changes in photosynthetic and time-lag parameters.

4.2. Reasons for Different Changes in the Magnitude of the Accuracy of the CWSI Inversions of Pn, Tr, and gs before and after Corrections of the Time-Lag Effect

After considering the time-lag effect, the magnitudes of correlations between the CWSI and Pn, Tr, and gs varied inconsistently, which might be related to the different major environmental factors affecting Pn, Tr, and gs [29], as well as their distinct critical soil moisture thresholds. When the crop was not subjected to water stress, environmental factors had a small and negligible effect on Pn, Tr was mainly limited by solar radiation, and gs was primarily limited by photosynthetically active radiation and crop canopy temperature. When crops were subjected to water stress, Pn was mainly limited by relative humidity and atmospheric temperature, Tr was chiefly limited by saturated water-vapor pressure difference, and gs was predominantly limited by saturated water-vapor pressure difference and wind speed [53]. Meanwhile, Pn, Tr, and gs showed different sensitivities to soil water deficit. The critical soil moisture thresholds for Pn, Tr, and gs were 62%, 60%, and 58% for maize at the seedling stage and 51%, 53%, and 48% at the nodulation stage, respectively. This indicated that crop photosynthetic parameters were sensitive to soil moisture in the order of gs > Tr > Pn [14], consistent with the magnitude of the CWSI correlation with Pn, Tr, and gs obtained in this study [18]. At the same time, this might result in a varying degree of improvement in the correlation between the CWSI and Pn, Tr, and gs before and after accounting for time-lag effects (gs > Tr > Pn).

4.3. Outlook

Physiological parameters of plants at different growth stages exhibit varying sensitivities to soil moisture [14]. This suggests that the photosynthetic parameters of winter wheat at different fertility stages show differential sensitivities to the CWSI under varying water stress conditions. The impact of water stress on the crop's gas exchange processes is minimal during the regrowth period, with little variation in Pn, Tr, and Gs across different water stress levels. However, the inhibitory effects of persistent water stress during the nodulation–irrigation period are more pronounced, indicating a more significant decrease in Pn, Tr, and gs in winter wheat under severe water stress [54]. Therefore, there are significant seasonal variations in the correlation of the CWSI with Pn, Tr, and gs in winter wheat subjected to different moisture treatments. In this study, the impact of the time-lag effect on the accuracy of the CWSI inversion of photosynthesis parameters is investigated only for the entire reproductive period. The influence of time lag between the Ta and Tc on the accuracy of the CWSI inversion of photosynthesis parameters during each reproductive phase is not discussed and requires further study.

5. Conclusions

In this study, we investigate the impact of the time-lag effect between the Tc and Ta on the correlation between the CWSI and photosynthetic parameters. The main conclusions are: (1) The magnitude of the time-lag parameter between the Tc and Ta in winter wheat, calculated by the four methods for the entire reproductive period, follows the order: gray time-lag correlation analysis > time-lag peak-seeking method > time-lag mutual information method > time-lag cross-correlation method. All time-lag parameters of severe water stress treatment experience a sudden decrease. (2) The CWSI empirical model is more sensitive to the time-lag effect than the theoretical model. Time-lag correction, particularly using the time-lag mutual information method, significantly improves the correlation between the CWSI and photosynthetic parameters. (3) The GA-SVM machine learning algorithm provides the highest prediction accuracy for daily changes in the CWSI empirical model corrected with the time-lag mutual information method (R^2 = 0.982, RMSE = 0.017).

Author Contributions: Y.W.: Field trials collecting data, analyzing data, conceptualizing, calculating, and writing first drafts; Y.L.: Assisted with calculating the data; N.Y.: Assisted with the experiments; J.C.: Assisted with the conceptualization; J.W. and Z.H.: Assisted with reviewing the manuscripts; Z.Z.: provided funding for experiments. All authors have read and agreed to the published version of the manuscript.

Funding: This study was financially supported by the National Natural Science Foundation of China (51979232, 52179044), and Natural Science Foundation in Shaanxi Province of China (2019JM-066).

Data Availability Statement: Data are contained within the article.

Conflicts of Interest: The authors declare no conflicts of interest.

References

1. Kifle, M.; Gebretsadikan, T.G. Yield and water use efficiency of furrow irrigated potato under regulated deficit irrigation, Atsibi-Wemberta, North Ethiopia. *Agric. Water Manag.* **2016**, *170*, 133–139. [CrossRef]
2. Liu, E.K.; Mei, X.R.; Yan, C.R.; Gong, D.Z.; Zhang, Y.Q. Effects of water stress on photosynthetic characteristics, dry matter translocation and WUE in two winter wheat genotypes. *Agric. Water Manag.* **2016**, *167*, 75–85. [CrossRef]
3. Saini, S.; Raj, K.; Saini, A.K.; Kumar, R.; Saini, A.; Khan, A.; Kumar, P.; Devi, G.; Bhambhu, M.K.; McKenzie, C.L.; et al. Unravelling the synergistic interaction of Thrips tabaci and newly recorded, Thrips parvispinus with Alternaria porri (Ellis.) Cif., inciting onion purple blotch. *Front. Microbiol.* **2024**, *15*, 1321921. [CrossRef]
4. Zhang, M.; Chen, W.; Jing, M.; Gao, Y.; Wang, Z. Canopy Structure, Light Intensity, Temperature and Photosynthetic Performance of Winter Wheat under Different Irrigation Conditions. *Plants* **2023**, *12*, 3482. [CrossRef]
5. Carrasco-Benavides, M.; Espinoza-Meza, S.; Umemura, K.; Ortega-Farías, S.; Baffico-Hernández, A.; Neira-Román, J.; Ávila-Sánchez, C.; Fuentes, S. Evaluation of thermal-based physiological indicators for determining water-stress thresholds in drip-irrigated 'Regina' cherry trees. *Irrig. Sci.* **2024**, *42*, 445–459. [CrossRef]
6. Desoky, E.-S.M.; Mansour, E.; El-Sobky, E.-S.E.A.; Abdul-Hamid, M.I.; Taha, T.F.; Elakkad, H.A.; Arnaout, S.M.A.I.; Eid, R.S.M.; El-Tarabily, K.A.; Yasin, M.A.T. Physio-Biochemical and Agronomic Responses of Faba Beans to Exogenously Applied Nano-Silicon Under Drought Stress Conditions. *Front. Plant Sci.* **2021**, *12*, 637783. [CrossRef]
7. Idso, S.B.; Jackson, R.D.; Pinter, P.J.; Reginato, R.J.; Hatfield, J.L. Normalizing the stress-degree-day parameter for environmental variability. *Agric. Meteorol.* **1981**, *24*, 45–55. [CrossRef]
8. Jackson, R.D.; Idso, S.; Reginato, R.; Pinter, P.J., Jr. Canopy temperature as a crop water stress indicator. *Water Resour. Res.* **1981**, *17*, 1133–1138. [CrossRef]
9. Chungloo, D.; Tisarum, R.; Samphumphuang, T.; Pipatsitee, P.; Sotesaritkul, T.; Cha-um, S. Exogenous glycine betaine alleviates water-deficit stress in Indian pennywort (*Centella asiatica*) under greenhouse conditions. *Protoplasma* **2024**. [CrossRef]
10. Pipatsitee, P.; Eiumnoh, A.; Praseartkul, P.; Taota, K.; Kongpugdee, S.; Sakulleerungroj, K.; Cha-um, S. Application of infrared thermography to assess cassava physiology under water deficit condition. *Plant Prod. Sci.* **2018**, *21*, 398–406. [CrossRef]
11. Ramos-Fernández, L.; Gonzales-Quiquia, M.; Huanuqueño-Murillo, J.; Tito-Quispe, D.; Heros-Aguilar, E.; Flores del Pino, L.; Torres-Rua, A. Water Stress Index and Stomatal Conductance under Different Irrigation Regimes with Thermal Sensors in Rice Fields on the Northern Coast of Peru. *Remote Sens.* **2024**, *16*, 796. [CrossRef]
12. Abdalla, M.; Carminati, A.; Cai, G.; Javaux, M.; Ahmed, M.A. Stomatal closure of tomato under drought is driven by an increase in soil–root hydraulic resistance. *Plant Cell Environ.* **2021**, *44*, 425–431. [CrossRef]
13. Rodriguez-Dominguez, C.M.; Brodribb, T.J. Declining root water transport drives stomatal closure in olive under moderate water stress. *New Phytol.* **2020**, *225*, 126–134. [CrossRef]

14. Ma, S.; Liu, S.; Gao, Z.; Wang, X.; Ma, S.; Wang, S. Water Deficit Diagnosis of Winter Wheat Based on Thermal Infrared Imaging. *Plants* **2024**, *13*, 361. [CrossRef]
15. Gonzalez-Dugo, V.; Zarco-Tejada, P.J. Assessing the impact of measurement errors in the calculation of CWSI for characterizing the water status of several crop species. *Irrig. Sci.* **2022**, *42*, 431–443. [CrossRef]
16. Huang, J.; Wang, S.; Guo, Y.; Chen, J.; Yao, Y.; Chen, D.; Liu, Q.; Zhang, Y.; Zhang, Z.; Xiang, Y. Hysteresis between winter wheat canopy temperature and atmospheric temperature and its driving factors. *Plant Soil* **2022**, *499*, 55–71. [CrossRef]
17. Zhang, Q.; Yang, X.; Liu, C.; Yang, N.; Yu, G.; Zhang, Z.; Chen, Y.; Yao, Y.; Hu, X. Monitoring soil moisture in winter wheat with crop water stress index based on canopy-air temperature time lag effect. *Int. J. Biometeorol.* **2024**, *68*, 647–659. [CrossRef]
18. Ma, X.; He, Q.; Zhou, G. Sequence of Changes in Maize Responding to Soil Water Deficit and Related Critical Thresholds. *Front. Plant Sci.* **2018**, *9*, 511. [CrossRef]
19. Farooq, M.; Kobayashi, N.; Ito, O.; Wahid, A.; Serraj, R. Broader leaves result in better performance of indica rice under drought stress. *J. Plant Physiol.* **2010**, *167*, 1066–1075. [CrossRef]
20. Cao, D.; Xu, N.; Chen, Y.; Zhang, H.; Li, Y.; Yuan, Z. Construction of a Pearson- and MIC-Based Co-expression Network to Identify Potential Cancer Genes. *Interdiscip. Sci. Comput. Life Sci.* **2022**, *14*, 245–257. [CrossRef]
21. Su, Z.-Y.; Wu, T.; Wang, Y.-T.; Huang, H.-Y. An investigation into the linear and nonlinear correlation of two music walk sequences. *Phys. D Nonlinear Phenom.* **2008**, *237*, 1815–1824. [CrossRef]
22. Wang, Y.; Huang, Y.; Zeng, X.; Wei, G.; Zhou, J.; Fang, T.; Chen, H. Faulty Feeder Detection of Single Phase-Earth Fault Using Grey Relation Degree in Resonant Grounding System. *IEEE Trans. Power Deliv.* **2017**, *32*, 55–61. [CrossRef]
23. Stockle, C.O.; Dugas, W.A. Evaluating canopy temperature-based indices for irrigation scheduling. *Irrig. Sci.* **1992**, *13*, 31–37. [CrossRef]
24. Ricotta, C.; Avena, G.C.; Teggi, S. Relation between vegetation canopy surface temperature and the Sun-surface geometry in a mountainous region of central Italy. *Int. J. Remote Sens.* **1997**, *18*, 3091–3096. [CrossRef]
25. Ding, Y.; Li, Z.; Peng, S. Global analysis of time-lag and -accumulation effects of climate on vegetation growth. *Int. J. Appl. Earth Obs. Geoinf.* **2020**, *92*, 102179. [CrossRef]
26. He, X.; Liu, A.; Tian, Z.; Wu, L.; Zhou, G. Response of Vegetation Phenology to Climate Change on the Tibetan Plateau Considering Time-Lag and Cumulative Effects. *Remote Sens.* **2024**, *16*, 49. [CrossRef]
27. Xu, S.; Wang, Y.; Liu, Y.; Li, J.; Qian, K.; Yang, X.; Ma, X. Evaluating the cumulative and time-lag effects of vegetation response to drought in Central Asia under changing environments. *J. Hydrol.* **2023**, *627*, 130455. [CrossRef]
28. Hatfield, J.L. The utilization of thermal infrared radiation measurements from grain sorghum crops as a method of assessing their irrigation requirements. *Irrig. Sci.* **1983**, *3*, 259–268. [CrossRef]
29. Maai, E.; Nishimura, K.; Takisawa, R.; Nakazaki, T.J.P.P.S. Light stress-induced chloroplast movement and midday depression of photosynthesis in sorghum leaves. *Plant Prod. Sci.* **2020**, *23*, 172–181. [CrossRef]
30. Liu, Y.; Dang, B.; Li, Y.; Lin, H.; Ma, H. Applications of Savitzky-Golay Filter for Seismic Random Noise Reduction. *Acta Geophys.* **2016**, *64*, 101–124. [CrossRef]
31. Guo, Y.; Mao, H.; Ding, H.; Wu, X.; Liu, Y.; Liu, H.; Zhou, S. Data-Driven Coordinated Development of the Digital Economy and Logistics Industry. *Sustainability* **2022**, *14*, 8963. [CrossRef]
32. Zeng, C.; Lai, W.; Lin, H.; Liu, G.; Qin, B.; Kang, Q.; Feng, X.; Yu, Y.; Gu, R.; Wu, J.; et al. Weak information extraction of gamma spectrum based on a two-dimensional wavelet transform. *Radiat. Phys. Chem.* **2023**, *208*, 110914. [CrossRef]
33. Feng, S.; Zhang, Z.; Zhao, S.; Guo, X.; Zhu, W.; Das, P. Time lag effect of vegetation response to seasonal precipitation in the Mara River Basin. *Ecol. Process.* **2023**, *12*, 49. [CrossRef]
34. Zhang, Z.; Wu, T. Time delay effect of summer maize canopy temperature change and its influence on soil moisture content monitoring. *Trans. Chin. Soc. Agric. Eng.* **2022**, *38*, 117–124. [CrossRef]
35. Zhang, B.; Huang, J.; Dai, T.; Jing, S.; Hua, Y.; Zhang, Q.; Liu, H.; Wu, Y.; Zhang, Z.; Chen, J. Assessing accuracy of crop water stress inversion of soil water content all day long. *Precis. Agric.* **2024**, 1–21. [CrossRef]
36. Konda, F.; Okamura, M.; Ali Akbar, M.; Yokoi, T. Fiber Speed and Yarn Tension in Friction Spinning. *Text. Res. J.* **1996**, *66*, 343–348. [CrossRef]
37. Ji, C.; Ma, F.; Wang, J.; Wang, J.; Sun, W. Real-Time Industrial Process Fault Diagnosis Based on Time Delayed Mutual Information Analysis. *Processes* **2021**, *9*, 1027. [CrossRef]
38. Yuan, G.; Luo, Y.; Tang, D.; Yu, Q.; Yu, L. Estimating Minimum Canopy Resistances of Winter Wheat at Different Development Stages. *Acta Ecol. Sin.* **2002**, *22*, 930–934.
39. Gontia, N.K.; Tiwari, K.N. Development of crop water stress index of wheat crop for scheduling irrigation using infrared thermometry. *Agric. Water Manag.* **2008**, *95*, 1144–1152. [CrossRef]
40. Mukoyama, T. Fitting of Gaussian peaks by simulated annealing. *X-Ray Spectrom.* **2017**, *46*, 63–68. [CrossRef]
41. Braswell, B.H.; Schimel, D.S.; Linder, E.; Moore, B. The Response of Global Terrestrial Ecosystems to Interannual Temperature Variability. *Science* **1997**, *278*, 870–873. [CrossRef]
42. Ma, Y.; Guan, Q.; Sun, Y.; Zhang, J.; Yang, L.; Yang, E.; Li, H.; Du, Q. Three-dimensional dynamic characteristics of vegetation and its response to climatic factors in the Qilian Mountains. *CATENA* **2022**, *208*, 105694. [CrossRef]
43. Pallotta, L.; Giunta, G. Accurate Delay Estimation for Multisensor Passive Locating Systems Exploiting the Cross-Correlation Between Signals Cross-Correlations. *IEEE Trans. Aerosp. Electron. Syst.* **2022**, *58*, 2568–2576. [CrossRef]

44. Zhang, Z.; Zhang, Q.; Yang, N.; Luo, L.; Huang, J.; Yao, Y. Time Lag Effect between Winter Wheat Canopy Temperature and Atmospheric Temperature and Its Influencing Factors. *Trans. Chin. Soc. Agric. Mach.* **2023**, *54*, 359–368. [CrossRef]
45. Albers, D.J.; Hripcsak, G. Estimation of time-delayed mutual information and bias for irregularly and sparsely sampled time-series. *Chaos Solitons Fractals* **2012**, *45*, 853–860. [CrossRef]
46. Liu, C.; Chen, Y.; Zou, L.; Cheng, B.; Huang, T. Time-Lag Effect: River Algal Blooms on Multiple Driving Factors. *Front. Earth Sci.* **2022**, *9*, 813287. [CrossRef]
47. Liu, J.; Zhang, J.; Shi, Q.; Liu, X.; Yang, Z.; Han, P.; Li, J.; Wei, Z.; Hu, T.; Liu, F. The Interactive Effects of Deficit Irrigation and Bacillus pumilus Inoculation on Growth and Physiology of Tomato Plant. *Plants* **2023**, *12*, 670. [CrossRef]
48. Egan, L.; Hofmann, R.; Nichols, S.; Hadipurnomo, J.; Hoyos-Villegas, V. Transpiration Rate of White Clover (*Trifolium repens* L.) Cultivars in Drying Soil. *Front. Plant Sci.* **2021**, *12*, 595030. [CrossRef]
49. Didion-Gency, M.; Gessler, A.; Buchmann, N.; Gisler, J.; Schaub, M.; Grossiord, C. Impact of warmer and drier conditions on tree photosynthetic properties and the role of species interactions. *New Phytol.* **2022**, *236*, 547–560. [CrossRef]
50. Wu, H.; Duan, A.; Yang, C. Physiological and Morphological Responses of Winter Wheat to Soil Moisture. *Acta Agric. Boreali-Sin.* **2000**, *15*, 92–96.
51. Struthers, R.; Ivanova, A.; Tits, L.; Swennen, R.; Coppin, P. Thermal infrared imaging of the temporal variability in stomatal conductance for fruit trees. *Int. J. Appl. Earth Obs. Geoinf.* **2015**, *39*, 9–17. [CrossRef]
52. Liu, J.; Wang, F.; Yu, Q.; Mao, F.; Bi, J.; Fan, G. Application of the leaf photosynthesis model for forecasting effect of drought on winter in north china plain. *J. Appl. Meteorol. Sci.* **2003**, *14*, 469–478.
53. Zhou, P.; Gao, P.; Chen, W.; Yang, M.; Ding, Z.; Liu, H.; Wang, W. Response of photosynthetic parameters of citrus trees to water and meteorological factors 53.108.under different water treatments. *Water Sav. Irrig.* **2022**, *6*, 90–95.
54. Zhao, W.; Liu, L.; Shen, Q.; Yang, J.; Han, X.; Tian, F.; Wu, J. Effects of Water Stress on Photosynthesis, Yield, and Water Use Efficiency in Winter Wheat. *Water* **2020**, *12*, 2127. [CrossRef]

plants

Article

A New Spectral Index for Monitoring Leaf Area Index of Winter Oilseed Rape (*Brassica napus* L.) under Different Coverage Methods and Nitrogen Treatments

Hao Liu [1,2], Youzhen Xiang [1,2,*], Junying Chen [1,2], Yuxiao Wu [1,2], Ruiqi Du [1,2], Zijun Tang [1,2], Ning Yang [1,2], Hongzhao Shi [1,2], Zhijun Li [1,2] and Fucang Zhang [1,2]

[1] Key Laboratory of Agricultural Soil and Water Engineering in Arid and Semiarid Areas, Ministry of Education, Northwest A&F University, Yangling 712100, China; hliunwafu@163.com (H.L.); junyingchen@nwafu.edu.cn (J.C.); phyllis12@nwafu.edu.cn (Y.W.); duruiqi@126.com (R.D.); tangzijun@nwsuaf.edu.cn (Z.T.); yangning1228@126.com (N.Y.); shihongzhao7@nwsuaf.edu.cn (H.S.); lizhij@nwsuaf.edu.cn (Z.L.); zhangfc@nwsuaf.edu.cn (F.Z.)

[2] College of Water Resources and Architectural Engineering, Northwest A&F University, Yangling 712100, China

* Correspondence: youzhenxiang@nwsuaf.edu.cn

Abstract: The leaf area index (LAI) is a crucial physiological indicator of crop growth. This paper introduces a new spectral index to overcome angle effects in estimating the LAI of crops. This study quantitatively analyzes the relationship between LAI and multi-angle hyperspectral reflectance from the canopy of winter oilseed rape (*Brassica napus* L.) at various growth stages, nitrogen application levels and coverage methods. The angular stability of 16 traditional vegetation indices (VIs) for monitoring the LAI was tested under nine view zenith angles (VZAs). These multi-angle VIs were input into machine learning models including support vector machine (SVM), eXtreme gradient boosting (XGBoost), and Random Forest (RF) to determine the optimal monitoring strategy. The results indicated that the back-scattering direction outperformed the vertical and forward-scattering direction in terms of monitoring the LAI. In the solar principal plane (SPP), EVI-1 and REP showed angle stability and high accuracy in monitoring the LAI. Nevertheless, this relationship was influenced by experimental conditions and growth stages. Compared with traditional VIs, the observation perspective insensitivity vegetation index (OPIVI) had the highest correlation with the LAI (r = 0.77–0.85). The linear regression model based on single-angle OPIVI was most accurate at $-15°$ (R^2 = 0.71). The LAI monitoring achieved using a multi-angle OPIVI-RF model had the higher accuracy, with an R^2 of 0.77 and with a root mean square error (RMSE) of 0.38 $cm^2 \cdot cm^{-2}$. This study provides valuable insights for selecting VIs that overcome the angle effect in future drone and satellite applications.

Keywords: leaf area index; multi-angle hyperspectral; machine learning; winter oilseed rape (*Brassica napus* L.)

Citation: Liu, H.; Xiang, Y.; Chen, J.; Wu, Y.; Du, R.; Tang, Z.; Yang, N.; Shi, H.; Li, Z.; Zhang, F. A New Spectral Index for Monitoring Leaf Area Index of Winter Oilseed Rape (*Brassica napus* L.) under Different Coverage Methods and Nitrogen Treatments. *Plants* **2024**, *13*, 1901. https://doi.org/10.3390/plants13141901

Academic Editor: Georgios Koubouris

Received: 14 May 2024
Revised: 3 July 2024
Accepted: 5 July 2024
Published: 10 July 2024

1. Introduction

The leaf area index (LAI) is a crucial parameter for characterizing plant canopy structure, influencing processes such as light interception, respiration, transpiration, and net primary productivity [1]. The LAI reflects the dynamic changes in growth characteristics, canopy light distribution, respiration, photosynthesis, water vapor release, and carbon cycling [2]. Traditional methods for monitoring crop LAI, such as length–width factor method, lattice method, paper weight method, and laser leaf area method, can provide accurate measurements for small areas. However, these methods are time consuming, destructive, and unsuitable for large-scale monitoring [3]. Hyperspectral remote sensing emerged as a prominent tool in precision agriculture for monitoring crop growth, offering robust spectral continuity and high spectral information content for real-time and extensive LAI monitoring [4].

The vegetation canopy exhibits significant vertical gradients in biochemical and biophysical properties, posing challenges for traditional remote sensing techniques that rely on vertical observations [5]. Multi-angle remote sensing addresses these challenges by capturing comprehensive geometrical and spatial distribution information from multiple directions. This approach is widely used to derive crop growth and nutritional status from remote sensing data [6,7]. Hasegawa et al. [8] improved the accuracy of LAI inversion and mitigated saturation effects associated with the normalized difference vegetation index (NDVI) by integrating traditional vertical angle vegetation indices with hotspot and dark point information from multi-angle remote sensing. Similarly, Stagakis et al. [9] compared the ability of different view zenith angles (VZAs) to invert the LAI using satellite imagery data and ground-measured LAI data from 64 vegetation indices (VIs). These studies collectively highlight that non-vertical observations offer more reliable insights into canopy structural information compared to vertical observations, thus reducing errors in LAI inversion associated with canopy structural features [10].

High reflectance in the near-infrared (NIR) band indicates multiple scattering within plant leaf blades and canopy leaves, making it a reliable indicator of the LAI [11]. Extensive research has compared the stability and accuracy of various VIs for LAI estimation, optimizing them to enhance their linear correlations [12–14]. For instance, the widely used NDVI has limitations such as saturation and non-linear relationships in dense canopies and vigorous growth conditions [15]. To account for soil background effects, soil-corrected VIs like the soil-adjusted vegetation index (SAVI), modified soil-adjusted vegetation index (MSAVI), and optimized soil-adjusted vegetation index (OSAVI) have been developed [16–18]. The meris terrestrial chlorophyll index (MTCI), involving green and red edges, can alleviate LAI saturation [19]. Introducing the blue band into the NDVI helps mitigate the impact of atmospheric and surface factors [20]. These established VIs maximize sensitivity to LAI changes while minimizing adverse effects from soil, light, and atmosphere. However, as VIs are derived from the reflectance of multiple bands through various combinations, they exhibit varying sensitivities to view zenith angles (VZAs). For example, the NDVI generally shows higher values at larger VZAs, whereas the enhanced vegetation index (EVI) displays stronger angle sensitivity [10,21]. Galvão et al. [22] proposed that the EVI and photochemical reflectance index (PRI) exhibit strong anisotropy, whereas the NDVI and the Vogelmann index (VOG) show weaker angle sensitivities. Verrelst et al. [23] suggested that VIs constructed from bands between 450 and 1050 nm display significantly different angle sensitivities depending on crop types. Hovi et al. [24] emphasized that canopy structure significantly influences reflectance variations at different VZAs. Most of these studies have focused on analyzing the angle sensitivity of VIs observed vertically, with limited research utilizing multi-angle remote sensing data for crop growth monitoring. Therefore, it is necessary to analyze the anisotropic characteristics of crop canopies to reduce the impact of VZAs on monitoring crop growth and to develop novel multi-angle VIs. He et al. [25] improved the prediction accuracy and stability of leaf nitrogen content (LNC) by introducing a new angle-insensitive vegetation index (AIVI). Li et al. [26] developed an index algorithm based on multi-angle reflectance factors for leaf surfaces, which is effective for measuring leaf water content and applicable to various plant species with different significant and moisture contents under different measurement conditions. Currently, there is a lack of systematic analysis of the relationship between multi-angle VIs and the LAI, particularly for oilseed crops like rapeseed. This gap severely limits the application of various spectral analysis methods in multi-angle remote sensing. Future studies should focus on analyzing the angle sensitivity of different spectral analysis methods to identify VIs and techniques most sensitive to the LAI while being least sensitive to VZAs.

Oilseed rape is a widely cultivated essential oilseed crop globally, including in China, and its growth and yield formation are significantly influenced by nitrogen nutrition. The real-time, accurate, and rapid estimation of rapeseed LAI is essential for diagnosing and managing its growth and predicting yields [27]. This study was conducted to design field experiments involving various coverage methods and levels of nitrogen fertilization, with the aim of analyzing parameter changes under different VZAs. The goal is to develop VIs that exhibit stable performance under different VZAs and experimental conditions. Additionally, the optimal VI is inputted into various machine learning models to compare the monitoring accuracy of each model. This research provides technical guidance and a theoretical foundation for enhancing ground-based technologies in multi-angle remote sensing applications.

2. Results

2.1. Changes in the Canopy Reflectance and VIs of Feature Bands at Different VZAs

In this study, we selected five representative feature spectral bands: blue (450 nm), green (550 nm), red (660 nm), red edge (720 nm), and NIR bands (780 nm). As illustrated in Figure 1, visible light exhibits a significantly asymmetrical shape within the SPP, while the red edge and NIR bands are almost symmetrical. For visible light, the reflectance of blue and red bands shows minimal change with the varying VZA, with amplitudes within 40%. In contrast, the green and NIR bands exhibit higher amplitudes, reaching up to 122%. Generally, in the back-scatter direction, the reflectance of all five feature bands increases with an increase in VZA, with the minimum reflectance occurring near the vertical angle.

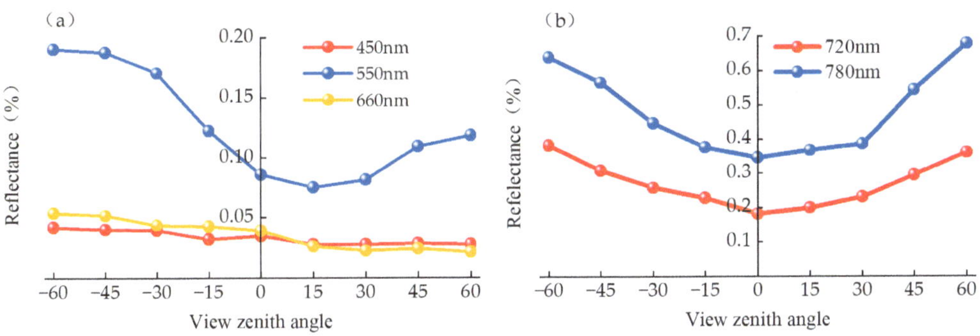

Figure 1. Reflectance changes in blue, green, red, red edge, and NIR bands with different VZAs at the stage of flowering of winter oilseed rape. (**a**) Visible wavelengths. (**b**) NIR wavelengths.

The changes in 16 VIs constructed from these five feature bands with respect to VZAs show fluctuations. Table 1 presents the change amplitude and F significance analysis of VIs relative to the vertical angle. A larger F-value indicates greater variation within that angular range. The results show that all the parameter variations exhibit significant differences at the $p < 0.05$ level in the range of −60° to +60°, while all the other parameters except the PRI, DVI, NDDA, VOG-2, DD, and CCII did not show significant differences at the $p < 0.05$ level in the range of +30° to +60° VZA. As shown in Figure 2, all parameters exhibit significant angular changes, with the NDDA showing the most prominent change between the forward- and back-scatter direction, with amplitudes reaching −29.77% and 95.42%. The F-test revealed significant difference at the $p < 0.01$ level in the range of −60° to +60° and +30° to +60°, indicating that the NDDA is the most sensitive parameter to VZAs. In the forward-scatter direction, most parameters show gradual stabilization after +30°, while the NDRE, NDDA, VOG-2, and CCII continue to increase with increasing VZAs. In the back-scatter direction, most parameters show a sharp increase or decrease after −15°, with the SAVI, NDDA, and DD being the most obvious. Additionally, the PRI, NDDA, and DD undergo sudden changes at a VZA of −45°. R1-dB, mND705, DDn, and DD exhibit a symmetrical shape in both the forward- and back-scatter direction, with their minimum values occurring at the vertical angle and increasing with VZAs. In summary, Vlopt, the OPIVI, and REP exhibit no significant change with the respect to VZAs and show a gentle angular change trend within the SPP.

Table 1. Percentages of change compared to nadir values for extreme VZAs and ANOVA F-ratio values.

Index	−60° vs. Nadir (%)	+60° vs. Nadir (%)	ANOVA F Values ($F_{8,1620}$) (−60°~+60°)	ANOVA F Value ($F_{2,360}$) (30°~60°)
Two bands				
PRI	1.74	−8.98	3.195 ***	2.805 **
R1-dB	2.33	8.72	7.129 ***	1.509
SAVI	12.52	5.47	3.248 ***	1.160
NDRE	3.23	21.18	5.723 ***	1.298
DVI	8.60	−6.87	2.722 ***	2.180 *
Vlopt	4.79	3.76	2.131 **	0.604
Three bands				
mND705	6.64	18.19	6.555 ***	1.226
NDDA	−29.77	95.42	3.901 ***	3.228 **
MTCI	−2.52	18.12	6.119 ***	1.297
EVI-1	11.81	3.90	1.819 *	1.597
DDn	20.22	17.21	1.947 *	1.751
OPIVI	1.60	−6.61	1.735 *	0.444
Four bands				
VOG-2	6.35	36.14	5.951 ***	3.447 **
DD	19.31	19.16	4.877 ***	2.062 *
REP	−2.36	3.31	1.877 *	0.760
CCII	−0.70	10.12	3.592 ***	2.499 *

* $p < 0.05$. ** $p < 0.01$. *** $p < 0.001$.

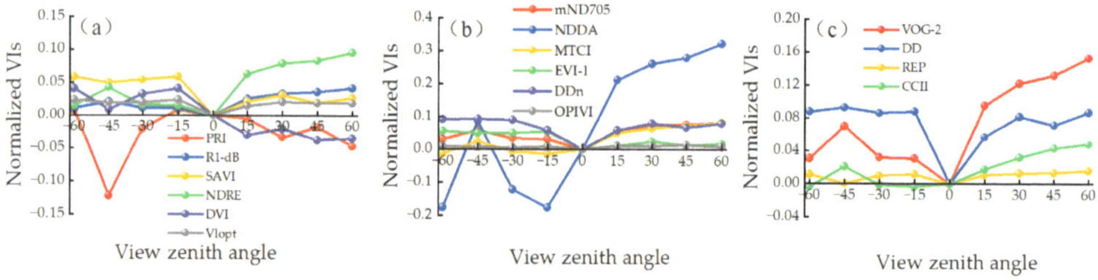

Figure 2. Normalized differences in VI values with respect to nadir between different VZAs within SPP in winter oilseed rape. *X*-axis represents the VZA (negative values represent back-scatter direction, positive values represent forward-scatter direction). *Y*-axis represents the normalized difference. (**a**) Two bands, (**b**) three bands, (**c**) four bands.

2.2. Relationship between LAI and VIs under Different VZAs

This study comprehensively analyzes the relationship between VIs and the LAI based on the performance of the same spectral parameter under different VZAs and the performance of different VIs under the same VZAs. A linear regression model was established to monitor the LAI using VIs under different VZAs. As shown in Figure 3, VIs have a closer relationship with the LAI in the back-scatter direction. Among the 16 parameters, MTCI exhibited the poorest regression potential (R^2 = 0.09–0.45) for all VZAs. Additionally, VIs such as the NDRE, NDDA, VOG-2, and CCII also showed poorer regression potential. For two-band VIs, although the PRI, SAVI, and DVI maintain a good correlation with the LAI near the vertical angle, their correlation coefficients decrease with increasing VZAs in both the forward- and back-scatter direction, indicating poor stability. In contrast, among the three-band VIs, mND705, EVI-1, and OPIVI exhibit strong correlations with the LAI. However, considering angular stability, the OPIVI performs the better. Among the four-band VIs, the REP shows high potential for monitoring crop LAI. Overall, most parameters are sensitive to changes in VZAs. As shown in Figure 4, EVI-1, the OPIVI, and the REP maintain high monitoring accuracy while demonstrating good angular stability. Notably, the OPIVI performed the best in the single-angle linear regression model at a VZA of −15°, with an R^2 of 0.71 and RMSE of 0.55 $cm^2 \cdot cm^{-2}$.

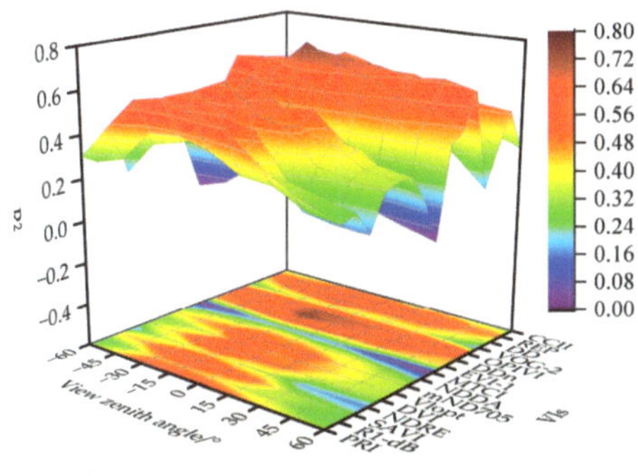

Figure 3. Accuracy of monitoring LAI with different VIs at different VZAs.

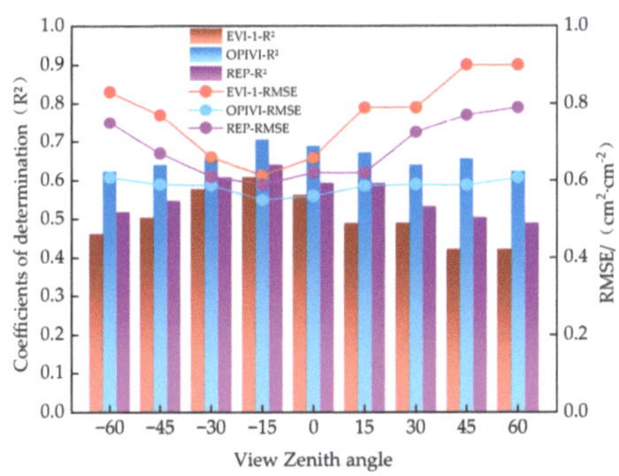

Figure 4. Quantitative relationship between VIs (EVI, OPIVI, and REP) and LAI at different VZAs.

2.3. Comparison of VIs in Different Experimental Conditions across All VZAs

Numerous studies have confirmed that different experimental conditions can affect the angular sensitivity of VIs, necessitating an analysis of how different reproductive stages and treatments impact VIs' angle sensitivity. As shown in Table 2, the correlation between different VIs and the LAI varies with different experimental factors. The R values under different experimental conditions show similar trends, with higher R values for vertical angles than for forward- and back-scatter direction. As shown in Figure 5, the parameters exhibit varying levels of correlation and angular stability under different experimental factors. During various growth stages, both the OPIVI and REP parameters demonstrate good angular stability while maintaining a high correlation with the LAI. However, there are significant differences in the correlation between EVI-1 and the LAI during the budding and flowering stages. Different treatments also results in significant differences in the correlation between the EVI-1 and REP parameters when comparing coverage method treatment and nitrogen fertilizer treatment.

Given the significant differences among experimental conditions (Table 2), the difference of R (DR, the difference in R among different experimental conditions) were calculated.

$$DR = \frac{R_{max} - R_{min}}{R_{average}} \quad (1)$$

Equation (1) demonstrates that the value of *DR* can reflect the angular stability of the correlation between VIs and VZAs. A smaller *DR* value indicates higher angular stability of the relationship between VIs and the LAI. The *DR* values were compared across all experimental factors. The results showed that the OPIVI had the strongest correlation with the LAI and exhibited the most stable change under different VZAs, followed by the REP and EVI-1. Specifically, the correlation coefficient for the OPIVI was slightly higher during flowering stage (0.85) compared to the seedling stages (0.82). Additionally, the correlation coefficient for cover treatment (0.86) was marginally higher than for nitrogen fertilizer treatment (0.82).

Table 2. Correlation coefficient (R) between three VIs and LAI at different VZAs. The highest r values for OPIVI are highlighted in bold.

Categories	Sub Datasets	VIs	−60°	−45°	−30°	−15°	0°	15°	30°	45°	60°
Growth stages	Bolting stage	EVI-1	0.70	0.75	0.81	0.85	0.80	0.74	0.72	0.69	0.69
		OPIVI	0.74	0.78	0.81	**0.82**	0.81	0.81	0.78	0.75	0.74
		REP	0.72	0.74	0.80	0.81	0.80	0.76	0.72	0.69	0.68
	Flowering stage	EVI-1	0.54	0.58	0.65	0.70	0.64	0.60	0.58	0.57	0.53
		OPIVI	0.78	0.79	0.83	**0.85**	0.83	0.80	0.78	0.75	0.76
		REP	0.68	0.71	0.75	0.78	0.76	0.74	0.72	0.69	0.67
Treatments	N rates	EVI-1	0.62	0.67	0.72	0.74	0.70	0.64	0.62	0.60	0.59
		OPIVI	0.77	0.78	**0.82**	**0.82**	**0.82**	0.81	0.76	0.77	0.76
		REP	0.61	0.64	0.71	0.74	0.70	0.65	0.62	0.61	0.60
	Overlay mode	EVI-1	0.72	0.76	0.81	0.82	0.81	0.75	0.72	0.70	0.69
		OPIVI	0.80	0.82	0.85	**0.86**	0.85	0.84	0.81	0.79	0.77
		REP	0.76	0.78	0.80	0.85	0.82	0.76	0.74	0.72	0.73

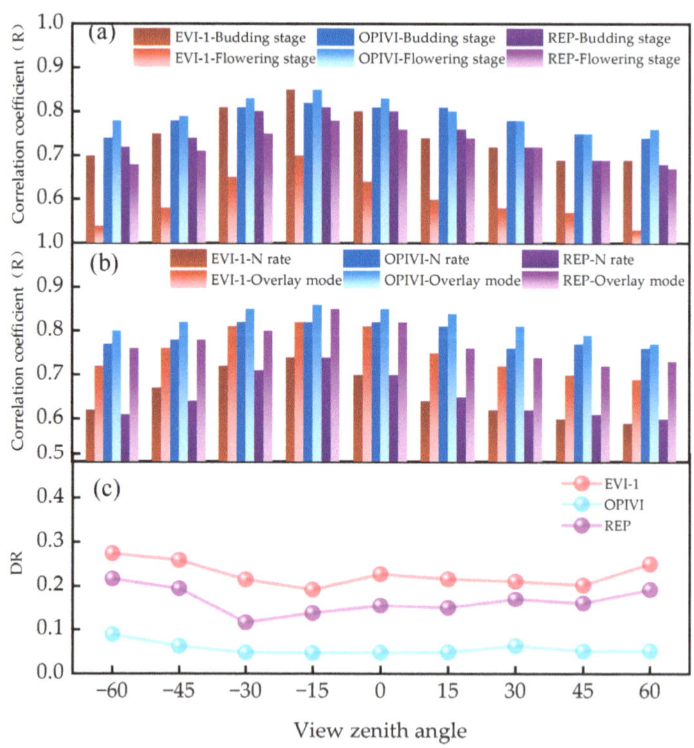

Figure 5. Influence of different experimental factors on the relationship between three VIs and LAI at different VZAs. (**a**) Changes in correlation coefficient of different growth stage VIs at different VZAs. (**b**) Changes in correlation coefficient of different experimental treatment VIs at different VZAs. (**c**) Changes in *DR* value of different VIs at different VZAs.

2.4. Estimating LAI by Different Machine Learning Algorithm

The three best angles (−30°, −15°, and 0°) of the OPIVI were selected as independent variables and input into three machine learning algorithms, SVM, XGBoost, and RF, for modeling. As shown in Figure 6, the results varied in monitoring accuracy among the algorithms. The RF algorithm performed the best, with an R^2 of 0.77 and an RMSE of 0.38 $cm^2{\cdot}cm^{-2}$, indicating that the Bagging model within the ensemble algorithm effectively utilized the information from the multi-angle VI data. The XGBoost algorithm achieved an R^2 of 0.73 and an RMSE of 0.41 $cm^2{\cdot}cm^{-2}$, slightly inferior to the RF algorithm. The SVM had the poorest performance in the multi-angle OPIVI monitoring of winter oilseed rape LAI (R^2 = 0.71, RMSE = 0.51 $cm^2{\cdot}cm^{-2}$). Overall, the RF algorithm, as an ensemble algorithm with integrated thinking, demonstrates strong comprehensive application capabilities and a high utilization rate of effective information in multi-angle VI data. It can serve as a robust algorithm support for monitoring crop nutrient parameters using multi-angle remote sensing data.

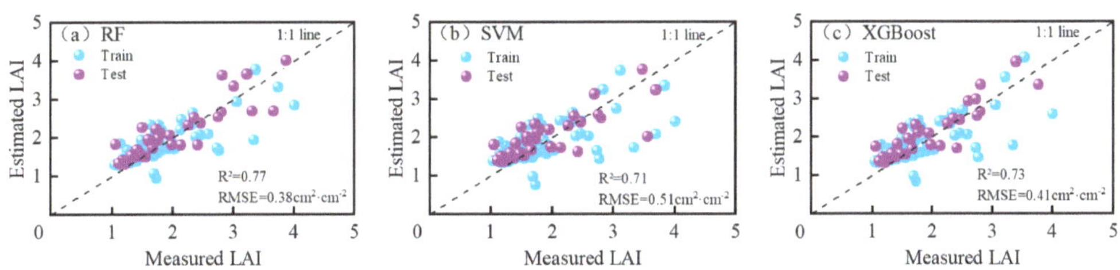

Figure 6. Quantitative relationships of OPIVI to LAI. (**a**) RF, (**b**) SVM, (**c**) XGBoost.

3. Materials and Methods

3.1. Experimental Design

This experiment was conducted at the Key Laboratory of Agricultural Soil and Water Engineering in Arid and Semiarid Areas of the Ministry of Education, Northwest A&F University, Yangling, from October 2022 to June 2023. The study area is a typical semi-humid and drought-prone region with a warm temperate semi-humid monsoon climate. The winter rapeseed variety used in this experiment is "Shaanyou No.18". A total of 45 plots were used for data collection in this experiment, each with a size of 4 m × 6 m (24 m^2) and arranged randomly. The experiment included five nitrogen (N) application rates: N0 (0), N1 (70 kg/hm^2), N2 (140 kg/hm^2), N3 (210 kg/hm^2), and N4 (280 kg/hm^2). Additionally, three types of mulching treatments were applied: straw mulching (SM) for flat crops, film mulching (FM) for ridges, and no mulching (NM) for flat crops. This resulted in a total of 15 treatments (Table 3), replicated three times. A 2 m wide protective belt surrounded the experimental area.

Table 3. The growth stage of the experiment, solar zenith angle, and azimuth angle.

Year	Date	Stage	Time	Solar Zenith Angle (°)	Solar Azimuth Angle (°)
2023	14 March	Budding	12:30–13:20	48.90–52.77	148.68–167.55
	18 March	Budding	12:30–13:20	50.23–54.23	147.68–167.10
	22 April	Flowering	12:30–13:20	61.52–67.06	135.46–160.89
	25 April	Flowering	12:20–13:10	62.32–68.01	134.18–160.15

3.2. Measuring Multi-Angular Spectra and LAI

The spectral reflectance of winter oilseed rape (*Brassica napus* L.) canopy was measured using an ASD field-spec 4 back-mounted field spectrometer (LICA United Technology Limited in Beijing, China) with the field gonimeter system (FIGOS) as the reference [28]. A multi-angle hyperspectral monitoring device was designed to meet specific requirements

(Figure 7a,b), ensuring the same target was observed at different VZAs. Measurements were taken between 11:00 and 13:00 under clear, windless conditions with good visibility.

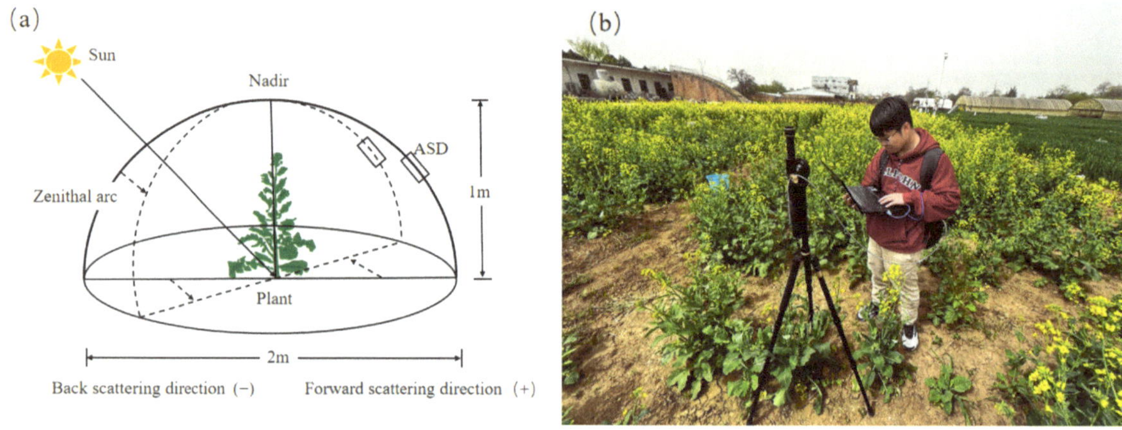

Figure 7. Schematic diagram and instrument for multi-angle hyperspectral measurement. (**a**) Schematic diagram of multi-angle observation at VZA of +60° and +45°. (**b**) Field measurement map.

For multi-angle hyperspectral data collection, the optical fiber is first fixed on the rocker arm of the bracket positioned at a height of 1.5 m. The real-time sun azimuth was obtained using an open-source sun azimuth calculator. The horizontal azimuth of the bracket was adjusted to ensure that the measurement plane was within the solar principal plane (SPP). Different VZAs were achieved by controlling the rocker arm, which moved back and forth to ensure consistent observations of the same area within the plot. The observation object was marked for accuracy. VZA was defined as 0° at vertical monitoring. The direction of sunlight opposite to the observation direction was defined as the forward-scatter direction (+), while the same side as the observation direction was defined as the back-scatter direction (−). VZAs were arranged as −60°, −45°, −30°, −15°, 0°, 15°, 30°, 45°, and 60° (Figure 7). Three measurements were taken at each VZA, and their average value was used as the spectral reflectance at that angle. Reference board calibration was performed immediately before and after each of VZA measurement (with the reflectance board having a reflectance of 1).

Following spectral data collection, three representative winter oilseed rape plants were randomly selected from each plot. These plants were separated into stems and leaves, and the leaf area was determined using the threshold segmentation of the photographs. The LAI was then calculated by multiplying the mean leaf area of the three plants by the number of single stems per unit area (obtained from field surveys conducted during critical growth stages).

3.3. Construction of the New Vegetation Index

Numerous studies have indicated that the NDVI is a frequently used and well-inverted spectral parameter. The saturation phenomenon of the parameters can be alleviated by introducing new bands or fixed coefficients based on the NDVI [29]. Considering that the blue and red bands are insensitive to changes in VZAs, and they are closely related to the leaf area index, it is proposed to introduce these bands into the NDVI. The specific formula is as follows:

$$OPIVI = \frac{R_{\lambda 1} - R_{\lambda 3}}{R_{\lambda 2} - R_{\lambda 3}} \tag{2}$$

Among them, $\lambda1$ refers to all red edge bands within the range of 700–760 nm. $\lambda2$ and $\lambda3$ represent red and blue bands, respectively. After screening and comparison, it is determined that $\lambda1$ = 720 nm, $\lambda2$ = 660 nm, and $\lambda3$ = 450 nm. The specific forms are as follows:

$$OPIVI = \frac{R_{720} - R_{450}}{R_{660} - R_{450}} \quad (3)$$

3.4. Data Analysis

3.4.1. Preprocessing of Spectral Data and Construction of VIs

Numerous studies have indicated that the wavelength range of 350–1300 nm is particularly sensitive and characteristic spectral for reflecting crop pigments, nutrients, and the overall growth and development status [30]. Consequently, this study utilizes this wavelength region to analyze the LAI of winter oilseed rape. To reduce the influence of background noise, baseline drift, and undesirable elements such as scattered light on hyperspectral reflectance, preprocessing techniques are employed. These techniques include Savitzky–Golay convolution smoothing and quadratic polynomial function fitting and filtering for denoising [31].

The VI inversion method is a well-established approach for parameter inversion. Table 4 lists the commonly used two-band, three-band, and four-band VIs for monitoring the LAI, leaf nitrogen concentration (LNC), and chlorophyll content.

Table 4. The selected vegetation indices used in this study.

Index	Formula	References
Two bands		
PRI (photochemical reflectance index)	(R570 − R531)/(R570 + R531)	[32]
RI-dB (redness index–decibels)	R735/R720	[33]
SAVI (soil-adjusted vegetation index)	1.5 × (R870 − R680)/(R870 + R680 + 0.5)	[34]
NDRE (normalized difference red edge)	(R790 − R720)/(R790 + R720)	[35]
DVI (difference vegetation index)	R860 − R560	[36]
Vlopt (variable light optical properties)	(1 + 0.45) × ($R800^2$ + 1)/(R670 + 0.45)	[37]
Three bands		
mND705 (modified normalized difference at 705 nm)	(R750 − R705)/(R750 + R705 − 2 × R445)	[38]
NDDA (normalized difference drought index)	(R680 + R756 – 2 × R718)/(R756 − R680)	[39]
MTCI (meris terrestrial chlorophyll index)	(R754 − R709)/(R709 − R681)	[40]
EVI-1 (enhanced vegetation index-1)	2.5 × (R860 − R645)/(1 + R860 + 6 × R645 − 7.5 × R470)	[20]
DDn (derivative difference normalized)	2.5 × R710 − R660 − R760	[41]
OPIVI (observation perspective insensitivity vegetation index)	(R720 − R450)/(R660 − R450)	This paper
Four bands		
VOG-2 (vogelmann red edge index 2)	(R734 − R747)/(R715 − R726)	[24]
DD (difference vegetation index)	(R749 − R720) − (R701 − R672)	[41]
REP (red edge position)	R700 + 40 × [(R670 + R780)/2 − R700]/(R740 − R700)	[42]
CCII (canopy chlorophyll index integrated)	TCARI/OSAVI	[43]

3.4.2. Training Dataset and Test Dataset

To ensure consistency across different viewing zenith angles (VZAs), we employed a uniform dataset partitioning method before modeling [39]. Specifically, the 180 samples were randomly divided into two sets: 70% (n = 126) for training and 30% (n = 54) for testing. This partitioning was maintained across all studies conducted at the same VZA.

3.4.3. Support Vector Machines (SVM)

Machine learning models can be categorized into single models and ensemble models. Among these models, support vector machine (SVM) is particularly effective for the inverse estimation of crop growth parameters [44]. The SVM transforms data into a high-dimensional feature space to establish a linear model and fits a regression function based on this model. SVM can largely overcome issues such as multiple discrete values and overfitting. By choosing an appropriate kernel function, the performance and accuracy of SVM can be enhanced. In this study, we compared the effects of linear, polynomial, and radial basis function (RBF) and Sigmoid kernel functions, concluding that the RBF performed best. The RBF kernel offers advantages such as wide mapping dimensions, fewer parameters to determine, and relatively simple operation [45].

3.4.4. eXtreme Gradient Boost (XGBoost)

Among the ensemble models, boosting and bagging are commonly used. XGBoost, a boosting algorithm, is frequently employed for the inverse estimation of agronomic parameters [46]. XGBoost iteratively combines weak base learners to form stronger learners. To control overfitting, it is crucial to manage model complexity and increase randomness. In this study, the max_depth is chosen as 5, the learning rate is 0.01, and the regular term coefficient is realized by adjusting the alpha and lambda values.

3.4.5. Random Forest (RF)

Random Forest, a popular bagging model in ensemble learning, is extensively used for the inverse estimation of crop growth parameters. This method involves sampling n samples from the original training set using bootstrap resampling, building decision trees for each sample to obtain n modeling results, and finalizing the prediction through voting on all decision tree results. Random Forest is effective for handling large datasets, estimating specific feature variables, managing noise, and providing fast computations [47]. After error analysis and repeated experiments, we selected ntree = 500 and mtry = 3 for model construction.

3.4.6. Evaluating Model Performance

Model performance was evaluated using the using the coefficient of determination (R^2) and the root mean square error (RMSE). The formulas for R^2 and RMSE are as follows:

$$R^2 = \frac{\sum_{i=1}^{n} (O_i - P_i)^2}{\sum_{i=1}^{n} (O_i - \overline{O_i})^2} \quad (4)$$

$$RMSE = \sqrt{\frac{1}{n}\sum_{i=1}^{n} (O_i - P_i)^2} \quad (5)$$

where P_i and O_i are the predicted values and observations, respectively, $\overline{O_i}$ represents the average of the observations, and n is the number of samples.

3.4.7. Flowchart

The workflow chart of this study is illustrated in Figure 8. The main stages include: (1) Data measurement: Ground measurements (LAI) and hyperspectral remote sensing data were collected at specified intervals. (2) Data analysis: The Angle sensitivity of different VI under different experimental conditions after SG smoothing was analyzed. (3) Model and result: The optimal index OPIVI was input into different machine learning models to get the best modeling decision.

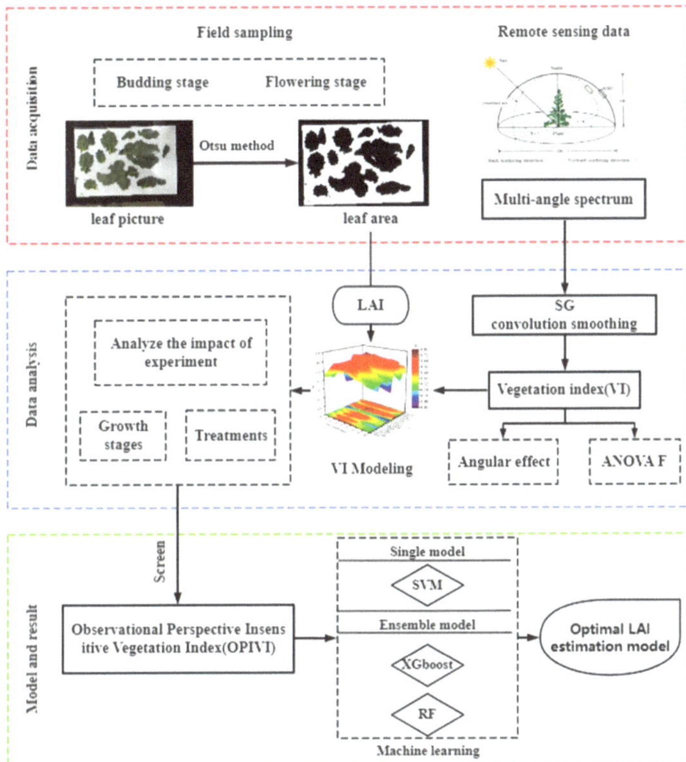

Figure 8. A brief flowchart of this study.

4. Discussion

4.1. The Impact of View Zenith Angle on Band Information and VI

He et al. [48] studied the impact of changes in VZAs on the crop spectral characteristics. They found significant differences in the reflectance values for green and NIR bands at varying VZAs. The reflectance values for green bands are primarily influenced by the absorption of pigments such as chlorophyll a/b, carotenoids, and lutein [49]. Discrepancies in these bands can be attributed to differences in pigment information acquisition at different VZAs. Conversely, NIR band reflectance is mainly influenced by canopy characteristics such as the LAI, biomass, the leaf tilt angle, and canopy distribution patterns [50]. Changes in VZAs influence these canopy characteristics, leading to variations in band emissivity.

In this study, most VIs were constructed at vertical angles, making them sensitive to changes in VZAs [39,43]. Among the nine VZAs, the LAI showed a strong correlation with VIs constructed at near-vertical angles. However, this correlation weakened as the viewing angle increased. This is because larger viewing angles capture canopy characteristics from the upper part of the canopy, while rapeseed plants typically have more leaf area in the lower parts of their canopies during the budding and flowering stages [51]. Thus, spectral information obtained at near-vertical angles is more closely related to the LAI. The angular effect of VIs can affect the stability of monitoring field crops. A new spectral index, the OPIVI, incorporates red, blue, and red edge reflectance into the structure of the NDVI, making it an effective parameter for monitoring the LAI. The advantages of the OPIVI lie in its angle insensitivity to blue and red reflectance and its ability to assess crop growth conditions using red edge parameters. Numerous studies have demonstrated strong relationships between the bands in the OPIVI formula and the LAI, nitrogen nutrient availability, contributing to the establishment of stable monitoring models within certain

VZA ranges [33,52]. Including blue band parameters reduces atmospheric correction effects and enhances the estimation of physiological indicators while mitigating the angle effect [20,38]. Consequently, the OPIVI maintains a stable and close relationship with the LAI within the SPP.

Different remote sensing monitoring directions provide different canopy information. Back-scattered signals primarily come from illuminated leaves or branches, while forward-scattered signals mainly come from shaded leaves or branches [9]. When monitoring plant water use efficiency (WUE) and crop LAI, back-scattered observations exhibit stronger robustness and information content than front-scattered ones [53]. In this study, most VIs demonstrated higher ability to monitor LAI at -15° than at vertical or frontal VZAs. Among the back-scattered viewing angles, -15° exhibited optimal performance. Huang et al. [51] and Ratutiainen et al. [54] also monitored leaf chlorophyll content and the LAI using similar VZAs, as spectral information at -15° viewing angles better reflected the vertical gradients of plant growth information. Furthermore, the relationship between different VIs, composed of various band numbers, and the LAI varies with changes in VZAs. Among these three VIs with high correlation and stability, namely EVI-1, the OPIVI, and the REP, all incorporate blue, red, and red edge parameters. This indicates that the stability advantage of OPIVI is not solely attributed to a single band but rather to the combined use of multiple bands, enhancing the accuracy and stability of the OPIVI.

4.2. The Impact of Experimental Factors on the Relationship between VIs and LAI

Different experimental treatments on crop canopy structure directly affects the angle sensitivity of VIs. Consequently, VIs exhibit varying angle sensitivities under varying experimental conditions. For example, He et al. [25] found that models performed better during the reproductive stage than the nutrient stage, likely due to stable canopy structure and reduced nitrogen dilution during reproduction. Additionally, the growth direction of leaves (horizontal or upright) influences canopy structure and vertical gradients [55]. In this study, the relationship between VIs and LAI varied with different experimental conditions. For instance, the relationship between VIs and LAI was stronger during the flowering stage compared to the budding stage, which may be attributed to the significant difference in canopy structure between these two stages. Crop nitrogen dilution effects were present from the bolting to flowering stages, and canopy structures reached a stable state during the flowering stage. Thus, there were significant differences in the relationship between VIs and the LAI during these stages. Among these three VIs, EVI-1, REP, and OPIVI, the red band parameter was crucial. Due to pigment absorption by rapeseed flowers during the two growth stages, there were significant differences in reflectance values for the red band, leading to underpinning the monitoring accuracy of EVI-1 depending on the growth stage. Nitrogen fertilization treatments influenced the population density and LAI of rapeseed plants, while cover treatments mainly affected soil water content changes in rapeseed populations [56,57]. Cover treatment had little impact on LAI changes during the late growth stage of rapeseed plants. As a result, correlations between VIs and the LAI were weak under nitrogen fertilization treatments but were stronger under cover treatments, as evidence by the REP parameter. The OPIVI maintained high stability across different growth stages and experimental treatments due to its balanced composition, allowing it to maintain consistent monitoring accuracy under varying experimental conditions. In conclusion, different growth stages, nitrogen fertilization treatments, and coverage methods significantly affect the relationship between the LAI and VIs. Therefore, it is essential to consider these experimental factors when selecting sensitive indices and constructing monitoring models.

4.3. The Suitable Algorithm for OPIVI to Monitor LAI

Different types of machine learning algorithms significantly impact the accuracy of monitoring VIs. He et al. [48] used neural network algorithms to estimate wheat LNC based on different VZAs, achieving a high accuracy with an R^2 value of 0.82. It is important to note that neural networks are single-model machine learning algorithms. In this study, the SVM algorithm, another single-model approach, performed poorly, with an R^2 value of 0.71. This may be because single models are prone to dataset fragmentation, which affects the monitoring accuracy [58]. Moreover, the main challenge of the SVM lies in determining the kernel functions and related parameters [59]. Due to limitations in parameter selection, such as kernel functions and penalty factors, its application is somewhat restricted. Ensemble algorithm-based machine learning models generally provide higher accuracy in monitoring crop nutrient parameters. For instance, Yuan et al. [60] developed a model to estimate the LAI of rice by combining single spectral and texture indicators, with the RF model achieving the highest R^2 value of 0.84. In this study, the LAI monitoring model developed by inputting the OPIVI data from three sensitive VZAs into the RF model achieved the highest accuracy, with an R^2 value of 0.77. This was the highest accuracy among the three machine learning algorithms tested and was significantly improved compared to the single-band linear regression models. This RF algorithm, as an ensemble learning method, demonstrates strong comprehensive application capabilities and high efficiency in utilizing effective information from multiple-angle VIs [61].

This study contributes to the application of multi-angle remote sensing in agriculture for monitoring crop growth and nutrient parameters. We aim to expand our dataset by including more experimental treatments, planting densities, locations, years, and rapeseed varieties to better analyze the angular effects of VIs under different conditions. To further validate the applicability of VIs, it is crucial to conduct analyses on other crops and utilize multiple datasets. Additionally, applying PROSAIL models to simulate additional VZAs and integrating them with various experimental conditions and algorithms will enable further investigation into the anisotropy of crop canopies. In summary, this study provides a more accurate and stable method for monitoring crop LAI from various perspectives.

5. Conclusions

The changes in reflectance and VIs with the change in the VZA highlight the importance of considering angle effects in remote sensing for crop LAI monitoring. Acknowledging the vertical distribution of the leaf area in winter oilseed rape, we developed a new index called the OPIVI that captures the dynamic changes in LAI using blue, red, and red-edge parameters. The OPIVI offers a simple and practical method for monitoring LAI in winter oilseed rape. The main conclusions are as follows:

(1) Multi-angle observations reveal that the relationship between the back-scatter direction and LAI is stronger than between the vertical and forward-scatter directions. Among the 16VIs tested, the OPIVI shows the highest potential for monitoring across different VZAs, performing best at an elevation angle of $-15°$.

(2) Different experimental factors, such as growth stage, nitrogen fertilizer application, and coverage method, have varying effects on different VIs and the LAI. Notably, the OPIVI maintains a high correlation and angular stability under various experimental conditions.

(3) For monitoring model selection, the combination of the RF model with clustering algorithms and multi-angle OPIVI provide optimal results in estimating winter oilseed rape LAI (R^2 = 0.77; RMSE = 0.38 $cm^2 \cdot cm^{-2}$). This approach significantly improves accuracy compared to single-angle inversion models.

Author Contributions: Data curation: H.L., Investigation: Y.X. and F.Z., Methodology: Z.L. and H.L., Project administration: Z.T. and H.S., Resources: Y.X. and F.Z., Software: R.D., Y.W., N.Y. and H.L., Supervision: J.C., Visualization: H.L., Writing—original draft: H.L., Writing—review and editing: H.L. All authors have read and agreed to the published version of the manuscript.

Funding: This research was supported by the National Natural Science Foundation of China (52179045, 52279047).

Data Availability Statement: The data that support the findings of this study are available from the corresponding author upon reasonable request.

Conflicts of Interest: The authors declare that they have no known competing financial interests or personal relationships that could have appeared to influence the work reported in this paper.

References

1. Guo, A.; Ye, H.C.; Huang, W.J.; Qian, B.X.; Wang, J.J.; Lan, Y.B.; Wang, S.Z. Inversion of maize leaf area index from UAV hyperspectral and multispectral imagery. *Comput. Electron. Agric.* **2023**, *212*, 108020. [CrossRef]
2. Viña, A.; Gitelson, A.A.; Nguy-Robertson, A.L.; Peng, Y. Comparison of different vegetation indices for the remote assessment of green leaf area index of crops. *Remote Sens. Environ.* **2011**, *115*, 3468–3478. [CrossRef]
3. Yang, N.; Zhang, Z.; Zhang, J.; Guo, Y.; Yang, X.; Yu, G.; Bai, X.; Chen, J.; Chen, Y.; Shi, L. Improving estimation of maize leaf area index by combining of UAV-based multispectral and thermal infrared data: The potential of new texture index. *Comput. Electron. Agric.* **2023**, *214*, 108294. [CrossRef]
4. Tang, H.; Brolly, M.; Zhao, F.; Strahler, A.H.; Schaaf, C.L.; Ganguly, S.; Zhang, G.; Dubayah, R. Deriving and validating Leaf Area Index (LAI) at multiple spatial scales through lidar remote sensing: A case study in Sierra National Forest, CA. *Remote Sens. Environ.* **2014**, *143*, 131–141. [CrossRef]
5. Li, L.; Jákli, B.; Lu, P.; Ren, T.; Ming, J.; Liu, S. Assessing leaf nitrogen concentration of winter oilseed rape with canopy hyperspectral technique considering a non-uniform vertical nitrogen distribution. *Ind. Crop. Prod.* **2018**, *116*, 1–14. [CrossRef]
6. He, L.; Zhang, H.-Y.; Zhang, Y.-S.; Song, X.; Feng, W.; Kang, G.-Z.; Wang, C.-Y.; Guo, T.-C. Estimating canopy leaf nitrogen concentration in winter wheat based on multi-angular hyperspectral remote sensing. *Eur. J. Agron.* **2016**, *73*, 170–185. [CrossRef]
7. Song, X.; Feng, W.; He, L.; Xu, D.; Zhang, H.-Y.; Li, X.; Wang, Z.-J.; Coburn, C.A.; Wang, C.-Y.; Guo, T.-C. Examining view angle effects on leaf N estimation in wheat using field reflectance spectroscopy. *Isprs J. Photogramm.* **2016**, *122*, 57–67. [CrossRef]
8. Hasegawa, K.; Matsuyama, H.; Tsuzuki, H.; Sweda, T. Improving the estimation of leaf area index by using remotely sensed NDVI with BRDF signatures. *Remote Sens. Environ.* **2010**, *114*, 514–519. [CrossRef]
9. Stagakis, S.; Markos, N.; Sykioti, O.; Kyparissis, A. Monitoring canopy biophysical and biochemical parameters in ecosystem scale using satellite hyperspectral imagery: An application on a Phlomis fruticosa Mediterranean ecosystem using multiangular CHRIS/PROBA observations. *Remote Sens. Environ.* **2010**, *114*, 977–994. [CrossRef]
10. Wang, J.; Wang, H.; Tian, T.; Cui, J.; Shi, X.; Song, J.; Li, T.; Li, W.; Zhong, M.; Zhang, W. Construction of spectral index based on multi-angle spectral data for estimating cotton leaf nitrogen concentration. *Comput. Electron. Agric.* **2022**, *201*, 107328. [CrossRef]
11. Liu, Y.; An, L.; Wang, N.; Tang, W.; Liu, M.; Liu, G.; Sun, H.; Li, M.; Ma, Y. Leaf area index estimation under wheat powdery mildew stress by integrating UAV-based spectral, textural and structural features. *Comput. Electron. Agric.* **2023**, *213*, 108169. [CrossRef]
12. Broge, N.H.; Leblanc, E. Comparing prediction power and stability of broadband and hyperspectral vegetation indices for estimation of green leaf area index and canopy chlorophyll density. *Remote Sens. Environ.* **2001**, *76*, 156–172. [CrossRef]
13. Brown, L.; Chen, J.M.; Leblanc, S.G.; Cihlar, J. A Shortwave Infrared Modification to the Simple Ratio for LAI Retrieval in Boreal Forests: An Image and Model Analysis. *Remote Sens. Environ.* **2000**, *71*, 16–25. [CrossRef]
14. Qiao, L.; Gao, D.; Zhao, R.; Tang, W.; An, L.; Li, M.; Sun, H. Improving estimation of LAI dynamic by fusion of morphological and vegetation indices based on UAV imagery. *Comput. Electron. Agric.* **2022**, *192*, 106603. [CrossRef]
15. RAUTIAINEN, M. Retrieval of leaf area index for a coniferous forest by inverting a forest reflectance model. *Remote Sens. Environ.* **2005**, *99*, 295–303. [CrossRef]
16. Almeida-Ñauñay, A.F.; Tarquis, A.M.; López-Herrera, J.; Pérez-Martín, E.; Pancorbo, J.L.; Raya-Sereno, M.D.; Quemada, M. Optimization of soil background removal to improve the prediction of wheat traits with UAV imagery. *Comput. Electron. Agric.* **2023**, *205*, 107559. [CrossRef]
17. Rondeaux, G.; Steven, M.; Baret, F. Optimization of soil-adjusted vegetation indices. *Remote Sens. Environ.* **1996**, *55*, 95–107. [CrossRef]
18. Su, J.; Liu, C.; Coombes, M.; Hu, X.; Wang, C.; Xu, Z.; Li, Q.; Guo, L.; Chen, W.H. Wheat yellow rust monitoring by learning from multispectral UAV aerial imagery. *Comput. Electron. Agric.* **2018**, *155*, 157–166. [CrossRef]
19. Haboudane, D. Hyperspectral vegetation indices and novel algorithms for predicting green LAI of crop canopies: Modeling and validation in the context of precision agriculture. *Remote Sens. Environ.* **2004**, *90*, 337–352. [CrossRef]
20. Huete, A.; Didan, K.; Miura, T.; Rodriguez, E.P.; Gao, X.; Ferreira, L.G. Overview of the radiometric and biophysical performance of the MODIS vegetation indices. *Remote Sens. Environ.* **2002**, *83*, 195–213. [CrossRef]
21. Galvão, L.S.; dos Santos, J.R.; Roberts, D.A.; Breunig, F.M.; Toomey, M.; de Moura, Y.M. On intra-annual EVI variability in the dry season of tropical forest: A case study with MODIS and hyperspectral data. *Remote Sens. Environ.* **2011**, *115*, 2350–2359. [CrossRef]
22. Galvão, L.S.; Breunig, F.M.; dos Santos, J.R.; de Moura, Y.M. View-illumination effects on hyperspectral vegetation indices in the Amazonian tropical forest. *Int. J. Appl. Earth Obs.* **2013**, *21*, 291–300. [CrossRef]
23. Verrelst, J.; Schaepman, M.E.; Koetz, B.; Kneubuehler, M. Angular sensitivity analysis of vegetation indices derived from CHRIS/PROBA data. *Remote Sens. Environ.* **2008**, *112*, 2341–2353. [CrossRef]

24. Hovi, A.; Schraik, D.; Kuusinen, N.; Fabiánek, T.; Hanuš, J.; Homolová, L.; Juola, J.; Lukeš, P.; Rautiainen, M. Synergistic use of multi- and hyperspectral remote sensing data and airborne LiDAR to retrieve forest floor reflectance. *Remote Sens. Environ.* **2023**, *293*, 113610. [CrossRef]
25. He, L.; Song, X.; Feng, W.; Guo, B.B.; Zhang, Y.S.; Wang, Y.H.; Wang, C.Y.; Guo, T.C. Improved remote sensing of leaf nitrogen concentration in winter wheat using multi-angular hyperspectral data. *Remote Sens. Environ.* **2016**, *174*, 122–133. [CrossRef]
26. Li, X.; Sun, Z.; Lu, S.; Omasa, K. A multi-angular invariant spectral index for the estimation of leaf water content across a wide range of plant species in different growth stages. *Remote Sens. Environ.* **2021**, *253*, 112230. [CrossRef]
27. Li, L.; Ren, T.; Ma, Y.; Wei, Q.; Wang, S.; Li, X.; Cong, R.; Liu, S.; Lu, J. Evaluating chlorophyll density in winter oilseed rape (*Brassica napus* L.) using canopy hyperspectral red-edge parameters. *Comput. Electron. Agric.* **2016**, *126*, 21–31. [CrossRef]
28. Sandmeier, S.R.; Itten, K.I. A field goniometer system (FIGOS) for acquisition of hyperspectral BRDF data. *IEEE Trans. Geosci. Remote* **1999**, *37*, 978–986. [CrossRef]
29. Peng, Y.; Gitelson, A.A. Application of chlorophyll-related vegetation indices for remote estimation of maize productivity. *Agric. For. Meteorol.* **2011**, *151*, 1267–1276. [CrossRef]
30. Li, L.; Lu, J.; Wang, S.; Ma, Y.; Wei, Q.; Li, X.; Cong, R.; Ren, T. Methods for estimating leaf nitrogen concentration of winter oilseed rape (*Brassica napus* L.) using in situ leaf spectroscopy. *Ind. Crop. Prod.* **2016**, *91*, 194–204. [CrossRef]
31. Flynn, K.C.; Baath, G.; Lee, T.O.; Gowda, P.; Northup, B. Hyperspectral reflectance and machine learning to monitor legume biomass and nitrogen accumulation. *Comput. Electron. Agric.* **2023**, *211*, 107991. [CrossRef]
32. Gamon, J.A.; Penuelas, J.; Field, C.B. A narrow-waveband spectral index that tracks diurnal changes in photosynthetic efficiency. *Remote Sens. Environ.* **1992**, *41*, 35–44. [CrossRef]
33. Gupta, R.K.; Vijayan, D.; Prasad, T.S. Comparative analysis of red-edge hyperspectral indices. *Adv. Space Res.* **2003**, *32*, 2217–2222. [CrossRef]
34. Yang, Z.; Rao, M.N.; Elliott, N.C.; Kindler, S.D.; Popham, T.W. Using ground-based multispectral radiometry to detect stress in wheat caused by greenbug (Homoptera: Aphididae) infestation. *Comput. Electron. Agric.* **2005**, *47*, 121–135. [CrossRef]
35. Fitzgerald, G.; Rodriguez, D.; Christensen, L.; Belford, R.; Sadras, V.; Clarke, T. Spectral and thermal sensing for nitrogen and water status in rainfed and irrigated wheat environments. *Precis. Agric.* **2006**, *7*, 233–248. [CrossRef]
36. Jensen, T.; Apan, A.; Young, F.; Zeller, L. Detecting the attributes of a wheat crop using digital imagery acquired from a low-altitude platform. *Comput. Electron. Agric.* **2007**, *59*, 66–77. [CrossRef]
37. Yang, H.; Li, F.; Hu, Y.; Yu, K. Hyperspectral indices optimization algorithms for estimating canopy nitrogen concentration in potato (*Solanum tuberosum* L.). *Int. J. Appl. Earth Obs.* **2021**, *102*, 102416. [CrossRef]
38. Sims, D.A.; Gamon, J.A. Relationships between leaf pigment content and spectral reflectance across a wide range of species, leaf structures and developmental stages. *Remote Sens. Environ.* **2002**, *81*, 337–354. [CrossRef]
39. Feng, W.; Guo, B.B.; Wang, Z.J.; He, L.; Song, X.; Wang, Y.H.; Guo, T.C. Measuring leaf nitrogen concentration in winter wheat using double-peak spectral reflection remote sensing data. *Field Crop. Res.* **2014**, *159*, 43–52. [CrossRef]
40. Dash, J.; Curran, P.J. The MERIS terrestrial chlorophyll index. *Int. J. Remote Sens.* **2004**, *25*, 5403–5413. [CrossRef]
41. le Maire, G.; François, C.; Dufrêne, E. Towards universal broad leaf chlorophyll indices using PROSPECT simulated database and hyperspectral reflectance measurements. *Remote Sens. Environ.* **2004**, *89*, 1–28. [CrossRef]
42. Ryckewaert, M.; Gorretta, N.; Henriot, F.; Gobrecht, A.; Héran, D.; Moura, D.; Bendoula, R.; Roger, J.-M. Potential of high-spectral resolution for field phenotyping in plant breeding: Application to maize under water stress. *Comput. Electron. Agric.* **2021**, *189*, 106385. [CrossRef]
43. Haboudane, D.; Miller, J.R.; Tremblay, N.; Zarco-Tejada, P.J.; Dextraze, L. Integrated narrow-band vegetation indices for prediction of crop chlorophyll content for application to precision agriculture. *Remote Sens. Environ.* **2002**, *81*, 416–426. [CrossRef]
44. Zhang, L.; Zhang, Z.; Wu, C.; Sun, L. Segmentation algorithm for overlap recognition of seedling lettuce and weeds based on SVM and image blocking. *Comput. Electron. Agric.* **2022**, *201*, 107284. [CrossRef]
45. Xu, J.; Zhou, S.; Xu, A.; Ye, J.; Zhao, A. Automatic scoring of postures in grouped pigs using depth image and CNN-SVM. *Comput. Electron. Agric.* **2022**, *194*, 106746. [CrossRef]
46. Han, L.; Yang, G.; Yang, X.; Song, X.; Xu, B.; Li, Z.; Wu, J.; Yang, H.; Wu, J. An explainable XGBoost model improved by SMOTE-ENN technique for maize lodging detection based on multi-source unmanned aerial vehicle images. *Comput. Electron. Agric.* **2022**, *194*, 106804. [CrossRef]
47. Xiao, C.; Ji, Q.; Chen, J.; Zhang, F.; Li, Y.; Fan, J.; Wang, H. Prediction of soil salinity parameters using machine learning models in an arid region of northwest China. *Comput. Electron. Agric.* **2023**, *204*, 107512. [CrossRef]
48. He, L.; Liu, M.; Guo, Y.; Wei, Y.; Zhang, H.; Song, X.; Feng, W.; Guo, T. Angular effect of algorithms for monitoring leaf nitrogen concentration of wheat using multi-angle remote sensing data. *Comput. Electron. Agric.* **2022**, *195*, 106815. [CrossRef]
49. Li, L.; Wang, S.; Ren, T.; Wei, Q.; Ming, J.; Li, J.; Li, X.; Cong, R.; Lu, J. Ability of models with effective wavelengths to monitor nitrogen and phosphorus status of winter oilseed rape leaves using in situ canopy spectroscopy. *Field Crop. Res.* **2018**, *215*, 173–186. [CrossRef]
50. Guerrero, A.; Javadi, S.H.; Mouazen, A.M. Automatic detection of quality soil spectra in an online vis-NIR soil sensor. *Comput. Electron. Agric.* **2022**, *196*, 106857. [CrossRef]
51. Rautiainen, M.; Lang, M.; Mõttus, M.; Kuusk, A.; Nilson, T.; Kuusk, J.; Lükk, T. Multi-angular reflectance properties of a hemiboreal forest: An analysis using CHRIS PROBA data. *Remote Sens. Environ.* **2008**, *112*, 2627–2642. [CrossRef]

52. Li, J.; Wijewardane, N.K.; Ge, Y.; Shi, Y. Improved chlorophyll and water content estimations at leaf level with a hybrid radiative transfer and machine learning model. *Comput. Electron. Agric.* **2023**, *206*, 107669. [CrossRef]
53. Zhang, H.-Y.; Liu, M.-R.; Feng, Z.-H.; Song, L.; Li, X.; Liu, W.D.; Wang, C.Y.; Feng, W. Estimations of Water Use Efficiency in Winter Wheat Based on Multi-Angle Remote Sensing. *Front. Plant Sci.* **2021**, *12*, 614417. [CrossRef]
54. Huang, W.; Yang, Q.; Pu, R.; Yang, S. Estimation of Nitrogen Vertical Distribution by Bi-Directional Canopy Reflectance in Winter Wheat. *Sensors* **2014**, *14*, 20347–20359. [CrossRef]
55. Bertheloot, J.; Martre, P.; Andrieu, B. Dynamics of Light and Nitrogen Distribution during Grain Filling within Wheat Canopy. *Plant Physiol.* **2008**, *148*, 1707–1720. [CrossRef]
56. Kamiji, Y.; Pang, J.; Milroy, S.; Palta, J.A. Shoot biomass in wheat is the driver for nitrogen uptake under low nitrogen supply, but not under high nitrogen supply. *Field Crop. Res.* **2014**, *165*, 92–98. [CrossRef]
57. Yao, X.; Ren, H.; Cao, Z.; Tian, Y.; Cao, W.; Zhu, Y.; Cheng, T. Detecting leaf nitrogen content in wheat with canopy hyperspectrum under different soil backgrounds. *Int. J. Appl. Earth Obs.* **2014**, *32*, 114–124. [CrossRef]
58. Mehdizadeh, S.; Behmanesh, J.; Khalili, K. Using MARS, SVM, GEP and empirical equations for estimation of monthly mean reference evapotranspiration. *Comput. Electron. Agric.* **2017**, *139*, 103–114. [CrossRef]
59. Zheng, H.; Lu, H. A least-squares support vector machine (LS-SVM) based on fractal analysis and CIELab parameters for the detection of browning degree on mango (*Mangifera indica* L.). *Comput. Electron. Agric.* **2012**, *83*, 47–51. [CrossRef]
60. Yuan, W.; Meng, Y.; Li, Y.; Ji, Z.; Kong, Q.; Gao, R.; Su, Z. Research on rice leaf area index estimation based on fusion of texture and spectral information. *Comput. Electron. Agric.* **2023**, *211*, 108016. [CrossRef]
61. Luo, K.; Lu, L.; Xie, Y.; Chen, F.; Yin, F.; Li, Q. Crop type mapping in the central part of the North China Plain using Sentinel-2 time series and machine learning. *Comput. Electron. Agric.* **2023**, *205*, 107577. [CrossRef]

plants

MDPI

Article

Study of the Spectral Characteristics of Crops of Winter Wheat Varieties Infected with Pathogens of Leaf Diseases

Roman Danilov [1,*], Oksana Kremneva [1,*], Igor Sereda [2], Ksenia Gasiyan [1], Mikhail Zimin [2], Dmitry Istomin [1] and Alexey Pachkin [1]

1 Federal State Budgetary Scientific Institution «Federal Research Center of Biological Plant Protection» (FSBSI FRCBPP), Krasnodar 350039, Russia; gasiyankkk@mail.ru (K.G.); istomin.dmitriy.94@mail.ru (D.I.); capricorn-53@yandex.ru (A.P.)

2 Department of Cartography and Geoinformatics, Faculty of Geography, M. V. Lomonosov Moscow State University, Moscow 119991, Russia; iisereda@mail.ru (I.S.); ziminmv@mail.ru (M.Z.)

* Correspondence: daniloff.roman2011@yandex.ru (R.D.); kremenoks@mail.ru (O.K.); Tel.: +7-(988)-529-9098 (R.D.)

Abstract: Studying the influence of the host plant genotype on the spectral reflectance of crops infected by a pathogen is one of the key directions in the development of precision methods for monitoring the phytosanitary state of wheat agrocenoses. The purpose of this research was to study the influence of varietal factors and disease development on the spectral characteristics of winter wheat varieties of different susceptibility to diseases during the growing seasons of 2021, 2022 and 2023. The studied winter wheat crops were represented by three varieties differing in susceptibility to phytopathogens: Grom, Svarog and Bezostaya 100. Over three years of research, a clear and pronounced influence of the varietal factor on the spectral characteristics of winter wheat crops was observed, which in most cases manifested itself as an immunological reaction of specific varieties to the influence of pathogen development. The nature of the influence of the pathogenic background and the spectral characteristics of winter wheat crops were determined by the complex interaction of the development of individual diseases under the conditions of a particular year of research. A uniform and clear division of the spectral characteristics of winter wheat according to the intensity of the disease was recorded only at a level of pathogen development of more than 5%. Moreover, this gradation was most clearly manifested in the spectral channels of the near-infrared range and at a wavelength of 720 nm.

Keywords: ground-based spectrometry; winter wheat; wheat diseases; spectral characteristics

Citation: Danilov, R.; Kremneva, O.; Sereda, I.; Gasiyan, K.; Zimin, M.; Istomin, D.; Pachkin, A. Study of the Spectral Characteristics of Crops of Winter Wheat Varieties Infected with Pathogens of Leaf Diseases. *Plants* **2024**, *13*, 1892. https://doi.org/10.3390/plants13141892

Academic Editor: Youzhen Xiang

Received: 7 June 2024
Revised: 6 July 2024
Accepted: 7 July 2024
Published: 9 July 2024

1. Introduction

Winter wheat is one of the leading agricultural crops grown worldwide. Economically important diseases of wheat include pathogens of septoria spp. and brown spot (*Pyrenophora tritici-repentis* (Died.) Drechsler), as well as pathogens that cause powdery mildew (*Blumeria graminis* (DC.) Speer), brown rust (*Puccinia triticina* Erikss.) and yellow rust (*Puccinia striiformis* West.). These diseases are widespread in the world [1,2]. In Russia, there are also cases of wheat affected by these diseases. This is especially pronounced in the southern regions [3–5]. According to FAO and the UN, wheat diseases cause annual losses. Thus, in developed countries, crop losses of up to 10% are observed and, in developing countries, up to 20–50% [6]. In Russia, losses of grain crops of up to 25–35% are caused by rust fungi, septoria, yellow spot, powdery mildew, fusarium and root rot [7,8].

One of the factors for successful plant protection is the ability to quickly monitor large areas of agricultural land. This approach can provide high quality data [9,10]. Vast cultivated areas make it difficult to conduct phytosanitary monitoring using traditional visual accounting methods. Consequently, there is a lack of proper control by specialists and, therefore, there is an urgent need for a fundamental scientific and methodological ground for early diagnosis of the main pathogens of wheat diseases based on aerospace information technology, as well as high-performance ground-based measurements. The current level of Earth remote-sensing equipment is characterized by the emergence of space- and aviation-based hyper-spectral equipment. Researchers believe that determining the spectral properties of an object in specific narrow wavelength ranges creates the opportunity to identify hidden special changes [11–17]. It is not possible to record such changes on the ground without special tools and special methods.

Many researchers study the physiological state of wheat crops through hyper-spectral data analysis [18–34]. The most significant results have been obtained by identifying diseases such as fusarium head blight (*F. graminearum* Schwabe) [19–22]. Further research is needed to determine the spectral properties of different wheat varieties since these properties may distort the overall picture of biotic and abiotic stress in the wheat canopy. Thus, Delwiche et al., 2000, studying the possibilities of detecting fusarium head blight in three different wheat varieties, established a significant influence of the variety factor on the adequacy of pathogen-detection models. Models developed for only one variety were found to be useless when applied to other varieties [19]. Zhang et al. studied the relationship between nutritional stress and yellow rust disease in three wheat varieties. They identified the only disease-sensitive plant growth index PhRI [24,25]. Others have established the possibility of using hyper-spectral data for phenotyping the resistance of different wheat varieties affected by septoria [32,33]. The results of studying the influence on the spectral properties of different varieties and variety mixtures of pathogens causing winter wheat yellow spot have also been obtained [34].

Thus, studying the influence of the host plant genotype on the spectral reflectance of crops, as well as on the biochemical and physiological characteristics of plants exposed to pathogen infection, is one of the key directions in the development of precision methods for monitoring the phytosanitary state of wheat agrocenoses. Currently, there is a growing trend in the number of scientific papers on this topic. However, a complete picture and unambiguous results on the spectral parameters of plants have not been obtained. This is attributed to the poor reproducibility of the results due to the lack of experimental data in the field of reflective properties of different varieties of the same species throughout years or seasons of research.

Here we aim to study the influence of variety factors and disease intensity on the spectral characteristics of winter wheat varieties of different susceptibility during the growing seasons of 2021, 2022 and 2023.

Accordingly, the following tasks were set:

(1) Study the nature of similarities and differences in the spectral characteristics of crops of winter wheat varieties in individual years of research;
(2) Study the relationship between the spectral characteristics of winter wheat crops and disease intensity in individual years of research;
(3) Study the relationship between spectral characteristics and disease intensity of each individual variety during three years of research;
(4) Assess the possibility of differentiating crops according to the degree of pathogen development.

2. Results

Using two-factor analysis of variance, how the spectral characteristics of winter wheat crops are affected by varietal differences, as well as the intensity of various diseases (Table 1), was established. However, the nature of their mutual influence was not entirely clear during the three years of research. Thus, in 2021, only a separate influence of crop varietal characteristics and disease development indicators was observed. In 2022 and 2023, observations were made of the influence of plant varietal characteristics and the development of disease on the plant.

Table 1. The influence of variety factors and disease development on the spectral brightness coefficient indicators of crops of the studied varieties of winter wheat at the time of the onset of intensive manifestation of all leaf diseases in the GS 60–70 "blooming" phase in the growing seasons 2021–2023.

Spectral Channels	Variety			Infectious Background			Variety * Infectious Background		
	2021	2022	2023	2021	2022	2023	2021	2022	2023
490	*	-	*	*	-	*	-	*	*
520	*	-	*	*	-	*	-	*	*
550	*	*	*	*	-	*	-	*	*
575	*	*	*	*	-	*	-	*	*
660	*	-	*	*	*	*	-	*	*
700	*	*	*	*	*	*	-	*	*
720	*	*	*	*	-	*	-	*	*
845	*	-	*	-	-	*	-	-	-
920	*	*	*	-	-	*	-	-	-
1085	*	*	*	-	-	*	-	-	-
1135	*	*	*	*	-	*	-	-	-
1215	*	*	*	*	-	*	-	-	-
1245	*	*	*	*	-	*	-	-	-
1285	*	*	*	*	-	*	-	-	-
1445	-	*	*	-	*	-	-	*	-
1675	*	*	*	*	-	*	-	*	*
1725	*	*	*	-	-	*	-	*	*
2005	*	*	-	-	-	-	-	-	-
2035	-	*	-	*	*	*	-	*	*
2295	*	*	-	-	*	-	-	*	-
2345	-	*	-	-	-	*	-	-	-

Notes: *—mathematically reliable influence of the factor on SBC indicators is confirmed.

According to 2022 data, the full influence of variety factors and disease development, as well as their combination, was manifested in the spectral channels of 575, 700, 1445, 2035 and 2295 nm. In the spectrum channels of 550, 660, 720 and 1725 nm, the influence of individual varieties and their combination with the disease development factor was observed. At wavelengths of 920, 1085, 1135, 1245, 1285 and 2345 nm, which belong to the near- and mid-infrared spectral ranges, only the influence of the variety factor was observed.

In 2023, the influence of varietal factors and the development of diseases on winter wheat crops had the greatest visibility. This effect was especially strong at wavelengths 490, 520, 550, 575, 660, 700, 720, 1675 and 1725 nm. A separate influence of the two factors, similar to in 2021, was observed at mid-infrared wavelengths of 845, 920, 1085, 1135, 1215, 1245, 1285 and 1445 nm.

A comparative post-hoc (Duncan) analysis showed (Figure 1, Table 2) that, during three years of research, despite the difference in the level of development and pathogenic composition, a clear and pronounced separation of control and infected crops was noticed. Moreover, in 2021, this division manifested itself in the form of an excess of the average values of the crops SBC in control plots over the infected ones in a number of spectral channels.

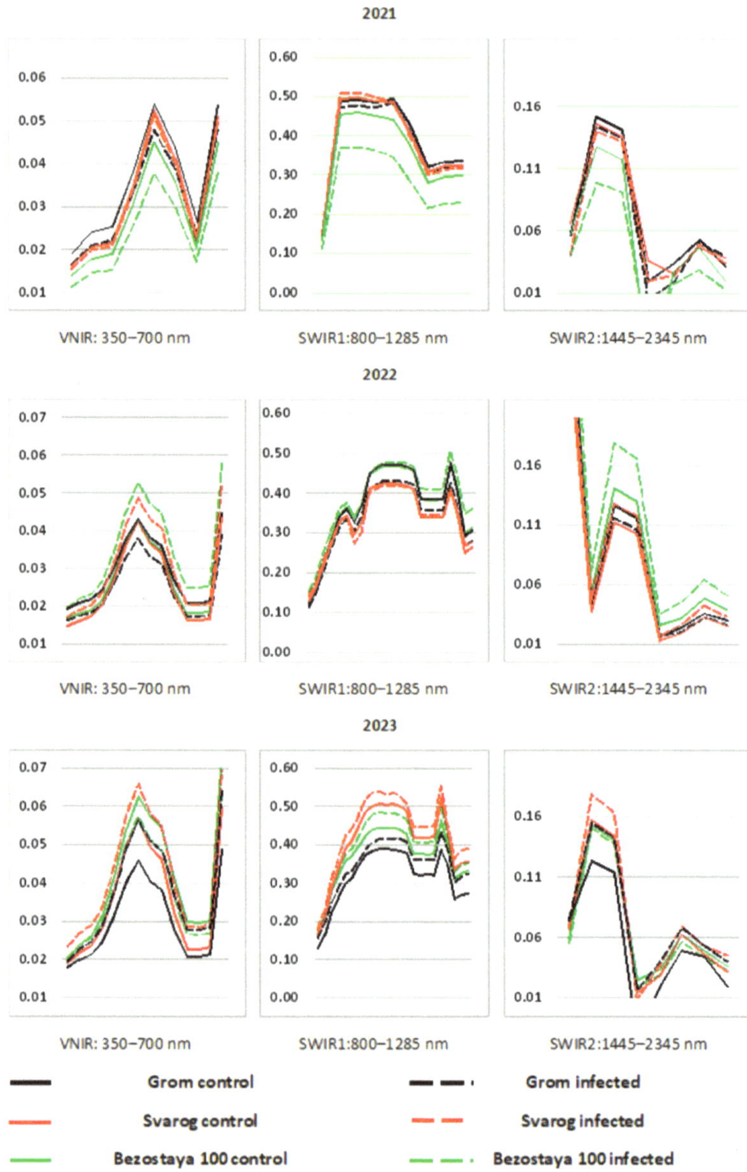

Figure 1. Spectral images of crops of studied varieties of winter wheat at the moment of the onset of intensive manifestation of all leaf diseases in the "flowering" phase of GS 60–70 in the growing season 2021–2023.

On the contrary, infected plots of the studied varieties of winter wheat exceeded the control ones in terms of reflectivity in 2022 and 2023. At the same time, the most informative spectral channels in which the difference between control and infected plots appeared are wavelengths of 575, 660, 700 and 2035 nm. A distinctive feature of the 2021 studies was the lowest development of pathogens compared to those in 2022 and 2023, as well as the simultaneous manifestation of yellow and brown rust on crops in infected plots. One could assume that this is what determined the nature of the differences between the control plots and the infected ones.

Table 2. A comparative post-hoc (Duncan) analysis of spectral brightness coefficient values of the studied winter wheat varieties in different spectral ranges at the time of the onset of intensive manifestation of all leaf diseases in the GS 60–70 "blooming" phase in the growing seasons 2021–2023.

Variety	Experience Option	Spectral Range, nm							
		490	550	660	720	845	1445	1675	2345
2021									
Bezostaya 100	Control	0.019 b	0.045 b	0.021 b	0.134 b	0.454 b	0.057 a	0.128 b	0.020 a
Svarog		0.021 bc	0.052 c	0.023 b	0.152 b	0.494 b	0.066 a	0.146 b	0.038 a
Grom		0.025 d	0.054 c	0.027 c	0.150 b	0.487 b	0.058 a	0.152 b	0.031 a
Bezostaya 100	Infected	0.015 a	0.038 a	0.017 a	0.111 a	0.367 a	0.039 a	0.098 a	0.012 a
Svarog		0.021 bc	0.051 bc	0.021 b	0.150 b	0.506 b	0.041 a	0.140 b	0.034 a
Grom		0.022 c	0.048 bc	0.023 b	0.136 b	0.472 b	0.056 a	0.144 b	0.039 a
2022									
Bezostaya 100	Control	0.018 ab	0.043 ab	0.018 a	0.128 ab	0.460 a	0.053 b	0.141 a	0.039 b
Svarog		0.016 a	0.042 ab	0.016 a	0.127 ab	0.419 a	0.037 a	0.112 a	0.025 a
Grom		0.021 b	0.043 ab	0.021 a	0.122 ab	0.463 a	0.043 ab	0.126 a	0.030 ab
Bezostaya 100	Infected	0.022 b	0.053 c	0.025 b	0.153 c	0.466 a	0.072 c	0.179 b	0.051 c
Svarog		0.019 ab	0.049 bc	0.021 a	0.140 bc	0.412 a	0.047 ab	0.127 a	0.033 ab
Grom		0.017 ab	0.038 a	0.017 a	0.113 ab	0.422 a	0.039 a	0.116 a	0.026 a
2023									
Bezostaya 100	Control	0.024 b	0.063 c	0.030 c	0.176 cd	0.440 b	0.059 a	0.153 b	0.036 a
Svarog		0.022 ab	0.057 b	0.023 a	0.164 bc	0.500 c	0.067 a	0.157 b	0.032 ab
Grom		0.020 a	0.046 a	0.021 a	0.128 a	0.383 a	0.074 a	0.123 a	0.020 ab
Bezostaya 100	Infected	0.024 b	0.057 b	0.027 b	0.164 bc	0.477 c	0.056 a	0.150 b	0.033 ab
Svarog		0.027 c	0.066 c	0.029 bc	0.180 d	0.536 d	0.069 a	0.178 c	0.045 c
Grom		0.023 b	0.056 b	0.028 bc	0.156 b	0.406 a	0.076 a	0.155 b	0.040 c

Notes: R—an indicator of the degree of progression of the disease; data represent the average mean value of the SBC and standard error in each column; the average values with the same letter do not differ significantly.

It is important to note that over the entire three-year study, the influence of varietal characteristics had an advantage in influencing the spectral properties of winter wheat crops. The development of the pathogenic background affected varieties only in accordance with the influence of varietal characteristics. In addition, it was found that the nature of the similarities and differences in the crops of compared winter wheat varieties according to spectral characteristics in the same growth phase, but in different years of research, was also ambiguous.

Differences only appeared in the mid-infrared spectral channel in 2021 and 2023. These years of research were characterized by leaf rust in the pathogenic background. In the spectral channels 490, 520, 550, 845, 1675, 1725 and 2345 nm, differences appeared only in 2023, when the maximum level of pathogen development was observed. In 2022, differences were noted in the 1445 nm spectral channel. This, in all likelihood, can be associated with the characteristics of composition and intensity of the pathogenic background for a given period of time.

The average level of development of the entire pathogenic background in control plots in 2021 was 0.33% and, in the infected ones, 0.72%. The difference between them was 0.54%. In 2022, this was 1.07 and 1.52%, respectively, with a difference of 0.45%. In 2023, the average level of development of all pathogens in the control background was 1.7%, and in the infected background 6.8%. The difference between them was 5.03%.

Thus, we can state that the factor of separability of winter wheat crops according to spectral characteristics with a minimum difference in the development of pathogens was 0.5%. In individual spectral channels of 490, 520, 845 and 2345 nm, variations appeared with a difference in the pathogenic background of 5%.

In 2021, the difference in the spectral characteristics of infected and control areas of the studied winter wheat crops was determined by the mutual influence of the development of yellow and brown rusts. According to the results of the correlation analysis for yellow and brown rusts, a negative relationship was revealed between the indicators of their development and the variable values of the spectral brightness coefficient of the spectral channels of the visible and near-infrared ranges (Table 3). Also, for both pathogens, a high and statistically significant correlation was established in the mid-infrared spectral channel of 1445 nm. Separately, for yellow rust, a high and statistically significant correlation was identified in the 2035 nm spectral channel.

Table 3. Results of assessing the correlation between the degree of disease development and variable values of the spectral brightness coefficient of spectral channels in the 2021–2023 research.

Pathogen	Spectral Range, nm												
	490	550	660	700	720	845	1245	1445	1675	2005	2035	2295	2345
						2021							
Powdery mildew	−0.39	0.13	−0.39	0.13	0.13	0.13	−0.13	0.13	−0.13	0.13	0.39	−0.39	−0.65
Yellow spot	0.46	0.33	0.46	0.33	0.21	0.58	0.27	−0.27	0.27	0.33	−0.09	0.7	0.52
Septoria	0.49	0.2	0.49	0.2	0.09	0.31	0.26	−0.37	0.26	0.26	−0.37	0.66	0.54
Yellow rust	−0.46	−0.62	−0.46	−0.62	−0.62	−0.31	−0.62	−0.93 *	−0.62	−0.22	−0.93 *	−0.31	−0.15
Brown rust	−0.68	−0.51	−0.68	−0.51	−0.51	−0.17	−0.68	−0.85 *	−0.68	−0.1	−0.68	−0.51	−0.51
Generalized categories	0.03	−0.35	0.03	−0.35	−0.43	−0.23	−0.23	−0.72	−0.23	−0.17	−0.75	0.17	0.17
						2022							
Powdery mildew	0.03	0.31	0.03	0.31	0.2	−0.54	−0.66	−0.31	−0.31	−0.66	−0.31	−0.14	−0.31
Yellow spot	-	-	-	-	-	-	-	-	-	-	-	-	-
Septoria	−0.2	−0.49	−0.2	−0.49	−0.77	0.09	−0.26	−0.66	−0.66	−0.26	−0.66	−0.77	−0.66
Yellow rust	0.43	0.77	0.43	0.77	0.94 *	0.03	0.14	0.54	0.54	0.14	0.54	0.71	0.54
Brown rust	-	-	-	-	-	-	-	-	-	-	-	-	-
Generalized categories	0.43	0.7	0.43	0.7	0.64	−0.12	−0.14	0.2	0.2	−0.14	0.2	0.32	0.2
						2023							
Powdery mildew	0.49	0.71	0.14	0.14	0.71	0.94 *	0.94 *	−0.49	0.6	−0.14	−0.09	−0.14	0.2
Yellow spot	0.2	−0.09	0.14	0.14	−0.09	−0.2	−0.2	0.31	−0.09	0.09	0.6	0.26	0.49
Septoria	0.49	0.54	0.14	0.14	0.54	0.77	0.77	−0.49	0.43	0.2	0.26	−0.14	0.37
Yellow rust	-	-	-	-	-	-	-	-	-	-	-	-	-
Brown rust	−0.54	−0.78	−0.3	−0.3	−0.78	−0.78	−0.78	0.85 *	−0.3	−0.3	0.17	0.34	−0.07
Generalized categories	0.6	0.49	0.43	0.43	0.49	0.6	0.6	0.14	0.71	0.03	0.77	0.54	0.83 *

Notes: *—statistical significance of data correlation is confirmed.

Tan spot and septoria were allocated to another group. A high and statistically significant correlation was also identified between these pathogens. The development indicators of powdery mildew were characterized by an average level of correlation with the development indicators of tan spot and septoria. No significant level of correlation was found with variable SBC values of spectral channels for tan spot and septoria.

A comparative post-hoc analysis showed that the difference in the spectral characteristics of the studied winter wheat crops with different levels of pathogen development was largely determined by the influence of the variety factor. Thus, in 2021, the main and most pronounced difference between control crops and infected spots was identified. It was determined by the presence of the joint manifestation of yellow and brown rusts on the latter. Despite the fact that the distribution and development of brown spot and septoria blight were high, the influence of these pathogens on the spectral response of the studied winter wheat crops was not revealed.

The pathogenic background of 2022 was characterized by the highest level of yellow rust manifestation over the entire research period, which negatively correlated with the development of septoria (Table 5). Correlation analysis revealed that yellow rust development indicators were characterized by a high and positively directed relationship with variable values of the spectral channels of the visible range.

Correlation analysis revealed that yellow rust was characterized by a positive correlation of the development degree with variable SBC values in the visible spectral channels, and the highest and statistically significant correlation in the 720 nm spectral channel. Septoria stood out with an average (0.5–0.6) level of correlation in the spectral channels of the mid-infrared range (1445, 1675, 1725, 2035, 2295 and 2345 nm), and powdery mildew in the near-infrared spectral range (Table 3).

A comparative post-hoc analysis indicated that the most pronounced and significant difference was characterized by infected crops of the Bezostaya 100 and Svarog varieties. This was manifested by the highest SBC values for almost all spectrum channels (Appendix A, Table A1). It is important that the spectral properties of two different varieties, when considered separately, are very different, despite the same level of pathogenic background. In general, the most pronounced differentiation of crops according to the degree of development of yellow rust appeared in the 720 nm spectral channel. Differentiation of crops according to the degree of septoria development in most of the spectral channels was subject to the influence of the variety factor. Categories of crops with pathogen development of 0.5–1% were identified only at wavelengths of 1445 and 2345 nm.

In 2023, the highest level of disease development was observed compared to 2021 and 2022, respectively, which was manifested in the strongest impact on the spectral properties of the studied winter wheat crops (Table 5). Against a pathogenic background, a statistically significant negative correlation was observed between the development of powdery mildew and leaf rust. In addition to these pathogens, a positive correlation of development was observed in septoria with powdery mildew, as well as a negative correlation in septoria with leaf rust.

The mutual influence of the three pathogens manifested in the correlation between their development and the variable SBC values of the near-infrared spectral channels (Table 3). This relationship was highest and statistically significant for powdery mildew. A statistically significant relationship between the developments of leaf rust and variable SBC values appeared at a wavelength of 1445 nm.

A comparative post-hoc analysis demonstrated that the best differentiation of crops according to disease intensity was also revealed in the spectral channels of the near-infrared range (Appendix A, Table A2). At the same time, a tendency for SBC values to increase along with the growing intensity of disease development was observed for powdery mildew and septoria. On the other hand, for leaf rust, a decrease in SBC values with an increase in the severity of disease development was observed.

Thus, according to the intensity of powdery mildew, the studied crops were clearly divided into groups with corresponding indicators of disease intensity of 0–3, 4, 10–12 and 30%, but according to the degree of development of septoria by 0, 2–3 and 4%. Solely crops with an indicator of 0% stood out according to the intensity of leaf rust. When grouping SBC values obtained at a wavelength of 1445 nm, no statistically significant differentiation of crops according to the degree of leaf rust intensity was revealed, despite the high level of correlation with the development of the disease in this wavelength range.

There was no significant correlation found in any of the tan spot spectral channels, although in 2023 it reached an average level of development of 18.37%. The division of crops according to disease intensity in agreement with the Duncan criterion was also extremely ambiguous. Apparently, this was due to the very specific reaction of each specific variety to the influence of a given pathogen.

The most pronounced differentiation of crops according to generalized groups appeared in the 2345 nm spectral channel. At the same time, crops of the first category with minimal pathogenic background, as well as of the fifth and sixth categories of crops with maximum indicators of disease development, were identified. The differentiation of the second and third categories was determined by the variety factor, which was superimposed by the influence of pathogens of various levels of development and composition. However, upon closer examination, it is possible to distinguish between these categories based on combinations of differences in SBC values in various spectral channels.

As a result, tables were created that compare spectral properties between individual varieties over a three-year study period (Table 4, Appendix A, Tables A3–A5).

Against the background of pathogenic development on crops of the Bezostaya 100 variety, a statistically significant positive correlation of the development of powdery mildew with yellow spot was revealed, which influenced the appearance for this variety of a statistically significant correlation between the external signs of powdery mildew damage and the SBC values of the visible and near-infrared wavelength ranges (Table 4).

Table 4. Results of the assessment of the correlation between the degree of disease development and variable values of the spectral brightness coefficient of winter wheat crops of the varieties Bezostaya 100, Svarog and Grom for the research period 2021–2023.

Pathogen	Spectral Range, nm												
	490	550	660	700	720	845	1245	1445	1675	2005	2035	2295	2345
						Bezostaya 100							
Powdery mildew	0.84 *	0.84 *	0.84 *	0.84 *	0.84 *	0.61	0.81 *	0.41	0.75	0.46	0.58	0.32	0.46
Yellow spot	0.78	0.78	0.78	0.78	0.78	0.34	0.44	0.14	0.37	−0.03	0.14	−0.1	−0.03
Septoria	0.31	0.31	0.31	0.31	0.31	0.03	0.14	−0.14	0.03	−0.26	−0.26	−0.49	−0.43
Yellow rust	−0.46	−0.46	−0.46	−0.46	−0.46	0.15	0.21	0.03	0.21	0.7	0.39	0.39	0.58
Brown rust	−0.65	−0.65	−0.65	−0.65	−0.65	−0.65	−0.65	−0.65	−0.65	−0.39	−0.65	−0.65	−0.65
Generalized categories	0.66	0.66	0.66	0.66	0.66	0.71	0.89 *	0.37	0.77	0.6	0.6	0.31	0.54
						Svarog							
Powdery mildew	0.52	0.52	0.52	0.81 *	0.52	0.26	0.52	0.52	−0.81 *	0.81 *	0.12	0.06	−0.31
Yellow spot	0.7	0.7	0.7	0.52	0.7	0.94 *	0.7	0.7	−0.39	0.52	0.82	0.52	-
Septoria	0.66	0.66	0.66	0.43	0.66	0.77	0.66	0.66	−0.49	0.43	0.43	0.03	−0.66
Yellow rust	−0.88 *	−0.88 *	−0.88 *	−0.52	−0.88 *	−0.7	−0.88 *	−0.88 *	0.03	−0.52	−0.64	−0.52	0.54
Brown rust	−0.13	−0.13	−0.13	−0.39	−0.13	0.39	−0.13	−0.13	0.39	−0.39	0.39	0.13	-
Generalized categories	0.49	0.49	0.49	0.77	0.49	0.31	0.49	0.49	−0.83 *	0.77	0.14	0.03	0.2
						Grom							
Powdery mildew	−0.58	−0.72	−0.67	−0.52	−0.72	−0.23	−0.64	−0.41	−0.67	0.17	−0.12	−0.72	−0.81 *
Yellow spot	0.38	0.75	0.55	0.81 *	0.75	−0.46	0.06	0.93 *	0.55	−0.64	0.12	0.75	0.41
Septoria	−0.09	−0.09	−0.03	−0.31	−0.09	0.26	0.26	−0.43	−0.03	0.09	−0.09	−0.09	0.37
Yellow rust	−0.52	−0.76	−0.52	−0.88 *	−0.76	0.21	−0.21	−0.88 *	−0.52	0.27	−0.27	−0.76	−0.21
Brown rust	0.1	0.51	0.34	0.68	0.51	−0.78	−0.3	0.85 *	0.34	−0.54	0.34	0.51	0.17
Generalized categories	−0.37	−0.14	−0.09	−0.09	−0.14	−0.66	−0.49	0.03	−0.09	−0.14	0.37	−0.14	0.09

Notes: *—statistical significance of data correlation is confirmed.

A comparative post-hoc analysis showed that crops of the Bezostaya 100 variety were well differentiated by the degree of development of powdery mildew, tan spot and septoria into groups with indicators of 0–1 and 2–10% in the spectral channels of the visible and near-infrared ranges (Appendix A, Table A3). At the same time, there was a tendency for SBC values to increase with the intensification of disease manifestation.

The pathogenic background of crops of the Svarog variety during three years of research was characterized by a positive interaction between the development of powdery mildew, tan spot and septoria, which negatively correlated with yellow rust manifestation. A high level of statistically significant and negatively directed correlation of variable SBC values of most spectral channels with symptoms of yellow rust development was revealed for crops of the Svarog variety (Table 4). A positive, statistically significant correlation of variable SBC values was revealed for powdery mildew in the 700 and 2035 nm spectral channels, for tan spot in the 845, 920 and 1085 nm spectral channels, and for septoria in the 920 nm spectral channel.

A comparative post-hoc analysis showed that the clearest and most pronounced division of crops was also revealed when grouping variable SBC values corresponding to the degree of intensity of yellow rust (Appendix A, Table A4). Crops of the Svarog variety were clearly divided according to the degree of development of yellow rust with a gradation of 0, 1–2 and 6%. At the same time, crops with a lower level of pathogen development were characterized by the highest average values. That is, the opposite trend was observed for powdery mildew, tan spot and septoria on crops of the Bezostaya 100 variety. Only groups with maximum pathogen development rates were identified when dividing crops by degree of intensity of powdery mildew, tan spot and septoria.

For the pathogenic background of the Grom variety, a positive and statistically confirmed relationship between the development of tan spot and brown rust was revealed, which in turn was negatively correlated with yellow rust. For indicators of brown spot and yellow rust on crops of the Grom variety, a characteristic was determined—a significant dependence at wavelengths of 700 and 1445 nm (Table 4). A statistically significant high correlation for leaf rust appeared only at a wavelength of 1445 nm, and for powdery mildew in the 2345 nm spectral channel.

A comparative post-hoc analysis showed that the differentiation of crops of the Grom variety was ambiguous according to the development degree of pathogens and the intensity of powdery mildew, tan spot and septoria (Appendix A, Table A5). Groups of crops with maximum pathogen development rates of 2–3% were identified for brown and yellow rust.

A cumulative comparison of the spectral characteristics of all three varieties over the entire period of research revealed a statistically significant relationship between only two pathogens: tan spot and yellow rust. A statistically significant correlation with the variable values of the visible and mid-infrared spectral channels was established for these two pathogens (Appendix A, Table A6).

3. Discussion

3.1. Assessment of the Mutual Influence of Variety Factors and Disease Development on the Spectral Characteristics of Winter Wheat Crops

Using two-factor analysis of variance, the influence of varietal factors and disease development on the spectral characteristics of the studied winter wheat crops was established (Table 1). In 2021, only a separate influence of crop varietal characteristics and disease development indicators was observed. In 2022 and 2023, an interaction between varietal factors and disease development was observed. This interaction of factors manifested itself in the form of differences in the spectral characteristics of the infected and control crops of each individual variety in the spectral channels of the visible and mid-infrared spectral ranges. In the near-infrared range, there was no interaction be-tween variety factors and disease development. Selected visible spectral channels (490, 520, 550, 575, 660 and 700 nm) are sensitive to pigment content (chlorophyll, carotenoids), phytomass, canopy structure, moisture content and plant stress. The mid-infrared range contains areas of water absorption, as well as areas sensitive to plant stress, lignin and starch content [12]. Thus, it can be assumed that the interaction of variety factors and disease development was a stress response of the host plant. This was accompanied by changes in the pigment composition of plants, disruption of the plant canopy structure, as well as disruption of water and temperature conditions [14,15].

A comparative post-hoc (Duncan) analysis showed (Table 2) that over three years of research, despite the difference in intensity and composition of pathogens, a clear and pronounced separation of control and infected crops was observed.

It was revealed that in the 575, 660, 700 and 2035 nm spectral channels, these differences appeared annually, regardless of the conditions and the specific year of research. At the same time, the manifestation of variability at other wavelengths of the spectral range depended on the composition and disease intensity in the pathogenic background. It is known that the 575 nm green spectral channel is sensitive to the content of plant pigments [11]. The 660 nm wavelength of the red part of the spectrum is associated with chlorophyll absorption and depends on many factors (phytomass, crop, canopy structure, nitrogen content, moisture content and stress) [12]. The 700 nm range is also associated with chlorophyll absorption and is most sensitive to changes in overall plant health. The light wavelength of 2035 nm corresponds to the maximum absorption of water in the mid-infrared region of the spectrum [13,14].

3.2. Comparison of the Spectral Characteristics of Crops of the Studied Varieties across the Time Period of Each Individual Year of Research

When considering the spectral properties of the studied winter wheat crops across the time period of each individual year, the following patterns can be identified:

- The nature of the influence of the pathogenic background on the spectral characteristics of winter wheat crops was determined by the complex interaction of disease development in a specific year of research. For instance, in 2021, the differences in the spectral characteristics of the control and infected backgrounds were determined by mutual manifestations of yellow and brown rusts in the latter. In 2022, the greatest impact was exerted by a negative correlation between the development of yellow rust and septoria, and in 2023 powdery mildew and brown rust;
- A clear and pronounced influence of varietal characteristics on the spectral properties of winter wheat was observed over three years. In most cases, this manifested itself as an immunological reaction of a particular variety to the influence of pathogen development. Different cultivars with similar pathogen indicators often exhibited strong differences in spectral response;
- A regular and clear division of the spectral properties of winter wheat crops according to the intensity of the disease was observed only at a level of pathogen development of more than 5%. Moreover, this gradation was clearest in the spectral channels of the near-infrared range and at a wavelength of 720 nm. The most pronounced differentiation of crops according to generalized groups appeared in the 2345 nm spectral channel.

However, upon closer examination, it was possible to identify distinctions in these categories based on combinations of differences in SBC values in various spectral channels.

- When the pathogenic background was below 5%, as in 2022 and 2023, the reflectivity of crops was largely determined by the influence of the variety factor or the interaction of the variety factor and disease intensity;
- Tan spot showed no significant correlations with SBC variables, even at its highest level in 2023. In all likelihood, this pathogen is the most variety-specific, i.e., it is determined most by the immunological reaction of a particular variety;
- A high level of correlation with variable SBC values of the 1445 nm spectral channel was revealed for leaf rust in 2021 and 2023.

3.3. Comparison of the Spectral Characteristics of Each Individual Variety over Three Years of Research

A comparison of the spectral properties of each individual variety over three years of research revealed the following patterns:

- A fairly clear correlation and differentiation of spectral characteristics according to the development degree of individual pathogens was revealed for crops of each individual variety;
- The spectral properties of each individual variety were determined by the different direction of the correlation relationship of disease intensity indicators in the general pathogenic background. Moreover, the nature of this relationship was different for the compared varieties, even with similar indicators of external disease manifestation. For instance, a pronounced and statistically significant correlation of external signs of powdery mildew manifestation with variable values of SBC spectral channels of the visible and near-infrared ranges appeared for the Bezostaya 100 variety. Such a relationship was not found in crops of the Svarog variety, though they were characterized by the highest rates of powdery mildew intensity. But a high level of statistically significant and negatively directed correlation of variable SBC values of most spectral channels with symptoms of yellow rust development was revealed. Moreover, the gradation of yellow rust development indicators in the Bezostaya 100 and Svarog varieties was almost identical;
- The clearest and most consistent differentiation of the spectral characteristics of winter wheat varieties was manifested by pathogen intensity. Thus, crops of the Bezostaya 100 variety were well differentiated by the intensity of powdery mildew, tan spot and septoria into groups with indicators of 0–1 and 2–10%. Crops of the Svarog variety were clearly divided according to the intensity of yellow rust with a gradation of 0, 1–2 and 6%. Crops with minimum and maximum indicators were differentiated according to general categories of disease intensity.

3.4. *Prospects for Further Development of Research*

The international scientific literature presents works aimed at identifying the development of individual pathogens in different wheat varieties [18–34]. These works proved the possibility of diagnosing diseases using hyperspectral analysis methods. However, results obtained even within the study of a single disease vary widely. This difference in results is potentially explained by the biochemical characteristics of different wheat varieties, climate, as well as a complex combination of the influence of abiotic and biotic stress factors [13–15]. Research on varietal and biochemical differences forms only a very small part of all studies devoted to spectral studies of vegetation. In addition, there is no reference methodology or database on which the authors of the works could rely in their research.

A feature of these studies is a detailed study of the nature of the interaction of varietal factors and complex development of the main economically significant diseases on the pathogenic background of crops of three varieties of winter wheat in 2021, 2022 and 2023.

The research results allow us to conclude that it is necessary to accumulate and systematize data over a significant period of time in relation to specific wheat varieties. A solution to this problem may be to create a model based on long-term data. Such a model should contain parameters of the mutual influence of pathogens on a specific variety, taking into account the limiting weather factors of a particular year. It is also possible to create a generalized model that allows extrapolation of data for many varieties based on studying a group of reference varieties [32,33]. In addition, the importance of studying biochemical changes in plant tissues under the influence of pathogens should be recognized [35,36].

4. Materials and Methods

4.1. *Organization of Test Plots and Experimental Design*

We conducted the research in the experimental fields of the Federal Research Center of Biological Plant Protection (FRCBPP), Krasnodar (45° 2.413′ 0″ N, 38° 58.5598′ 0″ E, 29 m above sea level) in 2021–2023 (Figure 2). The Köppen climate classification scheme assigns the climate of the study area as transitional from temperate continental to subtropical (Cfa) [37]. This region is characterized by long, hot summers and mild to moderately warm winters. Transitional seasons are poorly expressed. The average annual precipitation is

700–750 mm. The average annual air temperature is +13.4 °C and the average annual air humidity is 71%. The soil cover of the territory is represented by leached chernozems with low humus [38].

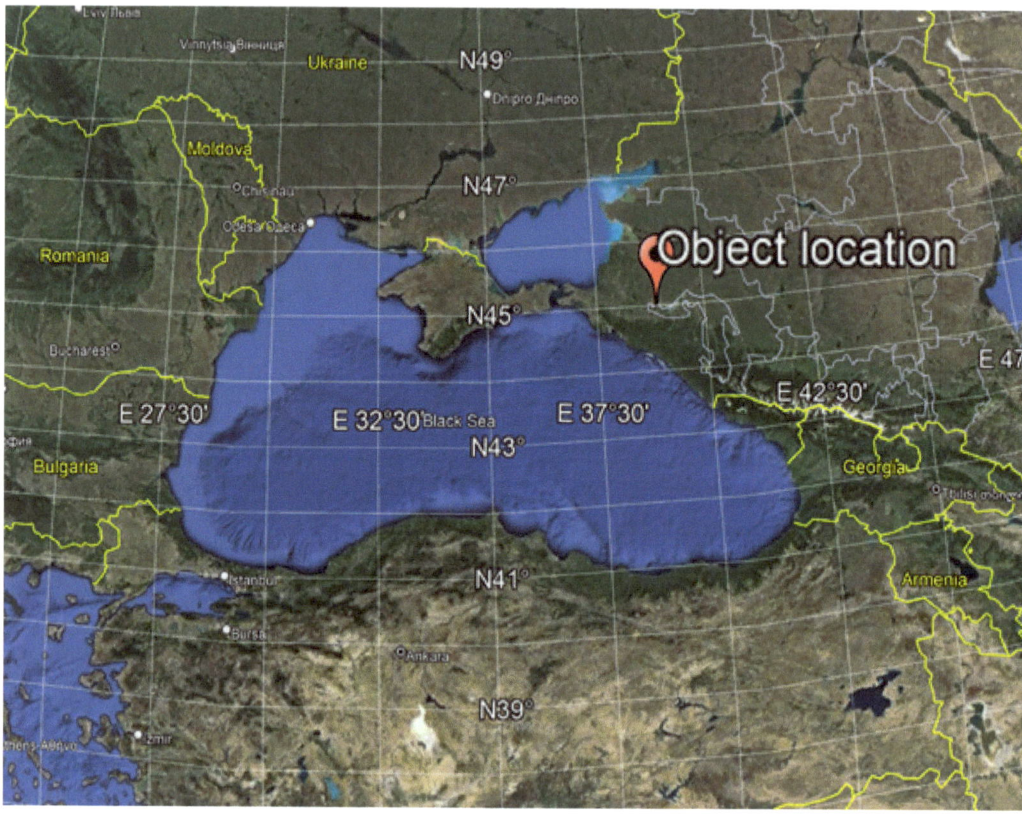

Figure 2. Geographical location of the research site.

The studied winter wheat crops were represented by three varieties bred by the National Grain Center named after. P. P. Lukyanenko (Krasnodar, Russia), which are susceptible to phytopathogens: Bezostaya 100, Svarog and Grom. Each plot was divided into two zones: 1—disease-protected by fungicides (clean background), 2—with an infectious background of pathogens. Artificial inoculation methods were used to develop brown and yellow rusts in the experimental area [39]. Inoculation of plants was carried out in the first ten days of April (HS phase 30–32). A mixture of urediniospores and talc in a ratio of 1:100 at a load of 5 mg spores/m^2 was used as an inoculating agent. The development of pathogens causing yellow spot, septoria and powdery mildew occurred against a natural infectious background. The creation of a control background (without diseases) was carried out by 2-fold treatment with the systemic fungicide Sokol, KS: 1st treatment on 25–31 April (flag-leaf phase), 2 May. 10–15 (phase "beginning of flowering" GS 61).

The research methodology was based on a comparative analysis of high-precision ground-based spectrometric measurements with the results of field phytopathological studies (Figure 3).

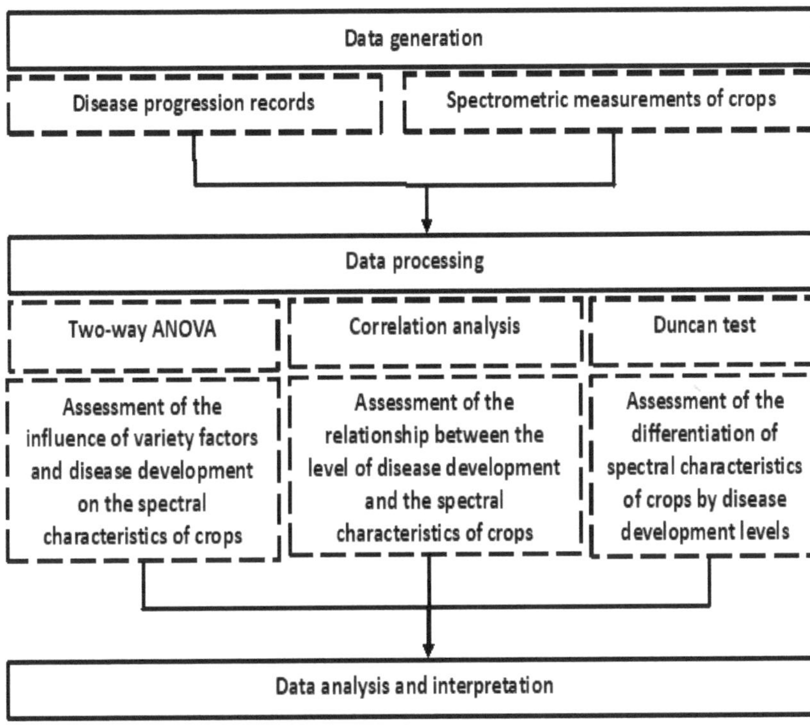

Figure 3. Research methodology diagram.

4.2. Field Experiments

The main period for conducting research was the beginning of the intensive appearance of leaf-stem diseases. This period fell in the second ten days of May (phase GS 61 "beginning of flowering"). This time period was the leading link for creating a predictive model of pathogen development since it allows for a comparative analysis of quantitative indicators of pathogen development. This analysis took into account the development of the pathogen from primary symptoms (after the incubation period) to intensive manifestation, taking into account the influence of varietal factors and weather conditions of a particular year.

The degree of development of the disease was assessed using the method of visual calculation of the ratio of the proportion of the affected area of the plant leaf blade to its total area (Figure 4). Visual observations of the development of winter wheat diseases were carried out while moving along the diagonal of each experimental plot with an area of 10 m^2. A total of 30 plants were selected for analysis. After this, for each tier (first, second leaf, etc.) a percentage assessment of leaf damage was given according to international scales. The degree of damage from rust diseases was assessed using the Peterson scale [40]; the degree of pyrenophorosis damage was assessed using the modified Saari–Prescott scale [41]; the degree of damage by powdery mildew and septoria was assessed using a special scale developed by CIMMYT [42].

Analyzing the test areas, the average degree of disease development was calculated using the following Formula (1):

$$R = \frac{1}{n}\sum_{i=1}^{n} r_i \quad (1)$$

where R is the average degree of disease development, %; r—degree of the disease development of an individual plant, %; n—total number of registered plants, pcs.

Areas with crops of the studied varieties were divided into general categories in accordance with the average indicators of pathogen development (Table 5).

Figure 4. Symptoms and degree of development of diseases. (**a**) Brown rust—10%; (**b**) Tan spot—25%; (**c**) Septoria—40%; (**d**) Brown rust—15%; (**e**) Brown rust—25%, Tan spot—5%; (**f**) Brown rust—10%, Tan spot—5%; (**g**) Brown rust—30%, Tan spot—10%.

4.3. Ground-Based Spectrometric Measurements

Ground-based spectrometry was carried out remotely at a height of 1.2–1.4 m from the earth's surface in the range of electromagnetic radiation from 350 to 2500 nm, with a spectral resolution of 1–10 nm. For this purpose, the ASD FieldSpec 3 Hi-Res spectroradiometer (Boulder, CO, USA) was used [43], which is designed for field remote sensing of the environment. The device has a non-removable fiber optic cable with factory calibration, thanks to which a high signal-to-noise ratio is achieved, which in turn ensures high accuracy of results for better identification and analysis of materials. To ensure comparability of the obtained data, measurements were carried out on days with clear sunny weather with a minimum amount of clouds. The sun's altitude was more than 35°.

Such analysis conditions were chosen due to the fact that, under such circumstances, lighting conditions change significantly less. This period of time and weather conditions reduce the possible error associated with the tilt of the sun. To analyze the spectral properties of the vegetation cover, two series of measurements of five repetitions were carried out. In the intervals between measurements, the panel reflecting light was calibrated. This decision was made due to the need to reduce the influence of uneven lighting. Vegetation cover was measured along the diagonal of the experimental plot, which corresponded to the methodology for conducting field surveys of plants for the presence of pathogens.

The results of ground-based spectrometric measurements are a set of spectral brightness coefficient (SBC) values. These values indicate the degree to which sunlight was reflected from plant surfaces at each wavelength.

Table 5. Average indicators of disease development in winter wheat crops of varieties Bezostaya 100, Svarog and Grom at the beginning of intensive manifestation of all leaf diseases in the GS 60–70 "blooming" phase in the growing seasons 2021–2023.

Variety	Experience Option	Powdery Mildew	Septoria	Yellow Tan	Yellow Rust	Brown Rust	General Categories
			2021				
Bezostaya 100	Control	0.01	0.8	0	0	0	1
Svarog		0.01	1	0	0	0	1
Grom		0.01	2.1	0.01	0	0	2
Bezostaya 100	Infected	0.01	1.9	0	0.3	0.3	4
Svarog		0.01	2.3	1.9	0.2	0.05	3
Grom		0	4.1	0.3	0.2	0	5
			2022				
Bezostaya 100	Control	0.8	0.26	-	1.48	-	1
Svarog		4.02	1.87	-	2.21	-	3
Grom		2.82	2.06	-	1.06	-	2
Bezostaya 100	Infected	2	0.94	-	7.72	-	5
Svarog		4.42	0.87	-	6.41	-	5
Grom		1.76	4.2	-	0.78	-	4
			2023				
Bezostaya 100	Control	3.43	1.23	1.87	-	0	2
Svarog		10.23	3.03	0.77	-	0	3
Grom		2.6	0	2.03	-	0.77	1
Bezostaya 100	Infected	12.93	3.77	10.03	-	0	4
Svarog		28.17	3.35	6.03	-	0	6
Grom		0	2.2	18.37	-	15.9	5

4.4. Data Processing

To identify specific spectral ranges indicating the manifestation of pathogenic changes, an analysis of changes in the morphology of reflective properties according to their actual state during field experiments was carried out.

Pre-processing of the analysis results and graphical visualization were carried out using the OriginPro 8.5.1 software package.

The pathogenic effect on the spectral properties of winter wheat plants in different wavelength ranges was assessed using two-way analysis of variance. To analyze the measurement results, the following wavelengths were selected: 490, 520, 550, 575, 660, 700, 720, 845, 1445, 1675 and 2345 nm. These spectral ranges are closely related to the biophysical characteristics of plants and are widely used in such studies [10]. Statistical processing of data from selected spectral channels was carried out with the calculation of the average value and standard deviation.

As a result of the analysis, the BCS values were grouped into categories corresponding to different degrees of the disease. The grouping of SBC values was carried out according to various parameters, including plant variety, damage from certain diseases, or compliance with three selected plant backgrounds. Correlation analysis of the relationship between the development of the disease and the number of detected pathogen spores was carried out on the basis of nonparametric statistical methods using Spearman's correlation at a high significance level of 95%. All methods of statistical analysis were performed in the Statistica 2010 program.

5. Conclusions

A regular and clear distinction in the spectral characteristics of winter wheat according to disease intensity was observed only when the pathogen development level exceeded 5%. Moreover, this gradation was clearest in the spectral channels of the near-infrared range and at a wavelength of 720 nm. During three years of research, a strong influence of varietal characteristics on the spectral properties of winter wheat crops was discovered. In most cases, this manifested itself as an immunological reaction of a particular variety to pathogens.

The features of the pathogenic influence on the spectral properties of winter wheat crops was characterized by a complex interaction between the manifestations of individual diseases in a specific year of research. The reflectivity of crops was largely determined by the influence of the variety factor or the interaction of the variety factor and disease intensity when the pathogenic background was below 5%, as in 2022 and 2023.

Author Contributions: Conceptualization methodology, field supervision, research, software, data processing—R.D. and O.K.; research, preparation of the initial draft, formal analysis, data processing—R.D., I.S., K.G. and D.I.; project management, funding, verification—O.K.; reviewing and editing, visualization—M.Z., O.K. and A.P. All authors have read and agreed to the published version of the manuscript.

Funding: The study was carried out within the framework of the State assignment of the Ministry of Education and Science of the Russian Federation on topic No. FGRN-2022-0001.

Data Availability Statement: Data is contained within the article.

Conflicts of Interest: The authors declare no conflicts of interest. The sponsors (Science Foundations) were not involved in the design of the study when collecting, analyzing or interpreting data, in the writing of the manuscript, or in the decision to publish the results.

Appendix A

Table A1. Results of a posteriori analysis of the spectral characteristics of winter wheat crops with different gradations of disease development according to the Duncan criterion in the growing season of 2022 research (GS 60–70 "flowering").

R, %	Spectral Range, nm							
	490	550	660	720	845	1445	1675	2345
				Septoria				
0.5	0.018 ± 0.001 ab	0.043 ± 0.003 a	0.018 ± 0.002 a	0.128 ± 0.006 a	0.460 ± 0.020 a	0.053 ± 0.005 bc	0.141 ± 0.012 ab	0.039 ± 0.004 bc
1	0.020 ± 0.001 b	0.051 ± 0.002 b	0.023 ± 0.001 b	0.147 ± 0.004 b	0.439 ± 0.014 a	0.060 ± 0.004 c	0.153 ± 0.008 b	0.042 ± 0.003 b
1.5	0.016 ± 0.001 a	0.042 ± 0.003 a	0.016 ± 0.002 a	0.127 ± 0.006 a	0.419 ± 0.020 a	0.037 ± 0.005 a	0.112 ± 0.012 a	0.025 ± 0.004 a
2	0.021 ± 0.001 b	0.043 ± 0.003 a	0.021 ± 0.002 ab	0.122 ± 0.006 a	0.463 ± 0.020 a	0.043 ± 0.005 ab	0.126 ± 0.012 ba	0.030 ± 0.004 ab
4	0.017 ± 0.001 ab	0.038 ± 0.003 a	0.017 ± 0.002 a	0.113 ± 0.006 a	0.422 ± 0.020 a	0.039 ± 0.005 ab	0.116 ± 0.012 a	0.026 ± 0.004 a

Table A1. *Cont.*

R, %	Spectral Range, nm							
	490	550	660	720	845	1445	1675	2345
				Yellow rust				
1	0.019 ± 0.001 ab	0.041 ± 0.002 a	0.019 ± 0.001 a	0.118 ± 0.004 a	0.442 ± 0.015 a	0.041 ± 0.004 ab	0.121 ± 0.008 a	0.028 ± 0.003 a
1.5	0.018 ± 0.001 ab	0.043 ± 0.003 a	0.018 ± 0.002 ab	0.128 ± 0.006 ab	0.460 ± 0.021 a	0.053 ± 0.005 bc	0.141 ± 0.012 ab	0.039 ± 0.004 b
2	0.016 ± 0.001 a	0.042 ± 0.003 a	0.016 ± 0.002 a	0.127 ± 0.006 ab	0.419 ± 0.021 a	0.037 ± 0.005 a	0.112 ± 0.012 a	0.025 ± 0.004 a
7	0.020 ± 0.001 b	0.051 ± 0.002 b	0.023 ± 0.001 b	0.147 ± 0.004 b	0.439 ± 0.015 a	0.060 ± 0.004 c	0.153 ± 0.008 b	0.042 ± 0.003 b
				Generalized categories				
1	0.018 ± 0.002 ab	0.043 ± 0.003 a	0.018 ± 0.002 ab	0.128 ± 0.005 a	0.460 ± 0.014 a	0.053 ± 0.004 bc	0.141 ± 0.008 ab	0.039 ± 0.003 bc
2	0.017 ± 0.001 ab	0.038 ± 0.002 a	0.017 ± 0.001 ab	0.113 ± 0.003 a	0.422 ± 0.013 a	0.039 ± 0.001 ab	0.116 ± 0.004 a	0.026 ± 0.001 a
3	0.021 ± 0.001 b	0.043 ± 0.001 a	0.021 ± 0.001 bc	0.122 ± 0.005 a	0.463 ± 0.016 a	0.043 ± 0.002 ab	0.126 ± 0.006 ab	0.030 ± 0.002 ab
4	0.016 ± 0.001 a	0.042 ± 0.002 a	0.016 ± 0.001 a	0.127 ± 0.005 a	0.419 ± 0.022 a	0.037 ± 0.002 a	0.112 ± 0.006 a	0.025 ± 0.001 a
5	0.020 ± 0.001 b	0.051 ± 0.002 b	0.023 ± 0.001 b	0.147 ± 0.006 b	0.439 ± 0.018 a	0.060 ± 0.006 c	0.153 ± 0.013 b	0.042 ± 0.005 c

Notes: R—an indicator of the degree of progression of the disease; data represent the average mean value of the SBC and standard error. In each column, the average values with the same letter do not differ significantly.

Table A2. Results of a posteriori analysis of the spectral characteristics of winter wheat crops with different gradations of disease development according to the Duncan criterion in the growing season of 2023 research (GS 60–70 "flowering").

R, %	Spectral Range, nm							
	490	550	660	720	845	1445	1675	2345
				Powdery mildew				
0	0.023 ± 0.001 b	0.056 ± 0.002 b	0.028 ± 0.001 bc	0.156 ± 0.005 b	0.406 ± 0.011 a	0.076 ± 0.005 a	0.155 ± 0.004 b	0.040 ± 0.005 ab
3	0.020 ± 0.000 a	0.046 ± 0.001 a	0.021 ± 0.000 a	0.128 ± 0.002 a	0.383 ± 0.008 a	0.074 ± 0.014 a	0.123 ± 0.003 a	0.020 ± 0.010 a
4	0.024 ± 0.001 b	0.063 ± 0.003 c	0.030 ± 0.002 c	0.176 ± 0.009 cd	0.440 ± 0.018 b	0.059 ± 0.003 a	0.153 ± 0.010 b	0.036 ± 0.003 ab
10	0.022 ± 0.001 ab	0.057 ± 0.001 b	0.023 ± 0.001 a	0.164 ± 0.003 bc	0.500 ± 0.009 c	0.067 ± 0.005 a	0.157 ± 0.004 b	0.032 ± 0.004 ab
12	0.024 ± 0.000 b	0.057 ± 0.000 b	0.027 ± 0.000 b	0.164 ± 0.001 bc	0.477 ± 0.002 c	0.056 ± 0.001 a	0.150 ± 0.002 b	0.033 ± 0.001 ab
30	0.027 ± 0.001 c	0.066 ± 0.002 c	0.029 ± 0.001 bc	0.180 ± 0.004 d	0.536 ± 0.015 d	0.069 ± 0.006 a	0.178 ± 0.006 c	0.045 ± 0.006 b

Table A2. *Cont.*

R, %	Spectral Range, nm							
	490	550	660	720	845	1445	1675	2345
				Septoria				
0	0.020 ± 0.000 a	0.046 ± 0.001 a	0.021 ± 0.000 a	0.128 ± 0.002 a	0.383 ± 0.008 a	0.074 ± 0.014 a	0.123 ± 0.003 a	0.020 ± 0.010 a
2	0.024 ± 0.001 c	0.063 ± 0.003 c	0.030 ± 0.002 c	0.176 ± 0.009 c	0.440 ± 0.018 b	0.059 ± 0.003 a	0.153 ± 0.010 b	0.036 ± 0.003 b
3	0.022 ± 0.000 bc	0.056 ± 0.001 b	0.025 ± 0.001 b	0.160 ± 0.003 b	0.453 ± 0.010 b	0.072 ± 0.004 a	0.156 ± 0.003 b	0.036 ± 0.003 b
4	0.025 ± 0.000 c	0.062 ± 0.001 c	0.028 ± 0.000 bc	0.172 ± 0.003 b	0.507 ± 0.009 c	0.062 ± 0.003 a	0.164 ± 0.004 b	0.039 ± 0.003 b
				Generalized categories				
1	0.020 ± 0.000 a	0.046 ± 0.001 a	0.021 ± 0.000 a	0.128 ± 0.002 a	0.383 ± 0.008 a	0.074 ± 0.014 a	0.123 ± 0.003 a	0.020 ± 0.010 a
2	0.024 ± 0.001 b	0.063 ± 0.003 c	0.030 ± 0.002 c	0.176 ± 0.009 cd	0.440 ± 0.018 b	0.059 ± 0.003 a	0.153 ± 0.010 b	0.036 ± 0.003 ab
3	0.022 ± 0.001 ab	0.057 ± 0.001 b	0.023 ± 0.001 a	0.164 ± 0.003 bc	0.500 ± 0.009 c	0.067 ± 0.005 a	0.157 ± 0.004 b	0.032 ± 0.004 ab
4	0.024 ± 0.000 b	0.057 ± 0.000 b	0.027 ± 0.000 b	0.164 ± 0.001 bc	0.477 ± 0.002 c	0.056 ± 0.001 a	0.150 ± 0.002 b	0.033 ± 0.001 ab
5	0.023 ± 0.001 b	0.056 ± 0.002 b	0.028 ± 0.001 bc	0.156 ± 0.005 b	0.406 ± 0.011 a	0.076 ± 0.005 a	0.155 ± 0.004 b	0.040 ± 0.005 b
6	0.027 ± 0.001 c	0.066 ± 0.002 c	0.029 ± 0.001 bc	0.180 ± 0.004 d	0.536 ± 0.015 d	0.069 ± 0.006 a	0.178 ± 0.006 c	0.045 ± 0.006 b

Notes: R—an indicator of the degree of progression of the disease; data represent the average mean value of the SBC and standard error. In each column, the average values with the same letter do not differ significantly.

Table A3. Results of a posteriori analysis of the spectral characteristics of winter wheat crops of the Bezostaya 100 variety with different gradations of powdery mildew development according to the Duncan criterion (GS 60–70 "flowering").

R, %	Spectral Range, nm							
	490	550	660	720	845	1445	1675	2345
				Powdery mildew				
0	0.017 ± 0.001 a	0.041 ± 0.002 a	0.019 ± 0.001 a	0.123 ± 0.005 a	0.411 ± 0.020 a	0.048 ± 0.007 a	0.113 ± 0.007 a	0.040 ± 0.005 a
1	0.018 ± 0.002 a	0.043 ± 0.003 a	0.018 ± 0.002 a	0.128 ± 0.005 a	0.460 ± 0.014 ab	0.053 ± 0.004 a	0.141 ± 0.008 b	0.020 ± 0.010 bc
2	0.022 ± 0.001 b	0.053 ± 0.002 b	0.025 ± 0.001 b	0.153 ± 0.003 b	0.466 ± 0.018 ab	0.072 ± 0.003 b	0.179 ± 0.006 c	0.036 ± 0.003 c
4	0.024 ± 0.001 b	0.063 ± 0.003 c	0.030 ± 0.002 c	0.176 ± 0.009 b	0.440 ± 0.018 ab	0.059 ± 0.003 bc	0.153 ± 0.010 bc	0.032 ± 0.004 bc
12	0.024 ± 0.000 b	0.057 ± 0.000 bc	0.027 ± 0.000 bc	0.164 ± 0.001 b	0.477 ± 0.002 b	0.056 ± 0.001 a	0.150 ± 0.002 b	0.033 ± 0.001 b

Table A3. *Cont.*

R, %	Spectral Range, nm							
	490	550	660	720	845	1445	1675	2345
				Yellow spot				
0	0.018 ± 0.001 a	0.027 ± 0.031 a	0.020 ± 0.001 a	0.130 ± 0.004 a	0.432 ± 0.013 a	0.054 ± 0.004 a	0.133 ± 0.007 a	0.028 ± 0.006 a
2	0.024 ± 0.001 b	0.037 ± 0.046 b	0.030 ± 0.002 c	0.176 ± 0.009 b	0.440 ± 0.018 ab	0.059 ± 0.003 a	0.153 ± 0.010 a	0.036 ± 0.003 a
10	0.024 ± 0.000 b	0.038 ± 0.039 b	0.027 ± 0.000 b	0.164 ± 0.001 b	0.477 ± 0.002 b	0.056 ± 0.001 a	0.150 ± 0.002 a	0.033 ± 0.001 a
				Septoria				
0	0.018 ± 0.002 a	0.043 ± 0.003 a	0.018 ± 0.002 a	0.113 ± 0.143 a	0.460 ± 0.014 ab	0.053 ± 0.004 a	0.141 ± 0.008 a	0.039 ± 0.003 a
1	0.020 ± 0.001 ab	0.048 ± 0.002 ab	0.022 ± 0.001 ab	0.133 ± 0.151 ab	0.459 ± 0.015 ab	0.063 ± 0.006 a	0.149 ± 0.009 a	0.033 ± 0.009 a
2	0.022 ± 0.001 ab	0.056 ± 0.003 b	0.027 ± 0.002 b	0.141 ± 0.177 b	0.421 ± 0.016 a	0.054 ± 0.004 a	0.139 ± 0.009 a	0.030 ± 0.004 a
4	0.024 ± 0.000 b	0.057 ± 0.000 b	0.027 ± 0.000 b	0.162 ± 0.166 b	0.477 ± 0.002 b	0.056 ± 0.001 a	0.150 ± 0.002 a	0.033 ± 0.001 a
				Generalized categories				
1	0.019 ± 0.001 ab	0.045 ± 0.002 ab	0.021 ± 0.001 ab	0.134 ± 0.005 ab	0.454 ± 0.022 b	0.057 ± 0.009 bc	0.128 ± 0.008 b	0.020 ± 0.013 ab
2	0.015 ± 0.001 a	0.038 ± 0.002 a	0.017 ± 0.001 a	0.111 ± 0.005 a	0.367 ± 0.024 a	0.039 ± 0.010 a	0.098 ± 0.008 a	0.012 ± 0.010 a
3	0.022 ± 0.001 bc	0.053 ± 0.002 bc	0.025 ± 0.001 bc	0.153 ± 0.003 bc	0.466 ± 0.018 b	0.072 ± 0.003 c	0.179 ± 0.006 c	0.051 ± 0.003 d
4	0.018 ± 0.002 a	0.043 ± 0.003 a	0.018 ± 0.002 a	0.128 ± 0.005 ab	0.460 ± 0.014 b	0.053 ± 0.004 ab	0.141 ± 0.008 b	0.039 ± 0.003 cd
5	0.024 ± 0.001 c	0.063 ± 0.003 d	0.030 ± 0.002 d	0.176 ± 0.009 c	0.440 ± 0.018 b	0.059 ± 0.003 bc	0.153 ± 0.010 bc	0.036 ± 0.003 bcd
6	0.024 ± 0.000 c	0.057 ± 0.000 cd	0.027 ± 0.000 d	0.164 ± 0.001 c	0.477 ± 0.002 b	0.056 ± 0.001 b	0.150 ± 0.002 bc	0.033 ± 0.001 bc

Notes: R—an indicator of the degree of progression of the disease; data represent the average mean value of the SBC and standard error. In each column, the average values with the same letter do not differ significantly.

Table A4. Results of a posteriori analysis of the spectral characteristics of winter wheat crops of the Svarog variety with different gradations of powdery mildew development according to the Duncan criterion (GS 60–70 "flowering").

R, %	Spectral Range, nm							
	490	550	660	720	845	1445	1675	2345
				Yellow rust				
0	0.024 ± 0.001 c	0.060 ± 0.001 b	0.025 ± 0.001 c	0.169 ± 0.003 c	0.514 ± 0.009 b	0.068 ± 0.004 b	0.164 ± 0.004 b	0.039 ± 0.004 a
1	0.021 ± 0.001 bc	0.051 ± 0.002 ab	0.021 ± 0.001 bc	0.150 ± 0.006 bc	0.506 ± 0.017 b	0.041 ± 0.008 ab	0.140 ± 0.007 ab	0.034 ± 0.009 a
2	0.019 ± 0.002 ab	0.049 ± 0.005 a	0.021 ± 0.002 ab	0.140 ± 0.011 ab	0.412 ± 0.029 a	0.047 ± 0.008 ab	0.127 ± 0.018 a	0.033 ± 0.007 a
6	0.016 ± 0.001 a	0.042 ± 0.002 a	0.016 ± 0.001 a	0.127 ± 0.005 a	0.419 ± 0.022 a	0.037 ± 0.002 a	0.112 ± 0.006 a	0.025 ± 0.001 a

Table A4. *Cont.*

R, %	Spectral Range, nm							
	490	550	660	720	845	1445	1675	2345
Generalized categories								
1	0.021 ± 0.001 b	0.052 ± 0.003 bc	0.023 ± 0.001 b	0.152 ± 0.008 bc	0.494 ± 0.030 b	0.066 ± 0.015 b	0.146 ± 0.011 bc	0.038 ± 0.011 a
2	0.021 ± 0.001 b	0.051 ± 0.002 bc	0.021 ± 0.001 b	0.150 ± 0.006 bc	0.506 ± 0.017 b	0.041 ± 0.008 ab	0.140 ± 0.007 bc	0.034 ± 0.009 a
3	0.019 ± 0.002 ab	0.049 ± 0.005 ab	0.021 ± 0.002 b	0.140 ± 0.011 ab	0.412 ± 0.029 a	0.047 ± 0.008 ab	0.127 ± 0.018 ab	0.033 ± 0.007 a
4	0.016 ± 0.001 a	0.042 ± 0.002 a	0.016 ± 0.001 a	0.127 ± 0.005 a	0.419 ± 0.022 a	0.037 ± 0.002 a	0.112 ± 0.006 a	0.025 ± 0.001 a
5	0.022 ± 0.001 b	0.057 ± 0.001 c	0.023 ± 0.001 b	0.164 ± 0.003 cd	0.500 ± 0.009 b	0.067 ± 0.005 b	0.157 ± 0.004 cd	0.032 ± 0.004 a
6	0.027 ± 0.001 c	0.066 ± 0.002 d	0.029 ± 0.001 c	0.180 ± 0.004 d	0.500 ± 0.009 b	0.069 ± 0.006 b	0.178 ± 0.006 d	0.045 ± 0.006 a

Notes: R—an indicator of the degree of progression of the disease; data represent the average mean value of the SBC and standard error. In each column, the average values with the same letter do not differ significantly.

Table A5. Results of a posteriori analysis of the spectral characteristics of winter wheat crops of the Grom variety with different gradations of powdery mildew development according to the Duncan criterion (GS 60–70 "flowering").

R, %	Spectral Range, nm							
	490	550	660	720	845	1445	1675	2345
Yellow rust								
0	0.022 ± 0.000 b	0.051 ± 0.001 b	0.025 ± 0.001 b	0.143 ± 0.003 b	0.408 ± 0.008 a	0.072 ± 0.006 a	0.141 ± 0.003 b	0.030 ± 0.005 a
2	0.022 ± 0.001 b	0.048 ± 0.002 b	0.023 ± 0.001 b	0.136 ± 0.007 b	0.472 ± 0.024 b	0.056 ± 0.005 a	0.144 ± 0.009 b	0.039 ± 0.009 a
3	0.019 ± 0.001 a	0.041 ± 0.001 a	0.019 ± 0.001 a	0.118 ± 0.003 a	0.442 ± 0.012 ab	0.041 ± 0.001 a	0.121 ± 0.004 a	0.028 ± 0.001 a
Brown rust								
0	0.022 ± 0.001 ab	0.047 ± 0.002 a	0.023 ± 0.001 a	0.133 ± 0.004 a	0.464 ± 0.011 a	0.050 ± 0.003 a	0.137 ± 0.005 b	0.032 ± 0.004 a
1	0.020 ± 0.000 a	0.046 ± 0.001 a	0.021 ± 0.000 a	0.129 ± 0.002 a	0.386 ± 0.008 a	0.073 ± 0.014 ab	0.124 ± 0.003 a	0.016 ± 0.010 ab
2	0.022 ± 0.001 b	0.056 ± 0.002 b	0.027 ± 0.001 b	0.153 ± 0.005 b	0.402 ± 0.011 b	0.076 ± 0.005 b	0.152 ± 0.005 c	0.042 ± 0.005 b
Generalized categories								
1	0.025 ± 0.001 c	0.054 ± 0.002 c	0.027 ± 0.001 c	0.150 ± 0.006 cd	0.487 ± 0.020 c	0.058 ± 0.007 a	0.152 ± 0.008 c	0.031 ± 0.010 a
2	0.022 ± 0.001 b	0.048 ± 0.002 b	0.023 ± 0.001 b	0.136 ± 0.007 bc	0.472 ± 0.024 c	0.056 ± 0.005 a	0.144 ± 0.009 bc	0.039 ± 0.009 a
3	0.017 ± 0.001 a	0.038 ± 0.002 a	0.017 ± 0.001 a	0.113 ± 0.003 a	0.422 ± 0.013 ab	0.039 ± 0.001 a	0.116 ± 0.004 a	0.026 ± 0.001 a

Table A5. *Cont.*

R, %	Spectral Range, nm							
	490	550	660	720	845	1445	1675	2345
4	0.021 ± 0.001 b	0.043 ± 0.001 b	0.021 ± 0.001 b	0.122 ± 0.005 ab	0.463 ± 0.016 bc	0.043 ± 0.002 a	0.126 ± 0.006 ab	0.030 ± 0.002 a
5	0.020 ± 0.000 ab	0.046 ± 0.001 ab	0.021 ± 0.000 b	0.128 ± 0.002 ab	0.383 ± 0.008 a	0.074 ± 0.014 a	0.123 ± 0.003 a	0.020 ± 0.010 a

Notes: R—an indicator of the degree of progression of the disease; data represent the average mean value of the SBC and standard error. In each column, the average values with the same letter do not differ significantly.

Table A6. Correlation between the degree of disease development and variable spectral brightness coefficient values of winter wheat crops of all varieties in 2021–2023.

Spectral Channels	Pathogen				
	Powdery Mildew	Tan Spot	Septoria	Yellow Rust	Brown Rust
490	0.14	0.66 *	0.35	−0.56 *	−0.17
520	0.22	0.74 *	0.25	−0.66 *	−0.07
550	0.32	0.70 *	0.24	−0.59 *	−0.13
575	0.34	0.68 *	0.22	−0.52 *	−0.12
660	0.09	0.69 *	0.30	−0.56 *	−0.07
700	0.32	0.68 *	0.15	−0.53 *	−0.06
720	0.31	0.63 *	0.21	−0.53 *	−0.18
845	0.15	0.23	0.42	−0.28	−0.44
920	0.19	0.22	0.40	−0.28	−0.46
1085	0.23	0.31	0.40	−0.27	−0.44
1135	0.19	0.28	0.33	−0.22	−0.46
1215	0.22	0.44	0.35	−0.39	−0.38
1245	0.22	0.44	0.35	−0.39	−0.38
1285	0.23	0.44	0.32	−0.40	−0.37
1445	−0.02	0.52 *	−0.18	−0.56 *	0.18
1675	0.13	0.48 *	0.19	−0.39	−0.22
1725	0.12	0.50 *	0.21	−0.40	−0.20
2005	0.06	−0.09	−0.07	0.16	−0.33
2035	0.20	0.31	−0.05	−0.20	−0.21
2295	−0.27	0.44	0.03	−0.29	0.00
2345	−0.03	0.27	0.11	0.04	−0.18

Notes: *—statistical significance of data correlation is confirmed.

References

1. Kokhmetova, A.; Atishova, M.; Sapakhova, Z.; Kremneva, O.Y.; Volkova, G.V. Evaluation of wheat cultivars growing in Kazakhstan and Russia for resistance to tan spot. *J. Plant Pathol.* **2017**, *99*, 161–167. [CrossRef]
2. Kolmer, J.A.; Hughes, M.E. Physiologic specialization of *Puccinia triticina* on wheat in the United States in 2012. *Plant Dis.* **2014**, *98*, 1145–1150. [CrossRef] [PubMed]
3. Volkova, G.V.; Shulyakovskaya, L.N.; Kudinova, O.A.; Matveeva, I.P. Yellow rust of wheat in the Kuban. *Plant Prot. Quar.* **2018**, *4*, 22. [CrossRef]
4. Kremneva, O.Y.; Volkova, G.V.; Kovalenko, N.M. The dynamics of the racial composition of *Pyrenophora tritici-repentis* in the North Caucasus region. *Mycol. Phytopathol.* **2019**, *53*, 246–253. [CrossRef]

5. Kremneva, O.Y.; Danilov, R.Y.; Gasiyan, K.E.; Ponomarev, A.V. Spore-trapping device: An efficient tool to manage fungal diseases in winter wheat crops. *Plants* **2023**, *12*, 391. [CrossRef]
6. FAO (Food and Agriculture Organization of the United Nations). Available online: https://www.fao.org/home/en/ (accessed on 12 June 2022).
7. Fedorenko, V.F.; Mishurov, N.P.; Nemenushchaya, L.A. *Promising Technologies for the Diagnosis of Pathogens of Agricultural Plants: Scientific, Analyte, Review*; FGBNU «Rosinformagrotekh»: Moscow, Russia, 2018; p. 68.
8. Sanin, S. Biological, agroecological and ecological aspects of phytosanitary monitoring. *Plant Prot. News* **1999**, *1*, 62–67.
9. Abrosimov, A.V.; Dvorkin, B.A. Prospects of application of remote sensing data from space to improve the efficiency of agriculture in Russia. *Geomatika* **2009**, *4*, 45–49.
10. Antonov, V.N.; Sladkikh, L.A. Monitoring of crop condition and forecasting of productivity of spring wheat by remote sensing data. *Geomatika* **2009**, *4*, 50–53.
11. Thenkabail, P.S.; Lyon, J.G.; Huete, A. *Hyperspectral Remote Sensing of Vegetation*; CRC Press: Boca Raton, FL, USA, 2011.
12. Lowe, A.; Harrison, N.; French, A.P. Hyperspectral image analysis techniques for the detection and classification of the early onset of plant disease and stress. *Plant Methods* **2017**, *13*, 80–91. [CrossRef]
13. Avinash, P.; Ramathilaga, A.; Valarmathi, P. Hyperspectral remote sensing for discrimination for plant disease forecasting: Review. *J. Pharmacogn. Phytochem.* **2022**, *11*, 208–215.
14. Cheshkova, A.F. A review of hyperspectral image analysis techniques for plant disease detection and identification. *Vavilov J. Genet. Breed.* **2022**, *26*, 202–213. [CrossRef] [PubMed]
15. Wan, L.; Li, H.; Li, C.; Wang, A.; Yang, Y.; Wang, P. Hyperspectral Sensing of Plant Diseases: Principle and Methods. *Agronomy* **2022**, *12*, 1451. [CrossRef]
16. Danilov, R.Y.; Kremneva, O.Y.; Ismailov, V.Y.; Tretyakov, V.A.; Rizvanov, A.A.; Krivoshein, V.; Pachkin, A.A. General methods and results of ground hyperspectral studies of seasonal changes in the reflective properties of crops and certain types of weeds. *Curr. Probl. Remote Sens. Earth Space* **2020**, *1*, 113–127. [CrossRef]
17. Danilov, R.Y.; Kremneva, O.Y.; Pachkin, A.A. Identification of the Spectral Patterns of Cultivated Plants and Weeds: Hyperspectral Vegetation Indices. *Agronomy* **2023**, *13*, 859. [CrossRef]
18. Delwiche, S.R.; Kim, M.S. *Biological Quality and Precision Agriculture. II*; DeShazer, J.A., Meyer, G.E., Eds.; SPIE: Bellingham, WA, USA, 2000; Volume 4203.
19. Barbedo, J.G.A.; Tibola, C.S.; Fernandes, J.M.C. Detecting Fusarium head blight in wheat kernels using hyperspectral imaging. *Biosyst. Eng.* **2015**, *131*, 65–76. [CrossRef]
20. Alisaac, E.; Behmann, J.; Kuska, M.T.; Dehne, H.-W.; Mahlein, A.-K. Hyperspectral quantification of wheat resistance to Fusariumhead blight: Comparison of two Fusarium species. *Eur. J. Plant Pathol.* **2018**, *152*, 869–884. [CrossRef]
21. Mahlein, A.-K.; Alisaac, E.; Al Masri, A.; Behmann, J.; Dehne, H.-W.; Oerke, E.-C. Comparison and Combination of Thermal, Fluorescence, and Hyperspectral Imaging for Monitoring Fusarium Head Blight of Wheat on Spikelet Scale. *Sensors* **2019**, *19*, 2281. [CrossRef] [PubMed]
22. Ma, H.; Huang, W.; Jing, Y.; Pignatti, S.; Laneve, G.; Dong, Y.; Ye, H.; Liu, L.; Guo, A.; Jiang, J. Identification of Fusarium Head Blight in Winter Wheat Ears Using Continuous Wavelet Analysis. *Sensors* **2019**, *20*, 20. [CrossRef] [PubMed]
23. Zhang, J.; Pu, R.; Huang, W.; Yuan, L.; Luo, J.; Wang, J. Using in-situ hyperspectral data for detecting and discriminating yellow rust disease from nutrient stresses. *Field Crops Res.* **2012**, *134*, 165–174. [CrossRef]
24. Zhang, N.; Pan, Y.; Feng, H.; Zhao, X.; Yang, X.; Ding, C.; Yang, G. Development of Fusarium head blight classification index using hyperspectral microscopy images of winter wheat spikelets. *Biosyst. Eng.* **2019**, *186*, 83–99. [CrossRef]
25. Zhang, X.; Han, L.; Dong, Y.; Shi, Y.; Huang, W.; Han, L.; González-Moreno, P.; Ma, H.; Ye, H.; Sobeih, T.A. Deep Learning-Based Approach for Automated Yellow Rust Disease Detection from High-Resolution Hyperspectral UAV Images. *Remote Sens.* **2019**, *11*, 1554. [CrossRef]
26. Krishna, G.; Sahoo, R.N.; Pargal, S.; Gupta, V.K.; Sinha, P.; Bhagat, S.; Sahara, M.S.; Singh, R.; Chattopadhyay, C. Assessing Wheat Yellow Rust Disease through Hyperspectral Remote Sensing. *ISPRS Int. Arch. Photogramm. Remote Sens. Spat. Inf. Sci.* **2014**, *XL-8*, 1413–1416. [CrossRef]
27. Guo, A.; Huang, W.; Ye, H.; Dong, Y.; Ma, H.; Ren, Y.; Ruan, C. Identification of Wheat Yellow Rust Using Spectral and Texture Features of Hyperspectral Images. *Remote Sens.* **2020**, *12*, 1419. [CrossRef]
28. Guo, A.; Huang, W.; Dong, Y.; Ye, H.; Ma, H.; Liu, B.; Wu, B.; Ren, Y.; Ruan, C.; Geng, Y. Wheat Yellow Rust Detection Using UAV-Based Hyperspectral Technology. *Remote Sens.* **2021**, *13*, 123. [CrossRef]
29. Kremneva, O.Y.; Tutubalina, O.V.; Sereda, I.I.; Danilov, R.Y.; Zimin, M.V.; Kurilov, A.A. Studies of changes in the spectral characteristics of winter wheat varieties depending on the degree of infection with pathogens. *Curr. Probl. Remote Sens. Earth Space* **2020**, *3*, 149–161. [CrossRef]
30. Sereda, I.I.; Danilov, R.Y.; Kremneva, O.Y.; Zimin, M.V.; Podushin, Y.V. Development of Methods for Remote Monitoring of Leaf Diseases in Wheat Agrocenoses. *Plants* **2023**, *12*, 3223. [CrossRef] [PubMed]
31. Bohnenkamp, D.; Behmann, J.; Mahlein, A.-K. In-Field Detection of Yellow Rust in Wheat on the Ground Canopy and UAV Scale. *Remote Sens.* **2019**, *11*, 2495. [CrossRef]
32. Kang, Y.; Anderegg, J.; Mikaberidze, A.; Karisto, P.; Masche, F.; McDonald, B.A.; Walter, A.; Hund, A. Hyperspectral canopy sensing of Wheat *Septoria tritici* Blotch Disease. *Front. Plant Sci.* **2018**, *9*, 1195. [CrossRef]

33. Odilbekov, F.; Armonienė, R.; Henriksson, T.; Chawade, A. Proximal phenotyping and machine learning methods to identify Septoria tritici blotch disease symptoms in wheat. *Front. Plant Sci.* **2018**, *9*, 685. [CrossRef]
34. Kremneva, O.Y.; Danilov, R.Y.; Sereda, I.I.; Tutubalina, O.V.; Pachkin, A.A.; Zimin, M.V. Spectral characteristics of winter wheat varieties depending on the development degree of *Pyrenophora tritici-repentis*. *Precis. Agric.* **2022**, *24*, 830–852. [CrossRef]
35. Carter, G.A.; Spiering, B.A. Optical Properties of Intact Leaves for Estimating Chlorophyll Concentration. *J. Environ. Qual.* **2002**, *31*, 1424. [CrossRef]
36. Stellacci, A.M.; Castrignanò, A.; Troccoli, A.; Basso, B.; Buttafuoco, G. Selecting optimal hyperspectral bands to discriminate nitrogen status in durum wheat: A comparison of statistical approaches. *Environ. Monit. Assess.* **2016**, *188*, 1–15. [CrossRef] [PubMed]
37. Weatherbase. Krasnodar. *Russia Köppen Climate Classification (Weatherbase).* Available online: https://www.weatherbase.com/ (accessed on 12 April 2024).
38. Valkov, V.F.; Shtompel, Y.A.; Trubilin, I.T. *Soils of the Krasnodar Region Their Utilisation and Protection*; SKNTs VSh: Rostov-on-Don, Russia, 1996; p. 192.
39. Anpilogova, L.K.; Volkova, G.V. *Methods for Creating Artificial Infectious Backgrounds and Assessing Wheat Varieties for Resistance to Harmful Diseases (Spike Fusarium, Rust, Powdery Mildew)*; VNIIBZR, RASKHN: Krasnodar, Russia, 2000; 28p.
40. Peterson, R.F.; Cambell, A.B.; Hannah, A.E. A diagrammatic scale for estimating rust intensity on leaves and stems of cereals. *Can. J. Res.* **1948**, *26*, 496–500. [CrossRef]
41. Roelfs, A.P.; Singh, R.P.; Saari, E.E. *Rust Diseases of Wheat: Concepts and Methods of Disease Management*; CIMMYT: Mexico City, Mexico, 1992; ISBN 968612747X.
42. Eyal, Z.; Scharen, A.L.; Prescott, J.M.; van Ginkel, M. *The Septoria Diseases of Wheat: Concepts and Methods of Disease Management*; CIMMYT: Mexico City, Mexico, 1987.
43. ASD. *Fieldspec® 3 User Manual*; ASD Document 600540 Rev. J; ASD Inc.: Boulder, CO, USA, 2010; 110p, Available online: http://www.geo-informatie.nl/courses/grs60312/material2017/manuals/600540-JFieldSpec3UserManual.pdf (accessed on 10 June 2023).

plants

MDPI

Article

Maize Leaf Disease Recognition Based on Improved Convolutional Neural Network ShuffleNetV2

Hanmi Zhou [1,*], Yumin Su [1], Jiageng Chen [1], Jichen Li [1], Linshuang Ma [1], Xingyi Liu [1], Sibo Lu [1] and Qi Wu [2,*]

[1] College of Agricultural Equipment Engineering, Henan University of Science and Technology, Luoyang 471003, China; symm1037@163.com (Y.S.); cjg980227171x@163.com (J.C.); lijichenli@163.com (J.L.); mls15036005926@163.com (L.M.); liuxingyi5742@163.com (X.L.); 20143300216@stu.haust.edu.cn (S.L.)

[2] College of Water Conservancy, Shenyang Agricultural University, Shenyang 110866, China

* Correspondence: zhouhm@163.com (H.Z.); qiwu0701@syau.edu.cn (Q.W.)

Abstract: The occurrence of maize diseases is frequent but challenging to manage. Traditional identification methods have low accuracy and complex model structures with numerous parameters, making them difficult to implement on mobile devices. To address these challenges, this paper proposes a corn leaf disease recognition model SNMPF based on convolutional neural network ShuffleNetV2. In the down-sampling module of the ShuffleNet model, the max pooling layer replaces the deep convolutional layer to perform down-sampling. This improvement helps to extract key features from images, reduce the overfitting of the model, and improve the model's generalization ability. In addition, to enhance the model's ability to express features in complex backgrounds, the Sim AM attention mechanism was introduced. This mechanism enables the model to adaptively adjust focus and pay more attention to local discriminative features. The results on a maize disease image dataset demonstrate that the SNMPF model achieves a recognition accuracy of 98.40%, representing a 4.1 percentage point improvement over the original model, while its size is only 1.56 MB. Compared with existing convolutional neural network models such as EfficientNet, MobileViT, EfficientNetV2, RegNet, and DenseNet, this model offers higher accuracy and a more compact size. As a result, it can automatically detect and classify maize leaf diseases under natural field conditions, boasting high-precision recognition capabilities. Its accurate identification results provide scientific guidance for preventing corn leaf disease and promote the development of precision agriculture.

Keywords: precision agriculture; deep learning; plant diseases; convolutional neural network

Citation: Zhou, H.; Su, Y.; Chen, J.; Li, J.; Ma, L.; Liu, X.; Lu, S.; Wu, Q. Maize Leaf Disease Recognition Based on Improved Convolutional Neural Network ShuffleNetV2. *Plants* **2024**, *13*, 1621. https://doi.org/10.3390/plants13121621

Academic Editor: Vittorio Rossi

Received: 13 May 2024
Revised: 7 June 2024
Accepted: 10 June 2024
Published: 12 June 2024

1. Introduction

Maize (*Zea mays* L.), commonly referred to as corn, stands as a fundamental crop with a rich history of sustaining human civilizations over millennia [1]. Originating from Mesoamerica, maize has evolved into a global staple, indispensable not only for human consumption but also for animal feed, biofuel production, and various industrial applications. According to the Food and Agriculture Organization Statistics (FAOSTAT), maize ranks among the most cultivated cereals globally, with a staggering production of approximately 1.14 billion tons in 2019 [2]. In the United States, maize holds the title of the most produced crop, occupying over 92 million acres of cultivated land [3]. This widespread cultivation underscores maize's pivotal role in global food security and agricultural economies. However, maize is frequently threatened by a range of pests and diseases, including maize leaf spot disease, pathogenic spot disease, maize aphids, and maize borers [4]. These challenges not only compromise yield quantity and quality but also jeopardize food security. To ensure successful maize cultivation, comprehensive measures must be taken to manage pests and diseases.

At present, traditional methods of pest and disease identification rely on manual observation and empirical judgment, processes that are time-consuming and prone to human error [5]. Professionals are tasked with identifying and categorizing maize leaves through the visual inspection or microscopic examination of features such as morphology, color, and texture [6]. Complex backgrounds pose a significant challenge in these traditional methods due to visual noise, similarity in color and texture between disease symptoms and background elements, and dynamic environmental changes that can obscure or alter the appearance of disease symptoms. Advancements in computer vision and image processing technologies have revolutionized crop disease identification, offering more efficient and objective avenues for disease detection through automated methods leveraging techniques like feature extraction, image segmentation, and machine learning [7,8]. The identification of pests and diseases using leaf images has emerged as a critical domain in plant health detection [9].

In the 21st century, convolutional neural networks (CNNs), a deep learning technique, have achieved remarkable success in image-based recognition tasks [10]. A CNN autonomously extracts image features and classifies images based on these features [11]. For maize leaf disease recognition, researchers commonly employ CNNs to train on maize leaf images, discerning maize leaf disease types by learning image features [12]. Specifically, CNN scrutinizes maize leaf image intricacies and characteristics, mapping them to maize leaf disease categories, thereby achieving rapid and precise disease recognition and classification. However, the presence of complex backgrounds in leaf images further complicates this task. Background elements such as soil, other plants, and varying lighting conditions can introduce noise that confounds the disease detection algorithms. Successful models must therefore effectively differentiate between disease symptoms and irrelevant background details. In recent years, an increasing number of researchers have embraced CNN techniques for crop leaf disease identification, yielding significant outcomes [13]. For instance, Hlaing et al. [14] characterized tomato images using the Johnson SB distribution model, achieving an average accuracy of 85.1%. Junde et al. [15] integrated channel attention into the lightweight neural network model MobileNetV2, enhancing pest recognition in complex backgrounds, with an average accuracy of 92.79%. Yun et al. [16] embedded an improved channel and spatial attention module into ResNet, achieving an average accuracy surpassing 95.37%. Hidayatuloh et al. [17] utilized the Keras deep learning framework to enhance the SqueezeNet model for automatic disease detection in tomato leaf images, boasting an average recognition accuracy of 86.92%. Agarwal et al. [18] proposed a model for tomato leaf disease recognition and detection, achieving an average accuracy of 91.2%. Bhujel et al. [19] devised a lightweight CNN integrating various attention modules to bolster overall accuracy, validated on a tomato leaf disease dataset. Bari et al. [20] employed Faster R-CNN for the real-time detection of rice leaf diseases, achieving accuracies of 98.09%, 98.85%, and 99.17%. Trivedi et al. [21] utilized Google Testbed on a dataset containing tomato leaf samples, attaining a prediction accuracy of 98.49%. Deepalakshmi et al. [22] developed a CNN model capable of recognizing various image types with an average accuracy of 94.5% and a recognition cost of 3.8 s. Sibiya et al. [23] devised a tomato leaf disease detection model based on ResNet50 using PyTorch, achieving 97% accuracy.

Leveraging CNN technology for maize leaf disease identification holds immense significance, substantially reducing diagnosis time and enhancing diagnosis accuracy and efficacy, thereby positively impacting maize yield and quality [24]. Therefore, this paper proposes a CNN model (SNMPF), which is based on the improved ShuffleNetV2 and aims to achieve the high-precision recognition of maize leaf disease. By optimizing the network structure, the SNMPF model can improve the recognition performance of corn leaf disease. At the same time, the lightweight design of the model enables it to be deployed on mobile devices, facilitating its practical application in the field. It provides farmers with a new technological means for the timely detection and prevention of diseases and pests, thereby reducing grain yield losses.

Therefore, the focus of this paper is to achieve high accuracy in maize leaf disease recognition, as well as a lightweight model. To accomplish this, the lightweight convolutional neural network ShuffleNetV2 was utilized, and the following studies and improvements were conducted:

- Augmentation of data using the Augmentor tool, employing random combinations of operations such as random rotation, vertical flip, morphological magnification, region erasure, and brightness transformation to bolster model training generalization ability;
- Division of data into training and validation sets in a 4:1 ratio, with 3200 images allocated for training and 800 sheets for validation;
- Comparative study of four models, ShuffleNetV2, ShuffleNetV2 + Max pooling (MP), ShuffleNetV2 + SimAM (SAM), and ShuffleNetV2 + Max pooling (MP) + SimAM (SAM), to assess the contributions of different improvement measures;
- Comparison between SimAM and the attention mechanisms, including SE, ECA, EMA, and CSAM. The suitability of each attention mechanism for the task of maize leaf disease recognition is being determined through comparison;
- Comparative evaluation with existing network models such as EfficientNet, Reg-Net, MobileViT, EfficientNetV2, and DenseNet showcases the advantages of the SNMPF lightweight model in maize leaf disease identification amidst complex backgrounds.

2. Materials and Methods

2.1. Data Selection and Preprocessing

2.1.1. Image Data

The target dataset of this paper's experiment is maize leaf diseases. To address the issue of poor robustness in the trained model caused by image data with simple backgrounds, this paper opted to utilize maize leaf images captured under field conditions with complex backgrounds. Data sourced from (https://osf.io/s6ru5/ (accessed on 31 July 2023)). These images were captured through the 12-megapixel camera sensor on the iPhone 11 Pro smartphone. These images were originally sized at 3000 × 3000 pixels and maintained a 1:1 aspect ratio [25]. During the model training process, to ensure that the input image meets the model requirements, the resolution of the image was unified to 224 × 224. The dataset comprised healthy maize leaves along with images depicting three common maize leaf diseases in complex backgrounds, namely Northern Leaf Blight (NLB), Gray Leaf Spot (GLS), and Northern Leaf Spot (NLS) of maize (Figure 1), totaling 1902 maize leaf images. From Table 1, the original data are both limited in quantity and unevenly distributed. This may lead to a decrease in the model's generalization ability, subsequently affecting the accuracy of evaluation results and hindering the model from being adequately trained. To address these issues, this study plans to employ data augmentation techniques to balance the quantity of samples across different categories, thereby enhancing the robustness of the model.

Table 1. Data distribution.

Image Type	Number of Original Image	Number of Enhanced Image	Label
Northern Leaf Blight	497	1000	NLB
Healthy Leaves	331	1000	Health
Gray Leaf Spot	523	1000	GLS
Northern Leaf Spot	551	1000	NLS

2.1.2. Data Augmentation

To ensure a relatively balanced number of samples across various categories in the dataset and enhance the performance and generalization ability of the model, a total of 4000 images were obtained through the random expansion and enhancement of the original 1902 images for training the research model. In this paper, the Augmentor tool in Python was employed for data augmentation, encompassing random rotation, vertical flipping, shape amplification, region erasure, and brightness transformation (Figure 2).

Figure 1. Maize leaves in a complex field background.

Figure 2. Data augmentation effect image. (**a**) Original image; (**b**) vertical flip: vertical flip probability set to 0.5; (**c**) random rotation: rotation probability set to 0.7, with a maximum angle of 10 degrees to the left and to the right; (**d**) region erasure: erasure probability set to 0.3, with the size of the erasure area being 0.3 of the image size; (**e**) brightness transformation: brightness adjustment probability set to 0.5, with the minimum value being 0.7 times and the maximum value being 1.3 times; (**f**) morphological amplification: magnification probability set to 0.3, with the minimum amplification factor being 1.1 times and the maximum being 1.6 times.

2.2. Construction of Identification Model of Maize Leaf Disease

2.2.1. ShufflenetV2

ShuffleNetV2, a lightweight convolutional neural network model proposed by MegVII [26], presents several advantages over traditional neural network models, including its lightweight, flexible, and highly efficient nature. The ShuffleNetV2 block introduces a Channel Split operation based on the ShuffleNetV1 block, dividing the input feature map into two blocks. In each block, half of the feature channels pass directly through the block and join the next block. This strategy reduces computations and parameters while increasing the number of feature channels, thereby enhancing network accuracy. Within the network module, some feature channels are passed directly to the next module without undergoing convolution calculations. In the subsampling layer of the network, instead of employing channel separation, each branch is created by copying the input, and subsampling operations with a step size of 2 are performed on each branch. Finally, the feature maps from all branches are concatenated, halving the spatial size of the feature maps while doubling the number of channels (Figure 3). This design effectively reduces computation while maintaining information richness and enhancing network expression capability [27]. Compared to the ShuffleNetV1 version, where ShuffleNetV2 adds a conv5 convolution before global pooling, ShuffleNetV2 enhances network performance and generalization. This model demonstrates strong feasibility, enabling rapid image processing and recognition on mobile devices while remaining lightweight. It holds promising prospects for applications in embedded computing, mobile terminals, and large-scale image recognition.

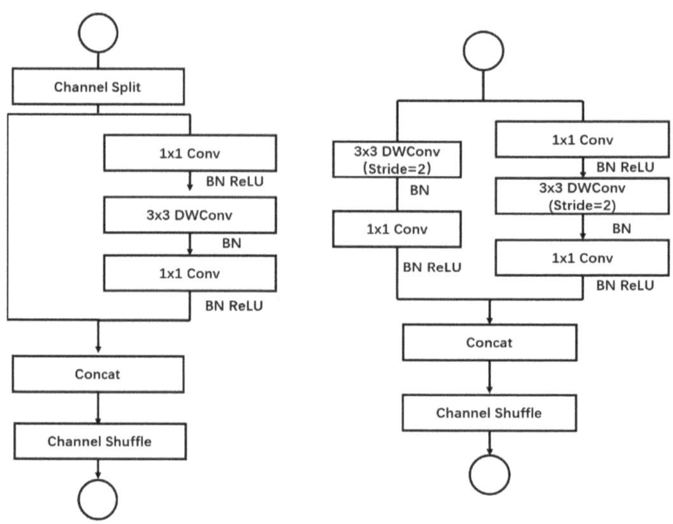

Figure 3. The basic unit of ShufflenetV2.

2.2.2. Max Pooling

Max Pooling is a down-sampling operation utilized to decrease computing and storage demands. This layer is commonly employed to diminish the size of a feature map while preserving crucial feature information. By partitioning the input feature map into fixed-size rectangular regions and selecting the maximum value within each region as the output, maximum pooling effectively diminishes the spatial dimension of the feature map while retaining features with the strongest response [28]. In the formula, the maximum pool selects the maximum value within the locally acceptable domain F, as demonstrated in Equation (1).

$$y = \mathrm{Max}(x_1, x_2, \cdots, x_i), (x_i \in F) \quad (1)$$

The maximum value serves as a significant feature extracted by the convolutional layer. By retaining these crucial features and discarding unimportant ones, the interference of irrelevant information is minimized (Figure 4). The incorporation of a maximum pooling layer can augment the characterization capability of the model. Furthermore, the maximum pooling layer also aids in reducing computation by diminishing the size of the feature map, thereby rendering the model more lightweight.

Figure 4. Illustration of Max Pooling.

In this paper, a Max Pooling layer is integrated into the ShuffleNetV2 network. This layer reduces the width and height of the feature map by half, gradually diminishing its size. It facilitates the extraction of key information from feature maps, the reduction in model overfitting, and the enhancement of the model's generalization capability.

2.2.3. SimAM Attention Module

Existing attention modules in computer vision typically focus on either the channel domain or spatial domain, corresponding to feature-based attention and spatial-based attention in the human brain, respectively. Traditional channel attention is one-dimensional, concentrating on the characteristics of different channels while treating all positions equally. Spatial attention, on the other hand, is two-dimensional, focusing on features at different locations while treating all channels equally. However, Yang et al. argue that computing three-dimensional values should be straightforward and allow modules to maintain lightweight properties [29]. Consequently, they proposed a simple yet effective attention mechanism called SimAM. SimAM can directly derive three values for the feature map without adding additional parameters, enabling the model to learn more discriminative neurons and improving the network's feature extraction ability (Figure 5). Additionally, based on neuroscience theory, SimAM optimizes the energy function to mine the importance of neurons, enhancing the ability to extract important features while suppressing the interference of non-important features. Furthermore, SimAM's operation is primarily based on the selection of an optimized energy function, avoiding excessive structural adjustments, and accelerating the calculation of attention weights. This allows the network to remain lightweight while better leveraging the effectiveness and flexibility of SimAM when integrated into our ShuffleNetV2 model.

In this paper, the SimAM attention mechanism is inserted into the basic network unit of ShuffleNet V2. This enhances the interaction between channels, enabling the model to better adapt to different input features, improve representation ability, and enhance classification performance. Additionally, the SimAM module assists the model in capturing important image features more efficiently and weighing these features to enhance recognition and classification accuracy. By introducing the SimAM module, ShuffleNetV2 can enhance its image processing and recognition capabilities while maintaining lightweight and efficient performance, thereby making the model more applicable to mobile devices and embedded systems.

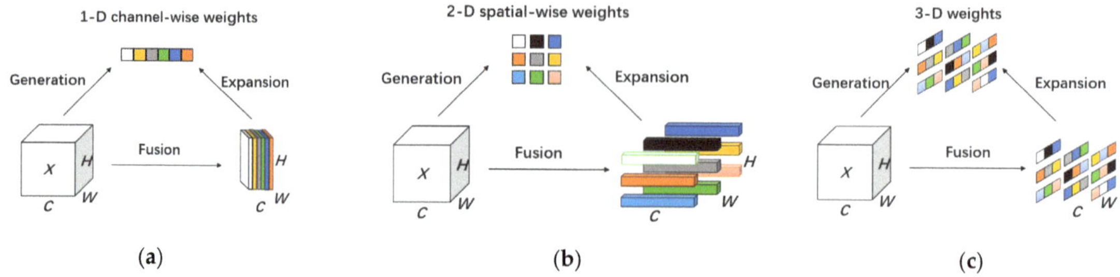

Figure 5. Comparisons of different attention steps. Most of the existing attention modules generate 1D or 2D weights from feature X and then expand the generated weights for channel (**a**) and spatial (**b**) attention. SimAM attention instead directly estimates 3D weights (**c**). In each subfigure, the same color denotes that a single scalar is employed for each channel, for the spatial location, or for each point on that feature.

2.2.4. Improved ShufflenetV2

Based on the characteristics of corn leaf disease images, ShuffleNetV2-0.5 was chosen as the baseline network and improved to create the SNMPF model. In the basic unit of the ShuffleNetV2 network, the stride in depth-wise (DW) convolution, which was originally set to 2 for down-sampling, is reduced to 1. Additionally, in the down-sampling module of the ShuffleNet model, the maximum pooling layer is employed instead of deep convolution for down-sampling. This enhancement aids in extracting crucial features from images, mitigating model overfitting, and enhancing the model's generalization ability (Figure 6).

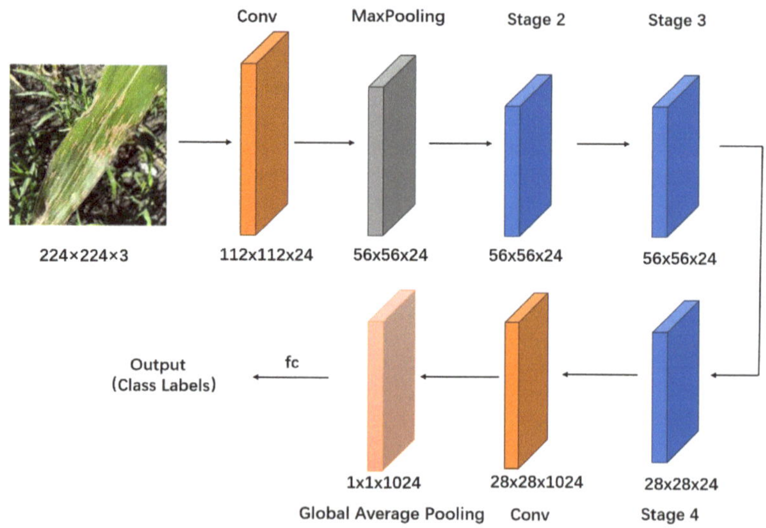

Figure 6. Structure of the SNMPF model. Input a 224 × 224 × 3 size image, first through a 3 × 3 convolution and then through the Max Pooling, followed by several Stage modules, and then through a 1 × 1 convolution. Finally, input through the full pooling layer (Global Pooling) and then connect to a fully connected layer to obtain the output.

Given that the dataset used in this paper consists of corn leaf images captured under complex backgrounds, non-important features such as environmental backgrounds may interfere with model recognition. Therefore, SimAM attention was integrated into the ShuffleNetV2 network (Figure 7) to alleviate the interference of non-important features, such as complex environmental backgrounds, on model recognition.

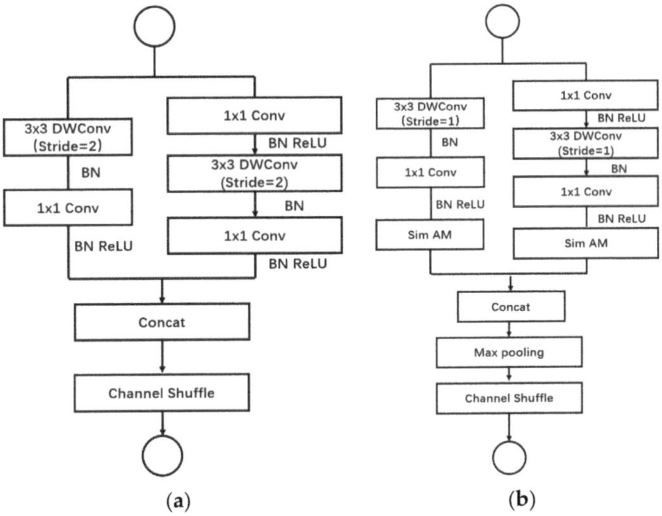

Figure 7. Comparison of the basic units of the initial network and the improved network. (**a**) Initial network; (**b**) improved network.

2.3. Experimental Setup

The experiments were carried out on a 64-bit Windows 10 operating system using the Python programming language. The PyTorch framework was employed for network construction, training, and testing purposes. The computer utilized for these experiments is equipped with an Intel(R) Core (TM) i7-10510U CPU and has 8 GB of RAM. Additionally, it is complemented by an NVIDIA GeForce MX250 GPU for accelerated processing.

2.4. Training Hyperparameter Settings

In our experiments, we use the SGD optimizer to iteratively update model parameters to minimize the loss function. This choice was made due to the advantages of the SGD optimizer in terms of computational efficiency and memory requirements. Additionally, we set the following hyperparameters: epoch = 30, learning rate = 0.01, and batch size = 4. These hyperparameters were carefully selected to maximize the effectiveness and accuracy of our model. By choosing these hyperparameters, we aimed to optimize our deep learning model and enhance its performance in recognition tasks.

2.5. Model Evaluation

In this paper, our main evaluation metrics include accuracy, model loss, and the size of the model. Accuracy (A) is calculated using Equation (2), while model loss is computed according to Equation (3). Accuracy is defined as the ratio of correctly recognized samples to the total number of samples.

$$A = \frac{TP + TN}{TP + TN + FP + FN} \times 100\% \tag{2}$$

TP (True Positive): The number of positive samples correctly predicted as positive by the model. In other words, the model successfully detected a positive sample.

FP (False Positive): The number of negative samples incorrectly predicted as positive by the model. In other words, the model incorrectly predicted negative samples as positive samples.

TN (True Negative): The number of negative samples correctly predicted as negative by the model. In other words, the model correctly determined negative samples.

FN (False Negative): The number of positive samples incorrectly predicted as negative by the model. In other words, the model incorrectly predicted positive samples as negative samples.

The loss function provided in Equation (3) is a formulation commonly used in margin-based methods, such as contrastive loss or triplet loss, for tasks like metric learning or Siamese network training.

$$\text{Loss} = \text{yd}^2 + (1 - \text{y})\cdot\max(0,\ \text{margin} - \text{d})^2 \quad (3)$$

In this equation, the following variables are defined:

d: Represents the Euclidean distance between two samples.

y: Represents the label indicating whether the two samples match. y = 1 indicates that the two samples belong to the same category, while y = 0 indicates that the two samples belong to different categories.

margin: Denotes the threshold.

The loss function encourages the model to learn embeddings such that the distance between samples from the same category is minimized while ensuring that the distance between samples from different categories is larger than the margin. This facilitates a better discrimination between classes in the learned feature space.

3. Results

3.1. Results on Data Enhanced Analytics

ShuffleNetV2 was utilized to train both the original maize leaf disease dataset and its augmented counterpart to assess the viability of data augmentation. It was observed that the performance of the ShuffleNetV2 model on the two datasets exhibited the following trend: the augmented data outperformed the raw data. When using an enhanced dataset, the accuracy of the model increased from 93.40% to 94.30%, an increase of 1.1 percentage points, while the loss slightly decreased from 0.416 to 0.414. This means that data augmentation effectively improves the performance of the model (Table 2). As the number of epochs increased, the recognition accuracy of the model trained on augmented data demonstrated improvement (Figure 8).

Table 2. Comparison before and after data augmentation.

Data Set	Accuracy	Loss	Model Size/MB
Original date	93.40%	0.416	1.50
Enhanced date	94.30%	0.414	1.50

It is worth noting that after data augmentation, the overall loss decreased compared to the original data, and the convergence speed accelerated (Figure 9). Furthermore, as epochs progressed, the enhanced data notably enhanced model accuracy, affirming the viability of data augmentation. Remarkable outcomes were achieved in enhancing model performance through data augmentation, as evidenced by the gradual reduction in the loss function and the progressive increase in accuracy. These findings underscore the positive impact of data augmentation on model training and performance.

3.2. Results of the Ablation Test

To validate the impact of various enhancements proposed in this paper on model performance, an ablation test was conducted for analysis. Throughout the experiments, consistency in test conditions was maintained, with only one enhancement altered in each experiment to assess its influence on model performance. Four models were evaluated: ShuffleNetV2, ShuffleNetV2 + Max Pooling (MP), ShuffleNetV2 + SimAM (SAM), and ShuffleNetV2 + Max Pooling (MP) + SimAM (SAM). Results revealed that incorporating Max Pooling into the ShuffleNetV2 network without increasing model complexity led to a 2.1 percentage point improvement in recognition accuracy for corn leaf diseases in complex environments, accompanied by a reduction in model loss by 0.135. The introduction of the SimAM module increased model size by only 0.06 MB, enhancing accuracy by 1.3 percentage points (Table 3). These findings indicate that the addition of SimAM resulted in improved accuracy and reduced loss, underscoring the effectiveness of model enhancements (Figure 10). The model incorporating both Max Pooling and SimAM achieved the highest performance (Figure 11). The synergistic effect of these enhancements did not adversely affect the model but rather boosted its accuracy. The enhanced model achieved an accuracy of 98.40%, surpassing the original model by 4.1 percentage points, with a loss of 0.228, 0.186 lower than that of the original model. Although the size of the improved model slightly increased, it remained lightweight.

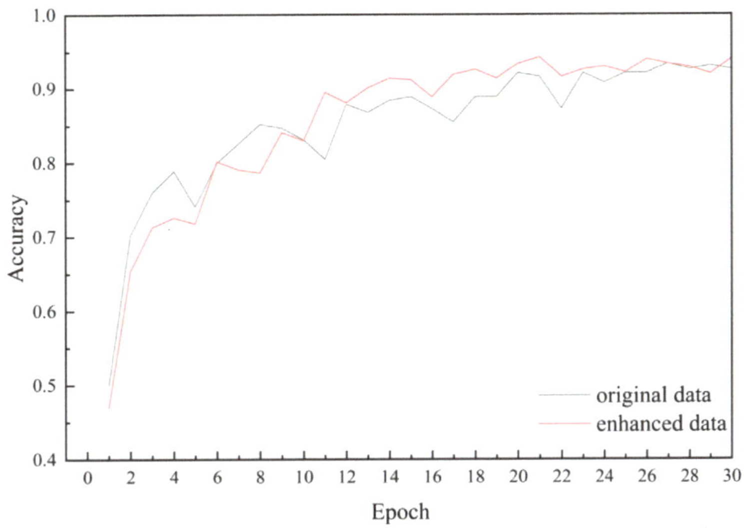

Figure 8. Accuracy curves before and after data augmentation.

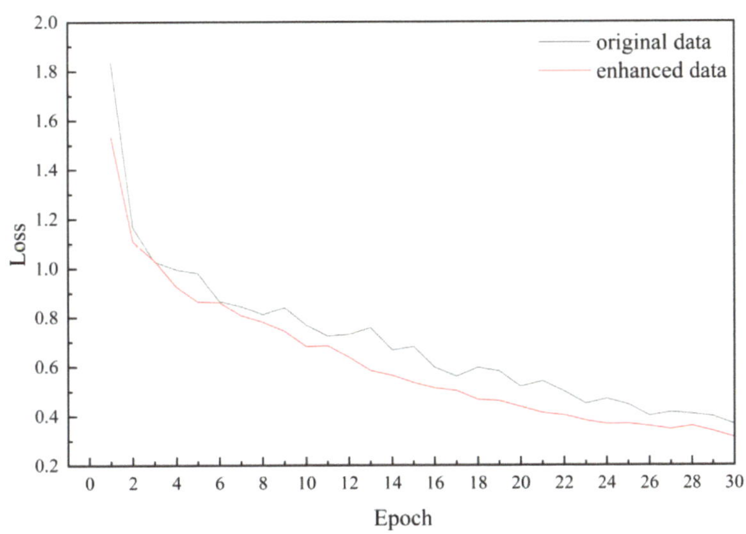

Figure 9. Loss curves before and after data augmentation.

Table 3. Comparison of ablation tests.

Model	Accuracy	Loss	Model Size/MB
ShuffleNetV2	94.30%	0.414	1.50
ShuffleNetV2 + MP	96.40%	0.279	1.50
ShuffleNetV2 + SAM	95.60%	0.316	1.56
ShuffleNetV2 + MP + SAM	98.40%	0.228	1.56

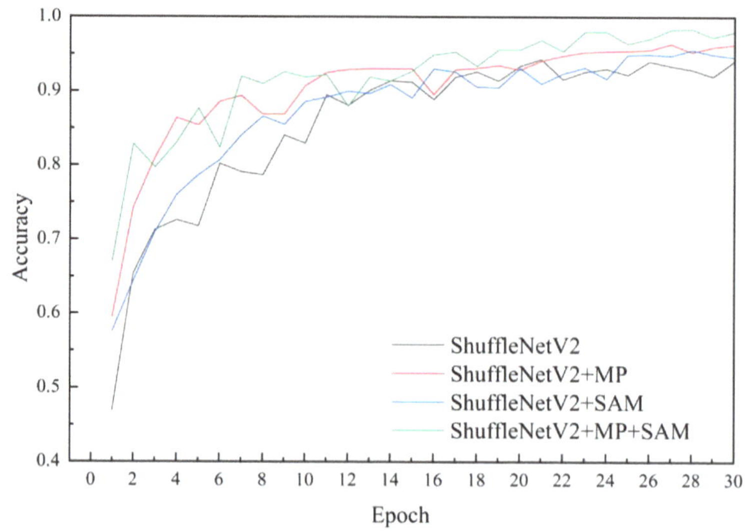

Figure 10. Accuracy curve of ablation test. This figure illustrates how adding different modules (MP, SAM, MP + SAM) affects the recognition accuracy of the model. The x-axis represents epoch, and the y-axis represents accuracy. The curve indicates that the model with the addition of MP + SAM has the best recognition performance.

3.3. Results of the Attention Test

To further validate the feasibility of incorporating SimAM attention in this paper, comparative experiments were conducted by adding SimAM, SE, ECA, EMA, and CSAM attention mechanisms to the ShuffleNetV2 model. These experiments aimed to analyze the effects of different attention mechanisms on the recognition and detection of maize leaf diseases amidst complex backgrounds. The results revealed that following the integration of SEM, ECA, EMA, CSAM, and SimAM attention mechanisms, the accuracy of maize leaf disease recognition by the ShuffleNetV2 model improved by 0.3%, 2.1%, 2.6%, 3.5%, and 4.1% (Table 4), respectively. Notably, the introduction of an attention mechanism effectively enhanced the model's capability to focus on crucial lesion features within complex backgrounds. Particularly, the introduction of the SimAM model exhibited the most significant enhancement in recognition performance.

Conversely, the introduction of the SE attention mechanism resulted in a decrease in model loss by 1.9 percentage points. Moreover, the integration of attention mechanisms led to an increase in the size of the model compared to the original ShuffleNetV2 model. Specifically, the ECA attention mechanism exhibited the smallest model size increase of 1.509 MB, followed by the SimAM attention mechanism with 1.556 MB, while the CSAM attention mechanism demonstrated the largest model size increase of 2.017 MB (Table 4). Consequently, SimAM attention emerged as a pivotal factor in enhancing the model's lightweight performance while achieving the highest accuracy gain, highlighting the significance of integrating attention mechanisms to enhance the performance of lightweight models in complex recognition tasks such as maize leaf disease identification.

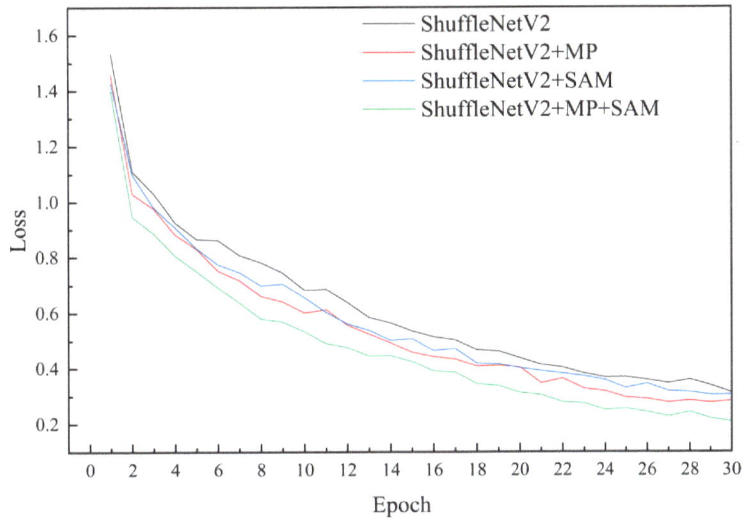

Figure 11. Loss curve of ablation test. This figure illustrates how adding different modules (MP, SAM, MP + SAM) affects the generalization ability of the model. The *x*-axis represents epoch, and the *y*-axis represents loss. The curve indicates that the model with the addition of MP + SAM fits the training data best.

Table 4. Comparison of different attentions.

Attention	Accuracy	Loss	Model Size/MB
No attention	94.30%	0.414	1.499
SE	94.60%	0.433	1.644
ECA	96.40%	0.298	1.509
EMA	96.90%	0.285	1.587
CSAM	97.80%	0.271	2.017
SimAM	98.40%	0.244	1.556

Subsequently, we draw the loss curve of the model to analyze the impact of different attention mechanisms on the robustness of the model. It was observed that the loss values of the models gradually decreased and stabilized with an increase in the number of model training iterations. By introducing ECA, EMA, CSAM, and SimAM attention mechanisms, the loss of the model was reduced. Especially after introducing the SimAM attention mechanism, the model exhibited the lowest loss value and significantly improved robustness. However, SE attention led to an increase in model loss (Figure 12).

3.4. Results of Model Comparison Test

To explore the advantages of the lightweight model developed in this paper for recognizing maize leaf diseases in complex scenarios, comparisons were drawn with contemporary network modeling algorithms. In the task of maize leaf disease recognition and detection, SNMPF achieved the highest recognition accuracy of 98.4%. DenseNet and EfficientNet performed similarly, while the MobileViT model exhibited the lowest recognition accuracy at only 64.7%. Furthermore, the model loss of MobileViT was notably high at 0.839 compared to several other models (Table 5). Although EfficientNetV2 is a widely optimized and efficient model, its performance in the recognition task of this paper was not as expected and even worse than EfficientNet. The main reason may be that EfficientNetV2 was originally designed to perform well on large and diverse datasets. However, our dataset size is relatively small, which may result in EfficientNetV2 not being able to fully leverage its advantages. In contrast, EfficientNet had better adaptability on the dataset in this paper.

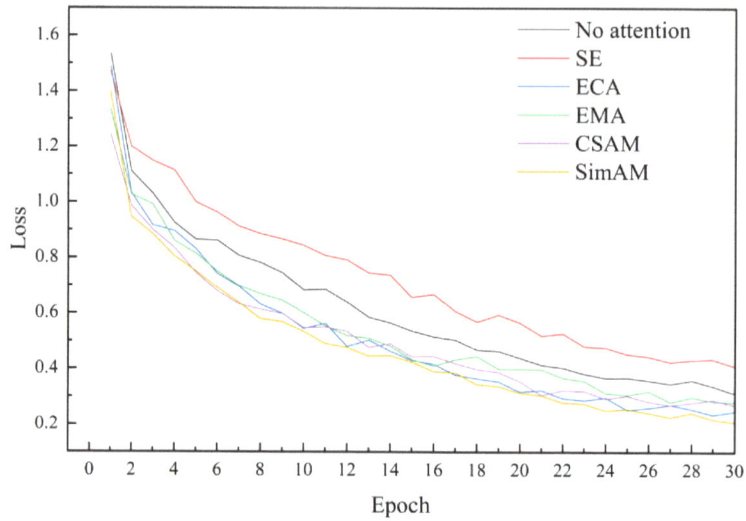

Figure 12. Attention test model loss curve. This figure depicts the loss of the model when adding different attention mechanisms (SE, ECA, EMA, CSAM, SimAM). The x-axis represents epoch, and the y-axis represents loss. The curve indicates that the model with added SimAM attention mechanism has the smallest loss.

Table 5. Comparison of different models.

Model	Accuracy	Loss	Model Size/MB
SNMPF	98.4%	0.228	1.56
EfficientNet	95.3%	0.369	15.98
RegNet	90.7%	0.504	15.49
MobileViT	64.7%	0.839	3.86
EfficientNetV2	83.9%	0.431	79.73
DenseNet	95.9%	0.290	27.78

The SNMPF model, built upon the enhanced ShuffleNetV2 architecture proposed in this paper, demonstrated superior performance. It achieved the highest recognition accuracy for maize leaf disease under complex backgrounds with minimal model loss (Figure 13). Moreover, the model size of SNMPF was a mere 1.56 MB, which was less than half of MobileViT's model size and even less than one-tenth of other models.

3.5. Visual Presentation of Model Prediction Results

The SNMPF model was used to predict the test image and then the activation heat map (Grad-CAM) technique was used to visualize the prediction of the model. Grad-CAM can display the recognition focus area of the model by evaluating the contribution of each region in the image to the prediction results. In this way, we can more accurately understand the features that the model focuses on, which can improve the understanding of the model's prediction results and the interpretability of the model.

The results revealed that for the four types of maize leaf images—GLS, NLB, NLF, and Health—the original ShuffleNetV2 model correctly predicted the disease with probabilities of 0.933, 0.922, 0.968, and 0.999, respectively. The SNMPF model developed in this paper achieved correct predictions with probabilities of 0.985, 0.995, 0.995, and 1.0, respectively (Figure 14). Even in the presence of complex background interference, SNMPF could still effectively capture the positions of maize leaf lesions and extract their features. It demonstrated relatively high recognition accuracy for the same type of leaf disease, enabling better focus on the identified lesion areas and thus reducing the impact of complex backgrounds.

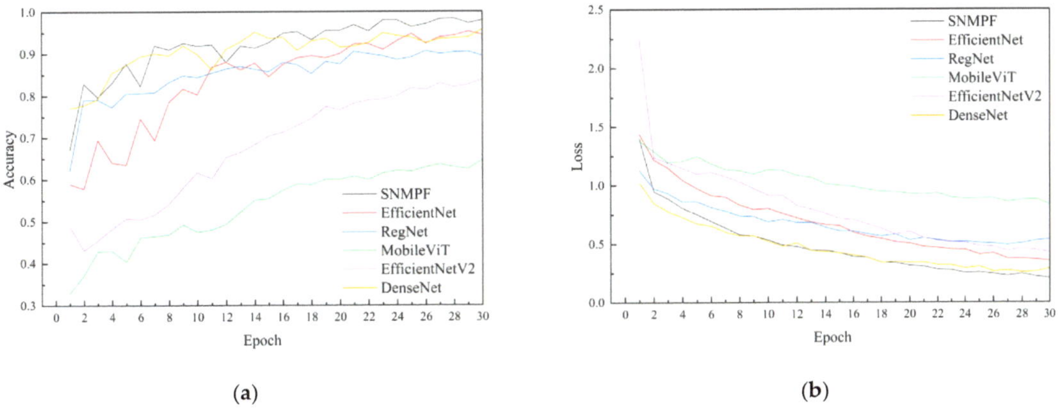

Figure 13. (**a**) Accuracy curves of different models; (**b**) loss curves of different models.

Figure 14. Visualization results of SNMPF in identifying maize leaf diseases. Different species of diseases have different thermogram characteristics, and the closer the color is to red, the more it plays a role in decision-making for the identification of that species.

This result further proves the effectiveness of the model improvement measures proposed in this paper, as well as the excellent performance of the SNMPF model in corn leaf disease recognition tasks in complex backgrounds, providing useful references for corn leaf disease recognition on mobile devices.

4. Discussion

After introducing the Maximum Pooling layer into ShuffleNetV2, the accuracy of the model in identifying corn leaf disease improved by 2.1 percentage points, and the model's loss was also reduced. This is because the max pooling layer can select the maximum value in each region, thereby reducing the size of the feature map while retaining the most significant feature information, thereby improving the model's generalization ability [30]. Introducing attention mechanisms enables the model to focus more sensitively on disease areas, thereby aiding in distinguishing various leaf disease images [31]. After introducing the SimAM attention mechanism in ShuffleNetV2, the model performance has been improved. SimAM can help models focus more on important features [32], thereby improving their ability to extract key information from images. Meanwhile, when both the max pooling layer and SimAM attention mechanism are added to the model, the performance of the model is further improved. This is because the max pooling layer effectively reduces the dimensionality of the feature map and preserves the most important information, while the SimAM attention mechanism further enhances the model's perception of key features, making it more accurate in identifying and classifying objects in the image. Overall, by combining these two mechanisms, the model's ability to extract and utilize features has been further enhanced, thereby improving overall performance.

To assess the feasibility of SimAM attention, a comparison was conducted with SE, ECA, EMA, and CSAM attention mechanisms within the ShuffleNetV2 model [33–36]. The results showed that after introducing attention mechanisms such as SE, ECA, EMA, CSAM, and SimAM, the recognition accuracy of the model was improved to varying degrees. These attention mechanisms help the model focus on the lesion area, thereby significantly improving the performance of the model [37]. However, after introducing the SE attention mechanism, the model's loss increases. This may be because SE introduces additional parameters, making the model more complex and leading to overfitting [38].

Furthermore, the recognition effects of SNMPF, EfficientNet, MobileViT, EfficientNetV2, RegNet, and DenseNet models on maize leaf diseases were compared [39–43]. Experimental results showed that EfficientNetV2 had the largest model size. However, large networks have high hardware requirements on computers and mobile devices, which is not conducive to a wide range of applications. Moreover, EfficientNetV2 had the worst recognition effect on maize leaf diseases, and the model loss value was also relatively large. This indicates that there is an overfitting problem [44], and EfficientNetV2 needs to be further optimized and adjusted for the identification and application of corn leaf diseases. The SNMPF model achieved the best recognition effect, and the highest recognition accuracy reaches 98.4%. Compared with the other models, the SNMPF model had the highest recognition accuracy and the smallest loss. The recognition accuracy is improved by 4.1%. This improvement was attributed to the addition of max pooling, which reduced model oversensitivity to feature locations, enhanced robustness [38], and minimized interference from complex backgrounds. Coupled with the SimAM attention mechanism, recognition accuracy was further improved.

Finally, comparing the prediction results of SNMPF and ShuffleNetV2 models for different types of maize leaf diseases, it can be observed that ShuffleNetV2 identifies some non-diseased areas as diseased areas, while the SNMPF model focuses more on diseased areas, reduces environmental impact, and improves the probability of accurate prediction [45]. These results further validate the effectiveness of the improvement measures in this article.

This paper verified the feasibility of ShuffleNetV2 in maize leaf disease recognition, providing technical support for deploying maize leaf disease recognition models on mobile devices. Due to limited public data on corn leaf disease, there are still limitations in this study, namely, the dataset used in this study is relatively limited. By comparing the model training results of the pre- and post-enhanced datasets, a large amount of data is beneficial for optimizing model performance. In the future, with the application of various image recognition technologies in the field, more types and quantities of corn leaf disease data images will be collected [46], which can increase the types of corn leaf disease recognition, further improve the recognition accuracy and performance of the model, and enhance the robustness and generalization ability of the model [47].

5. Conclusions

In this paper, we addressed the issues of low efficiency in traditional manual identification methods and the challenge of deploying existing recognition models on mobile devices due to their large size. We proposed a maize leaf disease recognition solution based on the improved lightweight convolutional neural network ShuffleNetV2. Focusing on maize leaf images in complex backgrounds, we improved the ShuffleNetV2 model by introducing max pooling layers to replace deep convolutional layers for down sampling, adding attention mechanisms, optimizing network structures, and we presented the SNMPF model.

The results indicate that the addition of max pooling layers is effective in improving the model's ability to recognize maize leaf diseases. Introducing attention mechanisms further enhances ShuffleNetV2's discriminative power in feature extraction and the classification of maize leaf diseases. After incorporating attention modules such as SE, ECA, EMA, CSAM, and SimAM, the accuracy of the model's recognition is enhanced, with the model achieving optimal recognition performance after introducing SimAM. In the task of maize leaf disease recognition under complex backgrounds, the SNMPF model proposed in this paper outperforms other traditional neural network models, achieving a recognition accuracy of 98.4%. Additionally, its model size is only 1.56 MB, making it suitable for deployment on small mobile devices.

In summary, this paper provides new technical means for the recognition and detection of maize leaf diseases, with potential applications in field environments in the future. Since this paper only focuses on recognizing three common types of maize leaf diseases, further improvements are needed in future research. Future studies will enrich image data to recognize more types of leaf diseases and deploy the SNMPF model on mobile devices to achieve the real-time recognition and detection of maize leaf diseases. Therefore, enabling the timely detection of maize leaf diseases can foster the development of precision agriculture.

Author Contributions: Conceptualization, H.Z. and Q.W.; methodology, H.Z., L.M. and S.L.; software, H.Z. and Y.S.; validation, J.C. and J.L.; formal analysis, J.C., J.L. and Q.W. resources, Y.S.; data curation, Y.S., J.C. and X.L.; writing—original draft preparation, Y.S.; writing—review and editing, L.M., H.Z. and Q.W.; visualization, X.L. and J.C.; supervision, Q.W.; funding acquisition, H.Z. All authors have read and agreed to the published version of the manuscript.

Funding: This research was funded by the National Natural Science Foundation of China (52379039); Science and Technology Specialist Program of Henan Province (FY2023); Young Backbone Teachers Program of Henan University of Science and Technology (13450001); and Experimental Technology Development Fund Program of Henan University of Science and Technology (SY2223003).

Data Availability Statement: Data supporting the findings of this paper are available from the corresponding authors.

Conflicts of Interest: The authors declare no conflicts of interest.

References

1. Solaimalai, A.; Anantharaju, P.; Irulandi, S.; Theradimani, M. *Maize Crop: Improvement, Production, Protection and Post Harvest Technology*; CRC Press: Boca Raton, FL, USA, 2020; ISBN 1000176959.
2. FAOSTAT Crops Processed. Available online: https://www.fao.org/faostat/zh/#data/QC/visualize (accessed on 17 January 2024).
3. USDA Corn and Other Feed Grains. Available online: https://www.ers.usda.gov/topics/crops/corn-and-other-feed-grains/ (accessed on 17 January 2024).
4. Langemeier, C.B.; Robertson, A.E.; Wang, D.; Jackson-Ziems, T.A.; Kruger, G.R. Factors Affecting the Development and Severity of Goss's Bacterial Wilt and Leaf Blight of Corn, Caused by *Clavibacter michiganensis* subsp. *nebraskensis*. *Plant Dis.* **2017**, *101*, 54–61. [CrossRef] [PubMed]
5. Sharma, S.; Kumar, V.; Sood, S. Pest Detection Using Random Forest. In Proceedings of the 2023 International Conference on IoT, Communication and Automation Technology (ICICAT), Gorakhpur, India, 23–24 June 2023; IEEE: New York, NY, USA, 2023; pp. 1–8.
6. Li, Z.; Zhou, G.; Hu, Y.; Chen, A.; Lu, C.; He, M.; Hu, Y.; Wang, Y. Maize Leaf Disease Identification Based on WG-MARNet. *PLoS ONE* **2022**, *17*, e0267650.

7. Li, L.; Zhang, S.; Wang, B. Plant Disease Detection and Classification by Deep Learning—A Review. *IEEE Access* **2021**, *9*, 56683–56698. [CrossRef]
8. Ouhami, M.; Hafiane, A.; Es-Saady, Y.; El Hajji, M.; Canals, R. Computer Vision, IoT and Data Fusion for Crop Disease Detection Using Machine Learning: A Survey and Ongoing Research. *Remote Sens.* **2021**, *13*, 2486. [CrossRef]
9. Khirade, S.D.; Patil, A.B. Plant Disease Detection Using Image Processing. In Proceedings of the 2015 International Conference on Computing Communication Control and Automation, Pune, India, 26–27 February 2015; IEEE: New York, NY, USA, 2015; pp. 768–771.
10. Pardede, H.F.; Suryawati, E.; Krisnandi, D.; Yuwana, R.S.; Zilvan, V. Machine Learning Based Plant Diseases Detection: A Review. In Proceedings of the 2020 International Conference on Radar, Antenna, Microwave, Electronics, and Telecommunications (ICRAMET), Tangerang, Indonesia, 18–20 November 2020; IEEE: New York, NY, USA, 2020; pp. 212–217.
11. Kattenborn, T.; Leitloff, J.; Schiefer, F.; Hinz, S. Review on Convolutional Neural Networks (CNN) in Vegetation Remote Sensing. *ISPRS J. Photogramm. Remote Sens.* **2021**, *173*, 24–49. [CrossRef]
12. Paymode, A.S.; Malode, V.B. Transfer Learning for Multi-Crop Leaf Disease Image Classification Using Convolutional Neural Network VGG. *Artif. Intell. Agric.* **2022**, *6*, 23–33. [CrossRef]
13. Saleem, M.H.; Potgieter, J.; Arif, K.M. Plant Disease Detection and Classification by Deep Learning. *Plants* **2019**, *8*, 468. [CrossRef] [PubMed]
14. Hlaing, C.S.; Zaw, S.M.M. Tomato Plant Diseases Classification Using Statistical Texture Feature and Color Feature. In Proceedings of the 2018 IEEE/ACIS 17th International Conference on Computer and Information Science (ICIS), Singapore, 6–8 June 2018; IEEE: New York, NY, USA, 2018; pp. 439–444.
15. Chen, J.; Chen, W.; Zeb, A.; Zhang, D.; Nanehkaran, Y.A. Crop Pest Recognition Using Attention-Embedded Lightweight Network under Field Conditions. *Appl. Entomol. Zool.* **2021**, *56*, 427–442. [CrossRef]
16. Zhao, Y.; Chen, J.; Xu, X.; Lei, J.; Zhou, W. SEV-Net: Residual Network Embedded with Attention Mechanism for Plant Disease Severity Detection. *Concurr. Comput.* **2021**, *33*, e6161. [CrossRef]
17. Hidayatuloh, A.; Nursalman, M.; Nugraha, E. Identification of Tomato Plant Diseases by Leaf Image Using Squeezenet Model. In Proceedings of the 2018 International Conference on Information Technology Systems and Innovation (ICITSI), Bandung, Indonesia, 22–26 October 2018; IEEE: New York, NY, USA, 2018; pp. 199–204.
18. Agarwal, M.; Singh, A.; Arjaria, S.; Sinha, A.; Gupta, S. ToLeD: Tomato Leaf Disease Detection Using Convolution Neural Network. *Procedia Comput. Sci.* **2020**, *167*, 293–301. [CrossRef]
19. Bhujel, A.; Kim, N.-E.; Arulmozhi, E.; Basak, J.K.; Kim, H.-T. A Lightweight Attention-Based Convolutional Neural Networks for Tomato Leaf Disease Classification. *Agriculture* **2022**, *12*, 228. [CrossRef]
20. Bari, B.S.; Islam, M.N.; Rashid, M.; Hasan, M.J.; Razman, M.A.M.; Musa, R.M.; Ab Nasir, A.F.; Majeed, A.P.P.A. A Real-Time Approach of Diagnosing Rice Leaf Disease Using Deep Learning-Based Faster R-CNN Framework. *PeerJ Comput. Sci.* **2021**, *7*, e432. [CrossRef] [PubMed]
21. Trivedi, N.K.; Gautam, V.; Anand, A.; Aljahdali, H.M.; Villar, S.G.; Anand, D.; Goyal, N.; Kadry, S. Early Detection and Classification of Tomato Leaf Disease Using High-Performance Deep Neural Network. *Sensors* **2021**, *21*, 7987. [CrossRef]
22. Deepalakshmi, P.; Lavanya, K.; Srinivasu, P.N. Plant Leaf Disease Detection Using CNN Algorithm. *Int. J. Inf. Syst. Model. Des.* **2021**, *12*, 1–21. [CrossRef]
23. Sibiya, M.; Sumbwanyambe, M. A Computational Procedure for the Recognition and Classification of Maize Leaf Diseases out of Healthy Leaves Using Convolutional Neural Networks. *AgriEngineering* **2019**, *1*, 119–131. [CrossRef]
24. Hu, R.; Zhang, S.; Wang, P.; Xu, G.; Wang, D.; Qian, Y. The Identification of Corn Leaf Diseases Based on Transfer Learning and Data Augmentation. In Proceedings of the 3rd International Conference on Computer Science and Software Engineering, Beijing, China, 22–24 May 2020; pp. 58–65.
25. Ahmad, A.; Saraswat, D.; El Gamal, A.; Johal, G. CD&S Dataset: Handheld Imagery Dataset Acquired under Field Conditions for Corn Disease Identification and Severity Estimation. *arXiv* **2021**, arXiv:2110.12084.
26. Lu, J.; Shi, R.; Tong, J.; Cheng, W.; Ma, X.; Liu, X. Lightweight Method for Plant Disease Identification Using Deep Learning. *Intell. Autom. Soft Comput.* **2023**, *37*, 525–544. [CrossRef]
27. Ma, N.; Zhang, X.; Zheng, H.-T.; Sun, J. Shufflenet v2: Practical Guidelines for Efficient Cnn Architecture Design. In Proceedings of the European Conference on Computer Vision (ECCV), Munich, Germany, 8–14 September 2018; pp. 116–131.
28. Murray, N.; Perronnin, F. Generalized Max Pooling. In Proceedings of the IEEE Conference on Computer Vision and Pattern Recognition, Columbus, OH, USA, 23–28 June 2014; pp. 2473–2480.
29. Yang, L.; Zhang, R.-Y.; Li, L.; Xie, X. Simam: A Simple, Parameter-Free Attention Module for Convolutional Neural Networks. In Proceedings of the International Conference on Machine Learning; PMLR, Virtual, 18–24 July 2021; pp. 11863–11874.
30. Chen, Y.; Jiang, H.; Li, C.; Jia, X.; Ghamisi, P. Deep Feature Extraction and Classification of Hyperspectral Images Based on Convolutional Neural Networks. *IEEE Trans. Geosci. Remote Sens.* **2016**, *54*, 6232–6251. [CrossRef]
31. Chen, Y.; Chen, X.; Lin, J.; Pan, R.; Cao, T.; Cai, J.; Yu, D.; Cernava, T.; Zhang, X. DFCANet: A Novel Lightweight Convolutional Neural Network Model for Corn Disease Identification. *Agriculture* **2022**, *12*, 2047. [CrossRef]
32. Jin, L.; Yu, Y.; Zhou, J.; Bai, D.; Lin, H.; Zhou, H. SWVR: A Lightweight Deep Learning Algorithm for Forest Fire Detection and Recognition. *Forests* **2024**, *15*, 204. [CrossRef]

33. Hou, Q.; Zhou, D.; Feng, J. Coordinate Attention for Efficient Mobile Network Design. In Proceedings of the IEEE/CVF Conference on Computer Vision and Pattern Recognition, Nashville, TN, USA, 20–25 June 2021; pp. 13713–13722.
34. Wang, Q.; Wu, B.; Zhu, P.; Li, P.; Zuo, W.; Hu, Q. ECA-Net: Efficient Channel Attention for Deep Convolutional Neural Networks. In Proceedings of the IEEE/CVF Conference on Computer Vision and Pattern Recognition, Seattle, WA, USA, 13–19 June 2020; pp. 11534–11542.
35. Ouyang, D.; He, S.; Zhang, G.; Luo, M.; Guo, H.; Zhan, J.; Huang, Z. Efficient Multi-Scale Attention Module with Cross-Spatial Learning. In Proceedings of the ICASSP 2023—2023 IEEE International Conference on Acoustics, Speech and Signal Processing (ICASSP), Rhodes Island, Greece, 4–10 June 2023; IEEE: New York, NY, USA, 2023; pp. 1–5.
36. Niu, B.; Wen, W.; Ren, W.; Zhang, X.; Yang, L.; Wang, S.; Zhang, K.; Cao, X.; Shen, H. Single Image Super-Resolution via a Holistic Attention Network. In *Proceedings of the Computer Vision–ECCV 2020: 16th European Conference, Glasgow, UK, 23–28 August 2020, Proceedings, Part XII 16*; Springer: Berlin/Heidelberg, Germany, 2020; pp. 191–207.
37. Daihui, L.; Shangyou, Z.; Wenhui, L.; Lei, Y. A New Cyclic Spatial Attention Module for Convolutional Neural Networks. In Proceedings of the 2019 IEEE 11th International Conference on Communication Software and Networks (ICCSN), Chongqing, China, 12–15 June 2019; IEEE: New York, NY, USA, 2019; pp. 607–611.
38. Chen, C.; Zhu, T.; Li, S.; Liu, B. Apple Leaf Disease Regcognition Method Base on Improved ShuffleNet V2. In Proceedings of the 2021 3rd International Conference on Advances in Computer Technology, Information Science and Communication (CTISC), Shanghai, China, 23–25 April 2021; IEEE: New York, NY, USA, 2021; pp. 276–282.
39. Tan, M.; Le, Q. Efficientnet: Rethinking Model Scaling for Convolutional Neural Networks. In Proceedings of the International Conference on Machine Learning; PMLR, Long Beach, CA, USA, 9–15 June 2019; pp. 6105–6114.
40. Mehta, S.; Rastegari, M. Mobilevit: Light-Weight, General-Purpose, and Mobile-Friendly Vision Transformer. *arXiv* **2021**, arXiv:2110.02178.
41. Evangelista, I.R.S.; Cabatuan, M.K.; Milagrosa, L.J.T.; Bandala, A.A.; Concepcion, R.; Dadios, E.P. Zea Mays Multi-Disease Classification and Severity Assessment with EfficientNetV2 Variants. In Proceedings of the 2023 IEEE Region 10 Symposium (TENSYMP), Canberra, Australia, 6–8 September 2023; IEEE: New York, NY, USA, 2023; pp. 1–6.
42. Schneider, N.; Piewak, F.; Stiller, C.; Franke, U. RegNet: Multimodal Sensor Registration Using Deep Neural Networks. In Proceedings of the 2017 IEEE Intelligent Vehicles Symposium (IV), Los Angeles, CA, USA, 11–14 June 2017; IEEE: New York, NY, USA, 2017; pp. 1803–1810.
43. Iandola, F.; Moskewicz, M.; Karayev, S.; Girshick, R.; Darrell, T.; Keutzer, K. Densenet: Implementing Efficient Convnet Descriptor Pyramids. *arXiv* **2014**, arXiv:1404.1869.
44. Zhang, H.; Zhang, L.; Jiang, Y. Overfitting and Underfitting Analysis for Deep Learning Based End-to-End Communication Systems. In Proceedings of the 2019 11th International Conference on Wireless Communications and Signal Processing (WCSP), Xi'an, China, 23–25 October 2019; IEEE: New York, NY, USA, 2019; pp. 1–6.
45. Liu, C.; Wu, Y.; Liu, J.; Sun, Z. Improved YOLOv3 Network for Insulator Detection in Aerial Images with Diverse Background Interference. *Electronics* **2021**, *10*, 771. [CrossRef]
46. Yu, Y. Research Progress of Crop Disease Image Recognition Based on Wireless Network Communication and Deep Learning. *Wirel. Commun. Mob. Comput.* **2021**, *2021*, 7577349. [CrossRef]
47. Atila, Ü.; Uçar, M.; Akyol, K.; Uçar, E. Plant Leaf Disease Classification Using EfficientNet Deep Learning Model. *Ecol. Inform.* **2021**, *61*, 101182. [CrossRef]

plants

MDPI

Article

Application of Unmanned Aerial Vehicle (UAV) Sensing for Water Status Estimation in Vineyards under Different Pruning Strategies

Juan C. Nowack [1,2], Luz K. Atencia-Payares [1,2,3], Ana M. Tarquis [1,4,*] and M. Gomez-del-Campo [1,2]

[1] CEIGRAM, ETSIAAB, Universidad Politécnica de Madrid (UPM), 28040 Madrid, Spain; j.nowack@hotmail.com (J.C.N.); lkatenciapayares@gmail.com (L.K.A.-P.); maria.gomezdelcampo@upm.es (M.G.-d.-C.)
[2] Departamento de Producción Agraria, ETSIAAB, Universidad Politécnica de Madrid (UPM), Av. Puerta de Hierro, n° 2–4, 28040 Madrid, Spain
[3] Unmanned Technical Works (UTW), Leganés, 28919 Madrid, Spain
[4] Grupo de Sistemas Complejos ETSIAAB, 28040 Madrid, Spain
* Correspondence: anamaria.tarquis@upm.es

Abstract: Pruning determines the plant water status due to its effects on the leaf area and thus the irrigation management. The primary aim of this study was to assess the use of high-resolution multispectral imagery to estimate the plant water status through different bands and vegetation indexes (VIs) and to evaluate which is most suitable under different pruning management strategies. This work was carried out in 2021 and 2022 in a commercial Merlot vineyard in an arid area of central Spain. Two different pruning strategies were carried out: mechanical pruning and no pruning. The stem water potential was measured with a pressure chamber (Ψ_{stem}) at two different solar times (9 h and 12 h). Multispectral information from unmanned aerial vehicles (UAVs) was obtained at the same time as the field Ψstem measurements and different vegetation indexes (VIs) were calculated. Pruning management significantly determined the Ψ_{stem}, bunch and berry weight, number of bunches, and plant yield. Linear regression between the Ψ_{stem} and NDVI presented the tightest correlation at 12 h solar time (R^2 = 0.58). The red and red-edge bands were included in a generalised multivariable linear regression and achieved higher accuracy (R^2 = 0.74) in predicting the Ψ_{stem}. Using high-resolution multispectral imagery has proven useful in predicting the vine water status independently of the pruning management strategy.

Keywords: canopy development; *Vitis vinifera*; production; vegetation index; chlorophyll

Citation: Nowack, J.C.; Atencia-Payares, L.K.; Tarquis, A.M.; Gomez-del-Campo, M. Application of Unmanned Aerial Vehicle (UAV) Sensing for Water Status Estimation in Vineyards under Different Pruning Strategies. *Plants* **2024**, *13*, 1350. https://doi.org/10.3390/plants13101350

Academic Editor: Youzhen Xiang

Received: 20 March 2024
Revised: 30 April 2024
Accepted: 4 May 2024
Published: 13 May 2024

1. Introduction

Pruning is usually performed to control the vine's vegetative development and generally implies a reduction in the leaf area, vigour and reserve accumulation compared to a non-pruned vine. A reduction in vegetative material can lead to more spaces in the canopy and exposed bunches [1]. Vinegrowers can adopt the no-pruning vineyard management practice to reduce operating costs and grape size [2], increasing the skin to pulp relation, an interesting feature for quality winemaking as it contributes to higher tannin and anthocyanin levels. However, it generally leads to an increment in the total leaf area [3], thus to greater transpiration levels [4] and water uptake demands, which, if not fulfilled, could lead to physiological water stress.

There is great interest in determining the plant water status in irrigated vineyards due to its relationship with the yield, fruit composition and wine quality [5,6], critical parameters for a profitable winemaking company. Water stress reduces photosynthetic activity and vegetative growth and limits berry ripening [7,8] The stem water potential (Ψ_{stem}) is a consistent and sensitive indicator of the plant water status in grapevines [9], and diverse authors have published optimal thresholds of Ψ_{stem} for different phenological stages [10]. It is an integrating measurement that can provide greater precision than the soil water content concerning irrigation management. However, determining the Ψ_{stem} in commercial vineyards has a series of downsides. Only small samples can be assessed rapidly because the Ψ_{stem} is a changing parameter throughout the day. The maximum Ψ_{stem} (higher vine hydration) occurs before dawn and starts descending as the plant transpires throughout the day [11]. To compare the Ψ_{stem} values from different treatments, they must be assessed in the shortest possible time.

Moreover, it is highly time-consuming, making it unpractical for large plantations where the intra-field variability is usually high. This evidences the necessity of developing a well-founded method for determining the grapevine water status in a cost-effective manner.

Remote sensing is a powerful tool increasingly being used at the commercial and research levels as it can obtain vast amounts of valuable and accurate geospatial data on a large-scale dimension. Some practical applications include the discrimination of plant species or vegetation types or the detection of diseased or physiologically stressed plants [12]. Precision viticulture, a recently acquired term, can be described as the precision agriculture sector focusing on the vineyard. Its main objective is to use diverse technologies to manage vineyard spatial heterogeneity to reduce environmental impacts while increasing profitability. This spatial and temporal variability can be expressed through productivity, vine development, water status, or exogenous factors such as the soil characteristics or microclimate conditions. Some common applications of remote sensing in vineyard management include the assessment or estimation of the chlorophyll and carotenoid concentrations [13,14], grape phenolic content [15] or colour [16], canopy structure [17,18] and water status [19–21].

With the use of vegetation indexes (VIs), a wide range of particular characteristics, like the vegetation biomass, productivity, biochemical properties or crop water status of a photosynthetically active plant, can be assessed based on the plant spectral response. Ref. [22] conducted an extensive review of the application of remote sensing-derived vegetation indexes (VIs) in viticulture, with 113 publications evaluated since 2000. They discovered that the most commonly used platforms are currently unmanned aerial vehicles (UAVs), aircraft, and Sentinel 2 satellites. The pursued objective and the imagery's price and resolution mainly conditioned each platform's utilisation. While commercial satellites can be suitable for regional-scale studies due to their extensive coverage [23], their generally low and inflexible spatial and temporal resolutions make them unattractive for managing at a vineyard or site-specific scope.

Concerning water variability management, ref. [24] used high-resolution multispectral and thermal sensors mounted on a UAV to estimate the water status in a rain-fed Tempranillo vineyard. They found that specific spectral indexes were significantly correlated to the Ψ_{stem} and stomatal conductance, both water status indicators, using 10 cm/pixel images. The same occurred with thermal indexes derived from 30 cm/pixel thermal images. They stated that thermal imagery could be helpful as a short-term water stress indicator, considering that the correlations changed throughout the season. Conversely, multispectral indexes can serve as long-term indicators as their correlations are more stable. Other related studies, like the one in [21], obtained similar results. They used 1 m/pixel multispectral images to prove that zones based on the Normalised Difference Vegetation Index (NDVI), one of the most well-known and -used VIs, values presented significant differences in the vine vegetative development, yield, and water status. Significant correlations between this index and the grapevine yield were also found by [5] using higher spatial resolution (2.6 cm/pixel) only 40 days before harvest. In general, the VIs are related to different agronomic parameters, such as the leaf area development, determined by pruning technique, which may modify vine water status and, therefore, the VI values. The leaf area is considered one of the most important agronomic parameters for evaluating the vegetative development of the plants; however, leaf area measurements are time-consuming and labour-intensive because of the inherent variability found within a vine and, on a larger scale, within a vineyard. Some methods are destructive and cannot be performed on a large scale [25]. On the other hand, the pruning weight is a parameter that encompasses the final performance of the plant. As the percentage of soil coverage, the pruning weight can differentiate between treatments, making it a good indicator in terms of the development of the vines. Other studies have observed a direct relationship between the yield through the pruning weight and the percentage of soil coverage [26–28]. The percentage of soil coverage appears to be an interesting tool as a proxy for the vegetative development and yield.

Other works have focused on using spectral proximal-sensing devices instead of UAVs. For example, ref. [29] evaluated hyperspectral reflectance indexes derived from a spectrometer (350–2500 nm region) to detect grapevine water status. Their study evidenced the existence of indexes capable of significantly correlating to water status parameters (both at canopy and leaf level), such as the total leaf water content, Ψ_{stem} or equivalent water thickness.

The above-mentioned studies and others will serve as a reference, considering that the availability of a varied set of spectral bands conditions the use of multiple VIs.

It must be emphasised that applying remote sensing to a discontinuous crop, like vineyards or fruit orchards, is technically more challenging than to a continuous herbaceous crop [30–32]. From an aerial perspective, the presence of the soil layer, which could be vegetated or not, does not allow for the use of widely accessible low-spatial-resolution satellite scenes (10 m/pixel for Sentinel-2). If the pixel size is excessive, the soil or intra-row vegetation reflectance deteriorates the data quality, which would not represent the canopy's status. In general terms, the pixel size should not be coarser than the targeted object or unit, an individual vine in this study. However, some studies have proven helpful in estimating the vine water status in Mediterranean climates with nanosatellite-based imagery (3 m/pixel) by compensating for the low spatial resolution with high temporal availability [33]. Until high-resolution satellite information is achieved and freely accessible, using UAVs with multispectral sensors remains one of the vineyard's best technical solutions for remote sensing, as pure canopy pixel information can be extracted. The pixel size may vary depending on the intended spatial coverage, but it can be down to 1.4 cm/pixel for monitoring grapevines.

The primary aim of this study was to assess the use of high-resolution multispectral imagery (12 cm/pixel) to estimate the plant water status through different bands and vegetation indexes (VIs) in vines under different pruning management. This work focuses on developing a well-founded indirect method for determining the grapevine water status that could be used to map the Ψ_{stem} in all the vineyard surfaces and aid in irrigation management.

2. Materials and Methods

2.1. The Study Vineyard

This study was carried out during the growing seasons of 2021 and 2022 on a 40 ha commercial vineyard in Yepes (39°56′26.2″ N, 3°42′49.7″ W), Spain, at 699 m above the mean sea level. The vineyard was planted in 2002, cv. Merlot (*Vitis vinifera* L.) over SO4 rootstock (*Vitis berlandieri* × *Vitis riparia*) and arranged on a trellis with a plantation frame of 2.6 m × 1.1 m (3500 vines/ha). The plants have been trained in a double-cordon system.

The climate of the area corresponds to a typical hot-summer Mediterranean climate. The region has an average daily temperature of 16 °C and an average annual rainfall of 394 mm, mainly concentrated at the end of autumn and the beginning of spring. Summers are characterised by a high atmospheric vapour demand derived from high temperatures (maximum temperatures > 40 °C) and low relative humidity. The temperature and relative humidity were measured in-field using a portable weather station OMEGAETTE model HH314A (OMEGA, Ltd., Bridgeport, NJ, USA, EEUU) that provided minute data. From these values, the saturation vapour pressure (e_s), actual vapour pressure (e_a) and vapour pressure deficit (VPD) were calculated for the measuring times.

Precipitation and reference evapotranspiration (ETo) data were extracted from the closest public weather station, 18 km from the study site, located in Magán (Toledo) (39°56′06.6″ N 3°56′32.4″ W). The rainfall in 2021 was 347 mm, while the ETo was 1284 mm. In 2022, these values were 348 and 1396 mm, respectively. The meteorological data are summarised in Figure 1.

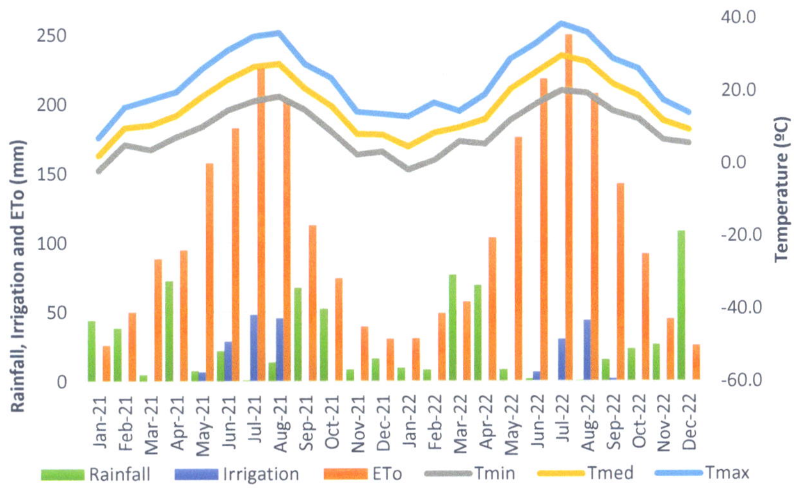

Figure 1. Monthly climatic variables during 2021 and 2022: rainfall, applied irrigation, reference evapotranspiration (ETo), average minimum temperature (Tmin), average temperature (Tmed) and average maximum temperature (Tmax). Data were obtained from the Magán public weather station (39°56′06.6″ N 3°56′32.4″ W).

2.2. Pruning Management

As in many other large-scale vineyards, this one is divided into smaller units for management purposes, designated as plots or parcels. Our study was executed in one of these plots (Figure 2), belonging to a zone classified as a coarse-loamy, gypsic, mesic Typic Calcixerepts soil [34]. In this 5 ha plot, two pruning strategies have been traditionally used, dividing it into two 2.5 ha areas with homogeneous conditions and management except for pruning. To a large extent, vineyard managers have done so over the last few years to assess its impact on qualitative performance. From this, two pruning treatments were selected for this study. The pruning treatment was developed over 10 years ago. The activity takes place in winter, around the second week of February. This plot area has a more intensive pruning approach. Firstly, mechanical pruning is performed with a horizontal trimmer 30 cm above the cordons to eliminate large amounts of wood in the most cost-effective way. Then, the vines are spur-pruned by hand, leaving approximately two buds per node. The no pruning treatment was developed over 10 years ago. Like the rest of the plots, the activity takes place in winter, around the second week of February. This treatment area was minimally hand-pruned, focusing on removing damaged or unnecessary parts. Therefore, the pruning magnitude is substantially smaller, considering the amount of wood removed after each productive campaign. Six vines out of each pruning system in adjacent lines were selected in order to take the different experimental measurements (Figure 2b).

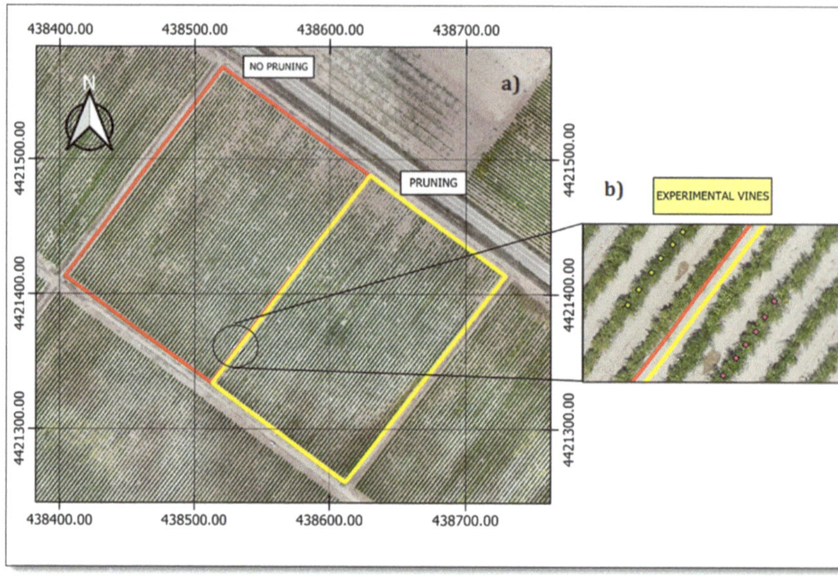

Figure 2. (**a**) Aerial RGB image of the commercial vineyard used for this study. Different pruning management plots (2.5 ha each) are outlined in different colours: red for no pruning and yellow for pruning. (**b**) Location details of the six experimental vines where measurements were taken.

2.3. Irrigation

Given the atmospheric conditions previously mentioned and the limited irrigation water availability in the growing area, the studied vines inevitably developed under a deficit irrigation strategy. During the irrigation season, from May to August in both years, the ETo was 769 in 2021 and 851 in 2022. Irrigation supplied 17% and 9% of the ETo in 2021 and 2022, respectively.

Irrigation was applied with one drip emitter per meter with a flow rate of 2 L h^{-1}. It was scheduled according to the standard practices followed by Bodegas Casa del Valle, i.e., with limited time intervals. The irrigation periods varied between the years of the study regarding the duration and total quantities applied. In 2021, the irrigation season went from mid-May to the beginning of September, and the irrigation in the study plot was 128 mm. In 2022, the irrigation season was shorter, going from the end of June to the beginning of September, and the applied amount was also reduced, with a total of 82 mm, 36% less than in the previous season.

2.4. Physiological and Agronomic Parameters

These measurements were conducted in six experimental vines for each pruning treatment in both study campaigns.

2.4.1. Stem Water Potential Measurements

The Ψ_{stem} (MPa) was measured at 9:00 and 12:00 h solar time on 9 different dates: 25 June 2021, 5 July 2021, 20 July 2021, 30 July 2021, 19 August 2021, 30 June 2022, 15 July 2022, 5 August 2022 and 12 August 2022. Measurements were performed on healthy and shaded leaves from the inner part of the canopy, where 6 leaves/plant were taken per treatment. They were covered with a plastic bag with aluminium foil one hour before the measurement, as standardised methods recommend attaining water status equilibrium between stem and leaves. For this measurement, a Scholander-type pressure chamber was used (Soil Moisture Equipment Corp., Santa Barbara, CA, USA).

2.4.2. Chlorophyll Measurements

The leaf chlorophyll level (μmol chlorophyll/m^2 leaf area) was measured on the same dates and times as the Ψ_{stem} using an Apogee MC-100 sensor (Apogee Instruments Inc., Logan, UT, USA). Measurements were performed on three healthy leaves per experimental vine.

2.4.3. Canopy Description

Measurements, necessary to describe the canopy development, were taken on 1 July 2021 and 20 June 2022 at an advanced stage in the season, ensuring vegetative growth had halted and maximum vegetative expression had been achieved.

In each experimental vine, three points were selected to measure the canopy contour (using a flexible tape) and distance between the highest and lowest leaves: trunk and 40 cm on either side. On the same points, the canopy width was noted at three heights (80, 110 and 120 cm from the ground). From these measurements, the canopy height and width were derived, and the canopy volume and external leaf area were calculated:

$$\text{Canopy volume}\left(m^3\right) = H * W * SV \tag{1}$$

$$\text{External canopy area}\left(m^2\right) = (2H + W) * SV \tag{2}$$

where: H: canopy height (m), W: canopy width (m), SV: spacing between vines (m).

The canopy soil coverage was calculated from a high-resolution (3 cm/pixel) RGB image using QGIS software (version 3.22.13 Białowieża) at the beginning of the irrigation season: 1 July 2021 and 30 June 2022. At this moment, the vegetative development stopped. The apex stops growing and does not develop more leaves (QGIS, Free Software Foundation, Boston, MA, USA). A non-supervised k-means classification with two classes was run to differentiate the vine canopy and soil pixels for a posterior percentage of covered soil calculation (Figure 3).

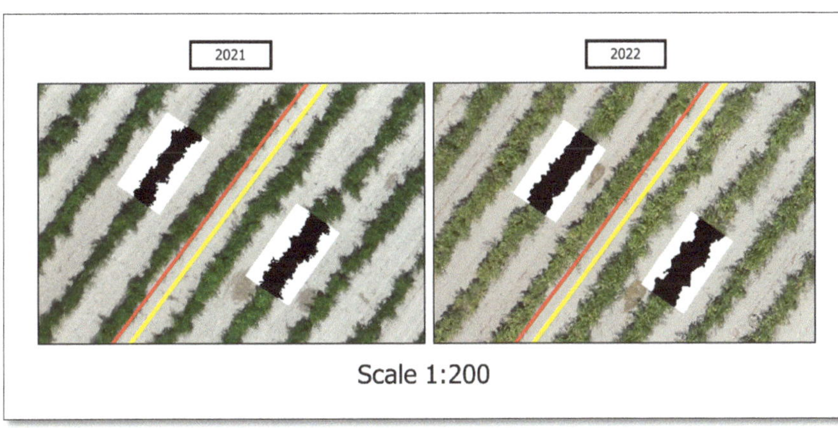

Figure 3. Details of the non-supervised classification performed to calculate the percentage of covered soil of the two pruning treatments in two consecutive seasons—a commercial vineyard in Yepes (Toledo).

2.4.4. Quantitative and Qualitative Analysis

The experimental vines were harvested by hand according to the commercial vineyard manager's decision concerning the optimal productive characteristics. The selected dates were 20 August 2021 and 16 August 2022. During the harvest, bunches were counted and production was weighed using a portable electronic scale. Between 150 and 200 berries of each vine were randomly selected, packaged, tagged, and stored in a cooler for subsequent analysis.

The selected berries were taken to the laboratory, counted, and weighed immediately. Then, they were processed to determine the total soluble solids (°Brix) with an Atago digital Brix refractometer (ATAGO CO., LTD., Tokyo, Japan), and the pH was measured with a pH Meter Hach sensION (Hach company., Loveland, CO, USA).

2.5. Multispectral Images and Vegetation Indexes Calculation

Aerial images were acquired by employing a UAV, model eBee (AgEagle Aerial Systems Inc., Wichita, Kansas), a commercial fixed-wing platform equipped with a Parrot Sequoia (Parrot© SA, 2017, Paris, France) multispectral sensor. Flights were carried out at 120 m of altitude with nadir mode. On the same days and times as the flights, the Ψ_{stem} measurements were carried out (9:00 and 12:00 solar time). The different high-resolution imagery consisted of RGB images (3 cm/pixel) and multi-band images (12 cm/pixel) that included the green (495–570 nm), red (620–750 nm), red-edge (670–760 nm) and near-infrared (NIR) (780–2500 nm) bands.

QGIS software was used to extract the band pixel information. Given the high spatial resolution of multi-band images (12 cm/pixel), pure canopy pixels could be selected, avoiding soil interference. A set of five points from the centremost area (top of the canopy) of each vine was selected to obtain each band reflectance value. These reflectance values were loaded on a spreadsheet where the VIs were calculated. Based on the available bands and the cited literature, five indexes (Table 1) were calculated: the NDVI, the Renormalized Difference Vegetation Index (RDVI), the Transformed Chlorophyll Absorption Ratio (TCARI), the Optimised Soil Adjusted Vegetation Index (OSAVI), and the Normalised Difference Red-Edge Index (NDRE).

Table 1. Formulas and references of the vegetation indexes used in this study.

Index	Formula	Reference
NDVI	$NDVI = \frac{R_{NIR} - R_{red}}{R_{NIR} + R_{red}}$	[35]
RDVI	$RDVI = \frac{R_{NIR} - R_{red}}{\sqrt{R_{NIR} + R_{red}}}$	[36]
TCARI	$TCARI = 3 * \left[\left(R_{RedEdge} - R_{red} \right) - 0.2 * \left(R_{RedEdge} - R_{green} \right) * \left(\frac{R_{RedEdge}}{R_{red}} \right) \right]$	[37]
OSAVI	$OSAVI = (1 + 0.16) * (R_{NIR} - R_{red}) / (R_{NIR} + R_{red} + 0.16)$	[38]
NDRE	$NDRE = \frac{R_{NIR} - R_{Red\ Edge}}{R_{NIR} + R_{Red\ Edge}}$	[39]

2.6. Statistical Analysis and Stem Water Potential Modelling

Data subjected to statistical variance analysis (ANOVA) were processed with Infostat Software version 1.5 (Universidad Nacional de Córdoba, Argentina). The mean values were classified using the LSD test ($p < 0.05$).

In order to estimate the Ψ_{stem} from the UAV-acquired multispectral information, different model approaches were evaluated using Statgraphics 19 software (Statgraphics Technologies, Inc. The Plains, Virginia, USA). Simple linear regressions between the Ψ_{stem} and all individual bands (green, red, red edge and NIR) and simple linear regressions between the Ψ_{stem} and Vis (NDVI, RDVI, TCARI, OSAVI and NDRE) were performed.

Multivariable linear regression models between the Ψ_{stem} and individual bands aim for a better correlation than simple linear regressions.

In the latter case, the Variance Inflation Factor (VIF), which measures multicollinearity, was used for this purpose, excluding solutions where VIF > 10 for any parameter. Moreover, the Durbin–Watson Statistic (DW) was used to assess the autocorrelation between the residuals. Models with DW < 1.5 were rejected. Other metrics we used to select which bands were best to include in the model are the Akaike Information Criterion (AIC), the Schwarz Information Criterion (SBIC), and the Hannan–Quinn Information Criterion (HQC). They are all indicators of the goodness of fit in a multivariable linear regression; the lower the value, the better.

3. Results

3.1. Climatic Characterisation

The climatic variables during the in-field measurements are summarised in Table 2, and the critical monthly parameters are presented in Figure 1. Our results reveal that the 2022 season was hotter (+2.0 °C and +1.8 °C at 9:00 and 12:00, respectively) and drier (10.4% and −3.7% RH lower at 9:00 and 12:00, respectively) than 2021. Moreover, the annual ETo (reference Evapotranspiration) was 109 mm higher in 2022, while a 36% reduced total irrigation was applied. The maximum monthly ETo occurred in July in both years and was 226.1 mm and 249.6 mm in 2021 and 2022, respectively.

Table 2. Temperature (°C), relative humidity (%), vapour pressure deficit (kPa) and daily ETo (mm/day) during the times and days that the Ψ_{stem} was evaluated and multispectral images were obtained for two hours of measurement and two campaigns in a commercial vineyard under two pruning treatments (mechanical and no pruning) in central Spain (Yepes, Toledo).

	9:00 Solar Time			12:00 Solar Time			ETo
Date	T (°C)	RH (%)	VPD (kPa)	T (°C)	RH (%)	VPD (kPa)	(mm/Day)
25 June 2021	27.4	43.9	2.05	33.5	29.8	3.63	6.6
5 July 2021	28.2	42.8	2.19	34.0	30.8	3.68	8.2
20 July 2021	28.8	39.6	2.39	35.9	19.8	4.74	7.9
30 July 2021	28.3	34.1	2.53	33.6	12.7	4.54	9.1
19 August 2021	28.7	41.0	2.32	33.1	22.8	3.91	6.3
Average 2021	28.3	40.3	2.30	34.0	23.2	4.10	7.6
30 June 2022	27.5	28.2	2.64	30.0	23.1	3.25	7.0
15 July 2022	35.2	22.6	4.40	40.2	15.4	6.32	8.1
5 August 2022	29.1	37.7	2.52	36.6	18.4	5.01	7.3
12 August 2022	29.3	31.0	2.81	36.6	21.2	4.85	6.3
Average 2022	30.3	29.8	3.09	35.8	19.5	4.86	7.2

The vapour pressure deficit (VPD) suffered an apparent increase between measuring hours due to a decrease in the RH and an increase in the temperature caused by solar radiation. This daily increase reached a topmost value of 98.8% on 5 August 2022. However, the average VPD increase between the hours was higher in 2021 (78.3% vs. 59.5%). Notable differences in climatic conditions between campaigns are also evidenced by the maximum VPD values in 2022, 73.9% and 33.3% higher than in 2022, at 9:00 and 12:00, respectively. These differences assuredly affected the vines' physiological and spectral behaviour. From our results, it can be noted that the VPD and daily ETo are tightly associated. The peak ETo values coincided with high VPD measures each season, although the ETo is a daily value and the VPD is an averaged punctual value.

3.2. Canopy Development

Characterisation of the different measurements and calculations relative to the canopy structure are reflected in Table 3. As expected, the pruning weight was notably higher in the pruning treatment vines (around double the amount in both years).

The height, width, volume, and external canopy area within the analysed geometry parameters did not show significant differences between the treatments in the two years of the study. However, the differences between the treatments were found to be influenced more by the effect of the year than by the treatments. On the other hand, the pruning weight and covered soil by the canopy show that the treatment and the year influence significant differences.

Lastly, remote sensing was applied to estimate the percentage of covered soil, which yielded statistically significant observations during both campaigns. This parameter is tightly related to the vine width, as reflected in the results. The pruned vines presented significantly higher soil coverage in both years, and both treatments experimented with an increase in this measure in 2022 (+6% and +9%, pruning and no pruning treatments, respectively).

Table 3. Vine canopy geometric parameters relative to the development and architecture over two campaigns (2021 and 2022), under mechanical pruning (P) and no pruning (NP), in a commercial vineyard in central Spain (Yepes, Toledo). Mean data and coefficient of variation (CV).

	2021					2022				
	P	CV (%)	NP	CV (%)		P	CV (%)	NP	CV (%)	
Pruning weight (g/plant)	177.5	27%	87.5	29%	*	216.9	30%	101.7	35%	*
Canopy height (cm)	69.75	21%	70.67	18%	ns	86.25	25%	106.08	11%	ns
Width (cm)	71.52	19%	58.9	15%	ns	96.07	10%	89.73	19%	ns
External canopy area (m^2)	2.34	8%	2.07	12%	ns	3.06	10%	3.14	9%	ns
Canopy volume (m^3)	0.53	13%	0.46	23%	ns	0.91	27%	1.04	14%	ns
Canopy contour (m)	1.64	11%	1.64	14%	ns	1.72	15%	1.73	6%	ns
Covered soil by the canopy (%)	43%	6%	36%	9%	*	49%	4%	45%	2%	*

ns: non-significant and * significant at $p < 0.05$.

3.3. Physiological Responses

The plant water status, determined by the Ψ_{stem} measurements, did not show a consistent trend for the different pruning treatments over the two campaigns (Table 4). In 2021, the no pruning treatment expressed lower values of Ψ_{stem} at both times and for almost all the measurement dates, with significant differences between the treatments on three dates and for both measuring times. The average Ψ_{stem} for the 2021 season was significantly lower in the pruning treatment, only at 9:00 solar time. The general tendency observed as the campaign advanced was a reduction in the Ψ_{stem} value, implying a decline in the plant water status for both treatments due to the depletion of water reservoirs in the soil.

Table 4. Ψ_{stem} (MPa) measured in two hours per day and over two campaigns (2021 and 2022), under mechanical pruning (P) and no pruning (NP), in a commercial vineyard in central Spain (Yepes, Toledo).

Date	9:00 Solar Time			12:00 Solar Time		
	P Ψ_{stem} (MPa)	NP Ψ_{stem} (MPa)		P Ψ_{stem} (MPa)	NP Ψ_{stem} (MPa)	
25 June 2021	−0.5	−0.6	ns	−0.8	−0.8	ns
5 July 2021	−0.6	−0.7	ns	−1.1	−0.9	ns
20 July 2021	−0.5	−0.8	*	−1.0	−1.1	*
30 July 2021	−0.6	−0.7	*	−1.0	−1.4	*
19 August 2021	−0.7	−1.1	*	−1.3	−1.6	*
Average 2021	−0.6	−0.8	*	−1.0	−1.2	ns
30 June 2022	−0.9	−0.9	ns	−1.1	−1.1	ns
15 July 2022	−1.1	−0.8	*	−1.1	−0.9	*
5 August 2022	−1.7	−1.5	*	−1.9	−1.7	ns
12 August 2022	−1.6	−1.3	*	−1.7	−1.7	ns
Average 2022	−1.3	−1.1	ns	−1.4	−1.3	ns

ns: non-significant and * significant at $p < 0.05$.

In 2022, the physiological behaviour of the pruning treatments reversed. In this campaign, the pruning treatment obtained lower Ψ_{stem} values, with significant differences on three dates and one date for the 9:00 and 12:00 measurements, respectively. The average campaign value was not significant for either of the two times. The progression throughout the season was also a decreasing trend for the Ψ_{stem}, for both times and treatments.

The chlorophyll concentration measurements (Table 5) did not show clear tendencies between the treatments, hours of measurement or campaigns, unlike the Ψ_{stem} values. In 2021, only three single-date significant differences between the pruning treatments were observed at solar noon. In 2022, this number was reduced to one date in the 9:00 measurement. The average campaign values did not show statistical differences for any measuring times.

Table 5. Chlorophyll concentration measurements (µmol chlorophyll/m^2 leaf area), two hours per day and over two campaigns (2021 and 2022), under mechanical pruning (P) and no pruning (NP), in a commercial vineyard in central Spain (Yepes, Toledo).

Date	9:00 Solar Time			12:00 Solar Time		
	P Chlorophyll µmol Chlorophyll/m^2 Leaf Area	NP Chlorophyll µmol Chlorophyll/m^2 Leaf Area		P Chlorophyll µmol Chlorophyll/m^2 Leaf Area	NP Chlorophyll µmol Chlorophyll/m^2 Leaf Area	
25 June 2021	15.00	17.05	ns	14.05	15.65	ns
5 July 2021	16.05	17.18	ns	19.75	14.65	*
20 July 2021	17.18	15.80	ns	20.40	17.08	ns
30 July 2021	14.45	14.75	ns	18.23	15.38	*
19 August 2021	18.98	16.32	ns	18.45	15.61	*
Average 2021	16.33	16.22	ns	18.18	15.67	ns
30 June 2022	14.9	14.7	ns	14.5	14.9	ns
15 July 2022	16.2	14.9	*	16.7	15.8	ns
5 August 2022	15.8	15.0	ns	15.6	15.5	ns
12 August 2022	14.1	14.2	ns	14.7	14.7	ns
Average 2022	15.3	14.7	ns	15.4	15.2	ns

ns: non-significant and * significant at $p < 0.05$.

3.4. Vine Production and Quality

Table 6 summarises the productive and qualitative parameters that were assessed. Only the bunch weight showed significant and consistent differences in both campaigns. The pruning treatment obtained statistically higher values in both years, with an increase of 104% and 34% compared to the no pruning in both years. In 2022, both treatments suffered a decrease in this parameter compared to the previous year: −46% and −19% for pruning and no pruning, respectively.

Table 6. Productive and qualitative parameters were evaluated for two campaigns (2021 and 2022), under mechanical pruning (P) and no pruning (NP), in a commercial vineyard in central Spain (Yepes, Toledo).

	2021			2022		
	P	NP		P	NP	
Production per plant (kg)	1.88	1.53	ns	1.27	1.60	ns
Berry weight (g)	0.77	0.51	*	0.39	0.34	ns
Bunch weight (g)	52.2	25.6	*	27.96	20.86	*
Number of bunches per plant	36.25	56.5	ns	37.17	62.0	*
TSS (°Brix)	28.48	28.88	ns	25.8	25.4	ns
pH	3.44	3.41	ns	3.65	3.81	*

ns: non-significant and * significant at $p < 0.05$.

The berry weight was higher in the pruning treatment but only significantly so in 2021. The pruned vines presented a higher berry weight than the non-pruned ones but showed the most significant decline between campaigns. The number of bunches per plant behaved contrarily to the previous parameters, as the non-pruned vines had higher numbers during both campaigns, with significant differences found only in the second campaign (+56% and +67% in 2021 and 2022, respectively).

The production per plant was not significantly different between the pruning treatments, but significant differences were observed between the mean values. In 2021, the pruning treatment had higher production than in 2022, caused by the significantly higher berry and bunch weight. In 2022, the non-pruned plants obtained higher production since the bunch and berry weights were balanced. Moreover, the non-pruned vines developed a very high number of bunches compared to 2021.

Concerning the qualitative parameters, the TSS (°Brix) did not show significant differences between the treatments in either campaign. the year factor influenced the performance, so in 2022, there was an apparent reduction in this parameter for both treatments, indicating less sugar content in the must. On the other hand, the pH measurements behaved contrarily to the TSS. The 2022 campaign reflected higher pH values for both treatments than the previous one, with significant differences between the treatments, which did not happen in 2021. As with the TSS, with the pH, the year factor influences the quality of the performance.

3.5. Vine Spectral Behaviour

In general terms, the pruning treatment obtained lower values for the NDVI, RDVI, OSAVI, and NDRE in the 2022 season than in 2021 for both flying times (Table 7). In 2021, the indexes average values were statistically higher in the pruning treatment at the 9:00 and 12:00 flights, except for the TCARI. In 2022, there were no significant differences between the two treatments for any of the indexes at both times, except for the NDRE at 9:00. The differences in values between the treatments decreased. In this season, the no pruning treatment reflected higher values for these indexes, although not significantly so (refer to Appendix A Figure A1). The VI values in 2022 were notably more alike between the treatments than in the previous campaign. As a result, statistically significant differences were reduced on individual dates and average seasonal values, as with the Ψstem values (refer to Appendix A Tables A1 and A2).

Table 7. Annual average and factorial analysis (FA) of vegetation indexes calculated from a high-resolution (12 cm/pixel) multispectral sensor mounted on board a UAV at two solar times (09:00 and 12:00), for two campaigns (2021 and 2022), under mechanical pruning (P) and no pruning (NP), in a commercial vineyard in central Spain (Yepes, Toledo).

9:00		**NDVI**		**RDVI**		**TCARI**		**OSAVI**		**NDRE**	
2021	P NP	0.78 0.67	*	0.61 0.47	*	0.03 0.17	*	0.71 0.58	*	0.23 0.21	*
2022	P NP	0.67 0.70	ns	0.56 0.56	ns	0.29 0.23	ns	0.63 0.65	ns	0.18 0.19	*
12:00											
2021	P NP	0.74 0.69	*	0.55 0.47	*	0.06 0.15	*	0.67 0.59	*	0.20 0.20	ns
2022	P NP	0.68 0.71	ns	0.51 0.51	ns	0.18 0.15	ns	0.61 0.63	ns	0.19 0.20	ns
FA											
Treatment	P NP	*		*		ns		*		ns	
Year	2021 2022	*		ns		*		ns		*	

ns: non-significant and * significant at $p < 0.05$.

The evolution of the different vegetation indexes during the year 2021 for the 9:00 and 12:00 flights is presented in Figure 4. In this year, the two treatments were significantly different for many indexes. The VI values showed a decline as the season advanced, which was more evident in the latest flights, and the same pattern followed in the 2022 season (refer to Appendix A Figure A1). However, the NDVI, RDVI, and OSAVI presented a more evident evolution in time than the rest (Figure 4). The evolution of these indexes shows that in the morning (Figure 4a,c,g), the separation between the treatments is more pronounced than at noon (Figure 4b,d,h).

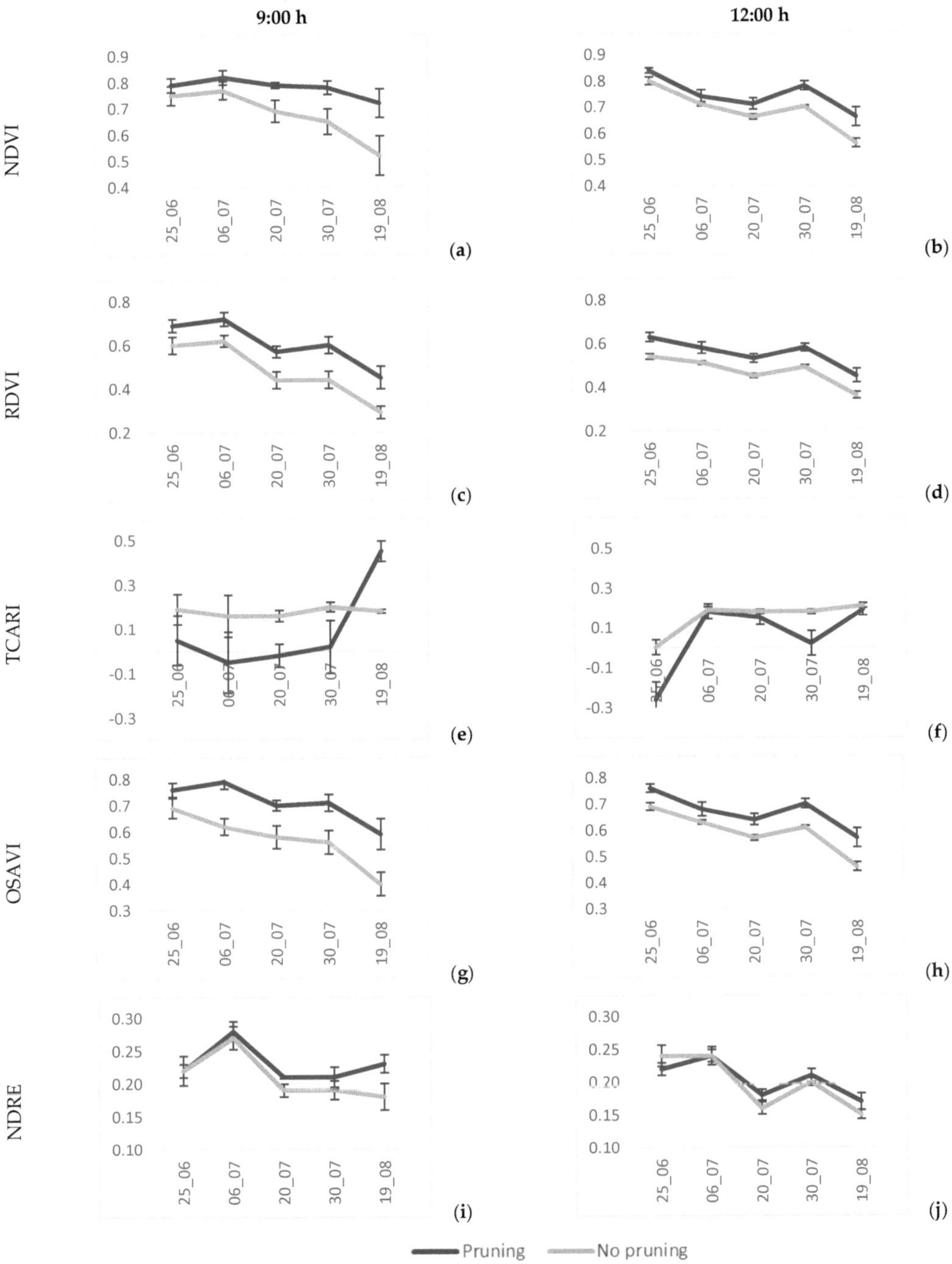

Figure 4. Evolution along the season of different vegetation indexes calculated from a high-resolution (12 cm/pixel) multispectral sensor mounted on board a UAV at 09:00 and 12:00 solar time in 2021 on a commercial vineyard under two pruning treatments (mechanical and no pruning) in central Spain (Yepes, Toledo).

The NDVI showed no differences in the first two measurement days at 9:00 (Figure 4a). The RDVI and OSAVI showed statistical differences between the treatments for all the dates and times in 2021(refer to Appendix A Tables A1 and A2). The evolution of the RDVI was constantly equidistant between the two treatments throughout the season at both times (Figure 4c,d). The NDRE did not exhibit a clear difference throughout the season between the two treatments (Figure 4i,j). The differences between the pruning treatments only showed statistical differences for the 9:00 flights in both campaigns (Table 7).

On the other hand, the TCARI presented no clear evolution along the season (Figure 4e,f), where this index was significantly higher in the non-pruned plants at both flying times. The TCARI reversed its response in 2022, and the pruned plants obtained slightly higher average values (Table 7).

A factorial analysis considering the year and treatment (Table 7) showed that the NDVI was influenced by the treatment and year; the RDVI and OSAVI were not influenced by the year but by the treatment. The TCARI was only influenced by the treatment, and the NDRE was influenced by only the year.

Figure 5 shows the NDVI performance between the treatments at two time (9:00 and 12:00) for the last flight date of the 2021 season (19 August). It is observed that the pruning treatment exhibits higher values of the NDVI than the non-pruning treatment at both times of day.

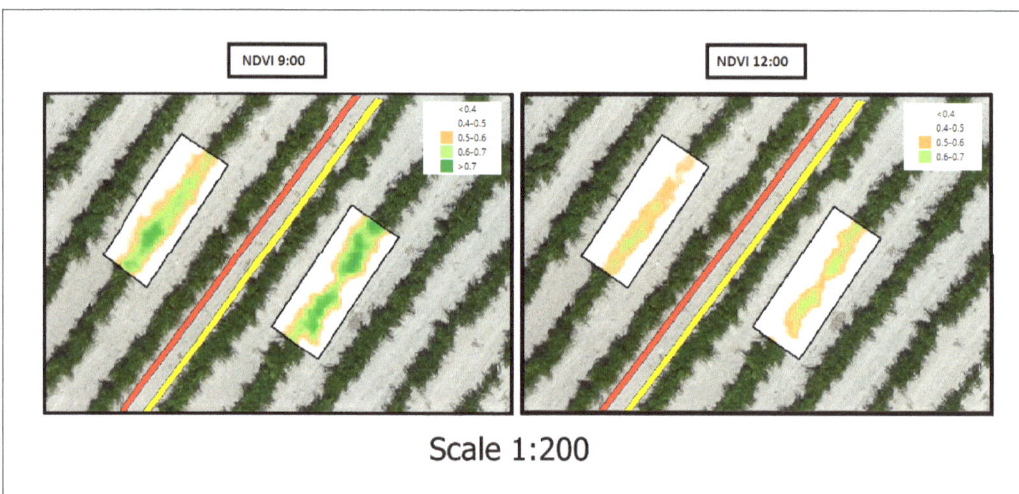

Figure 5. Details of the NDVI of two pruning treatments at two times of day. Left imagen corresponding to values of the NDVI at 9:00. Right imagen corresponding to values of the NDVI at 12:00. The no pruning treatment is located near the red line; the pruning treatment is located near the yellow line.

3.6. *Stem Water Potential Estimation*

3.6.1. Simple Linear Regression Models

The results of the calculated R^2 values for the simple linear regressions are depicted in Table 8. Different approaches were developed concerning the data included in the models. Seven different single-variable regression models were performed for each VI value and band reflection (four single-season and single-time, two multi-seasonal and single-time, and one multi-seasonal and multi-time). Even though our objective in this work was to develop a multi-seasonal robust model, the Ψ_{stem} was also predicted for individual seasons and flying times to better understand the results.

Table 8. Coefficients of determination (R^2) for the simple linear regression models where the Ψ_{stem} is predicted using the five calculated VIs (NDVI, RDVI, TCARI, OSAVI and NDRE) and the four individual bands (red, green, red edge and NIR) in different model-type scenarios.

Model Type	**Single-Season and Single-Time**				**Multi-Season 2021–2022**		
	2021 *		**2022 ****		**Single Time *****		**Both Times**
Indexes	**9:00**	**12:00**	**9:00**	**12:00**	**9:00**	**12:00**	**9:00–12:00**
NDVI	0.5	0.65	0.62	0.65	0.44	0.58	0.43
RDVI	0.49	0.6	0.37	0.48	0.12	0.39	0.23
TCARI	0.13	0.40	0.53	0.56	0.56	0.48	0.36
OSAVI	0.73	0.65	0.52	0.58	0.25	0.49	0.33
NDRE	0.23	0.41	0.30	0.12	0.40	0.00	0.10
Bands							
RED	0.02	0.47	0.71	0.68	0.69	0.64	0.38
GREEN	0.27	0.21	0.62	0.72	0.50	0.62	0.18
RED EDGE	0.51	0.25	0.0008	0.12	0.03	0.08	0.00
NIR	0.46	0.42	0.03	0.02	0.00	0.08	0.04

* n = 40, ** n = 48, *** n = 88.

In most cases, the correlations between the Ψ_{stem} and the different VIs were stronger at solar midday. The NDVI and OSAVI were, on average, the tightest-fitting VIs (R^2: 0.55 and 0.51, respectively). Not surprisingly, when modelling the Ψ_{stem} with the combined VI data from 2021 and 2022 or when integrating the measuring times, the R^2 values decreased due to higher variability being added to the models.

Concerning the use of individual bands to estimate the Ψ_{stem}, the red and green bands obtained the best results (average R^2: 0.51 and 0.45, respectively) and were the least variable between the models. The red-edge and NIR bands only obtained good correlations in 2021, with exceptionally low values when data from 2022 were included.

Simple linear regression models with the vegetation indexes proved more useful to estimate the Ψ_{stem} when data from both seasons and times were included (n = 176) than individual bands, with the NDVI and the red band providing the best fit in each category (R^2: 0.43 and 0.38, respectively).

3.6.2. Multiple-Variable Regression Modelling with Spectral Bands

A multivariable linear regression model was developed to estimate the Ψstem to improve the quality of the simple regression models (Table 8), especially concerning the multi-seasonal and multi-temporal approach (n = 176). Only individual bands were considered independent variables, as the VIs are a linear combination of individual band reflectance values.

The initial procedure included two categorical factors: measurement time and pruning treatment. However, the pruning treatment was not statistically significant at the 95% confidence level. Therefore, it was removed from the categorical factors, and only the measuring time was included. The difference in the average Ψ_{stem} values at 9:00 and 12:00 over the two seasons is statistically significant (−0.92 and −1.32 MPa, respectively), dividing the measuring time into two homogeneous groups to include it as a categorical factor. On the other hand, the Ψ_{stem} values divided by the pruning treatment could not be used to identify two homogeneous groups.

Of all the possible combinations of bands, the red and red edge yielded the best quality results (highest R^2 adjusted for degrees of freedom and lowest AIC, HQC, and SBIC). The equation for the fitted model is presented below, and Table 9 summarises the statistical metrics used to evaluate the model performance:

$$\Psi stem\ (MPa) = 0.199484 * T + [-0.365892 + 1.4701 *\ RED\ EDGE - 14.8059 * RED] \quad (3)$$

where:

$$\begin{cases} T = 1\ if\ time\ is\ 9:00 \\ T = -1\ if\ time\ is\ 12:00 \end{cases}$$

where T is the categorical factor for the time of day the data were obtained.

Table 9. Performance statistical metrics of the multivariable linear regression model used to predict the Ψ_{stem} with data from 2021 and 2022 and two flying times, 9:00 and 12:00 solar time: n, R^2 and R^2 adjusted for degrees of freedom (R^2 adj.), standard error of estimates (S_e), mean square error (MSE), mean absolute error (MAE) and Durbin–Watson Statistic (DW).

Metric	Value
n	176
R^2	0.72
R^2 adj.	0.72
S_e	0.215
MSE	0.05
MAE	0.17
DW	1.53

This model meets the imposed conditions stated beforehand: Durbin–Watson Statistic >1.5 and VIF for every parameter <10. Therefore, it is accepted as a viable and quality-fitting model. A comparison of the observed and predicted Ψ_{stem} using this model is represented in Figure 6. Noticeably, its coefficient of determination is impressively improved compared to the single-variable regression models. The model, using the red and red-edge bands as independent variables, can explain 72% (R^2 adj.) of the variance of the dependent variable Ψ_{stem}, independently of the pruning management carried out.

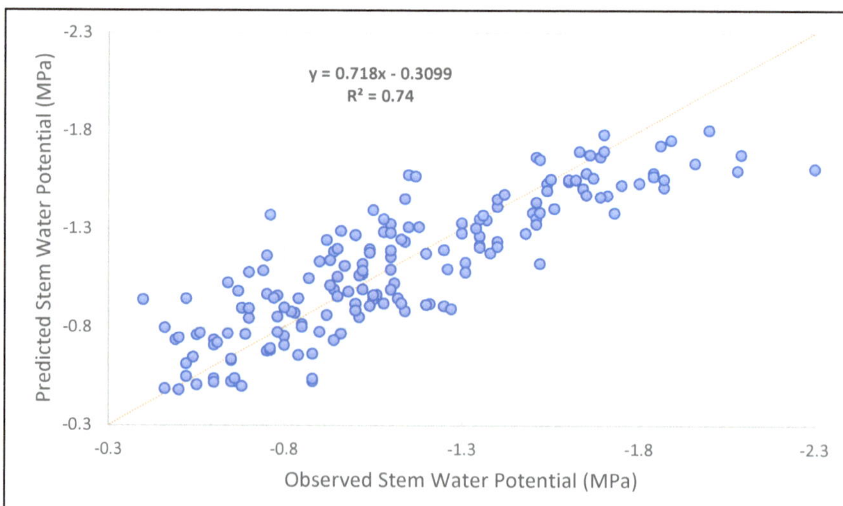

Figure 6. Plot of the observed vs. predicted Ψ_{stem} using the developed multivariable regression model with the red and red-edge bands as independent variables (n = 176). The dashed line indicates a 1:1 slope.

A visual representation of the model for the different measuring times is possible, given that it only has two dependent variables (Figures 7 and 8, for 9:00 and 12:00 solar time, respectively). As observed, lower Ψ_{stem} values are associated with lower red-edge and higher red reflection values. Conceptually, both models are parallel planes separated by the TIME factor, 0.40 MPa.

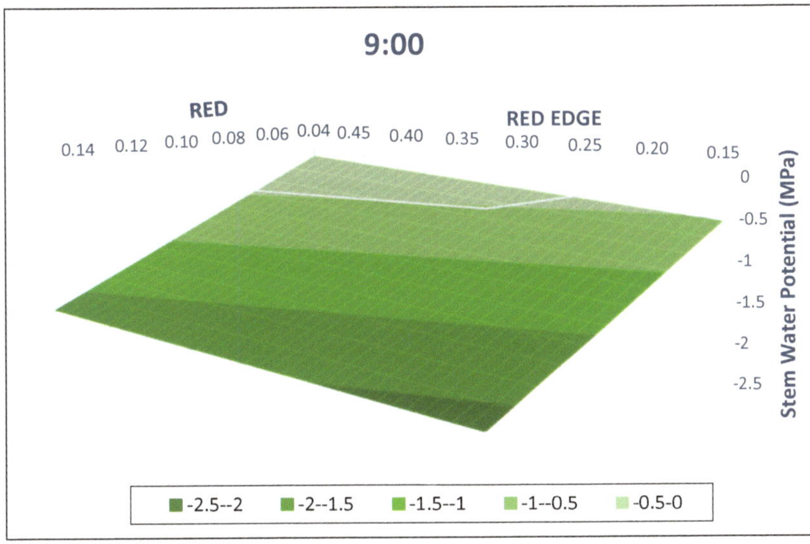

Figure 7. Visual representation of the multivariable linear regression model developed at 9:00. Horizontal axes correspond to the red and red-edge reflection values, and the vertical axis to the stem water potential (MPa). The model developed includes data from 2021 and 2022 and two pruning treatments for commercial vineyards in Yepes (Spain).

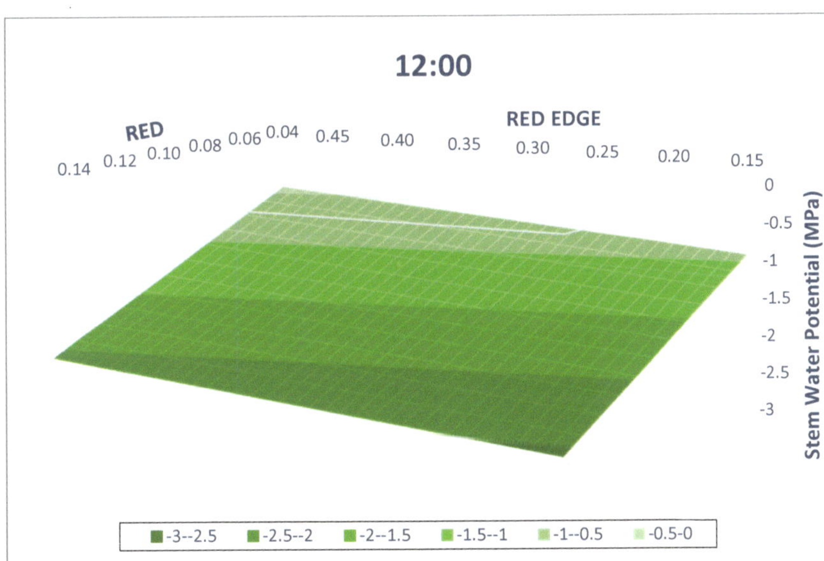

Figure 8. Visual representation of the multivariable linear regression model developed at 12:00. Horizontal axes correspond to the red and red-edge reflection values, and the vertical axis to the stem water potential (MPa). The model developed includes data from 2021 and 2022 and two pruning treatments commercial vineyard in Yepes (Spain).

4. Discussion

Both years show that climatic conditions in the studied area can be extremely harsh (>40 °C and high evaporative demand). Therefore, optimal water management would be necessary to avoid water stress incidences. The daily ETo values ranged from 6.3 to

9.1 (mm/day) (Table 2). However, it must be noted that the irrigation and rainfall in this commercial vineyard do not fulfil the evaporative demands, and therefore, the vines are grown under a deficit irrigation regime. Both seasons registered similar rainfall (around 348 mm), but in 2021, the annual ETo was lower than in 2022 (1284 vs. 1393 mm), and the irrigation supply was higher (128 vs. 82 mm).

Our results indicate that the vine water status presented a wide range of values in both seasons. In 2021, the measured solar midday Ψ_{stem} values oscillated between −0.8 MPa (beginning of the measuring campaign) and −1.6 MPa (last measurement day). According to the thresholds described by [40], these values correspond to non-stressed vines and intense water stress, respectively. In 2022, the solar midday Ψ_{stem} readings were noticeably lower, ranging from −0.9 MPa to −1.9 MPa, which can be interpreted as non-stressed and severely stressed vines, respectively.

Vine water use is majorly determined by atmospheric conditions (VPD) and plant structural characteristics like the canopy size or disposition. The VPD is an integrating climatic parameter more tightly associated with physiological variables such as the stomatal conductance or Ψ_{stem}. Ref. [41] concluded that around 71% of the variability in the Ψ_{stem} could be explained by the VPD, and this correlation was not affected by the location or cultivar. Given that the climatic and irrigation conditions were equal for both pruning treatments, it could be assumed that the differences in their water status were caused almost solely by canopy management. However, climatic data can be used to explain the general behaviour of the crop during the day and throughout the campaign. The atmospheric conditions were clearly reflected in the Ψ_{stem} values, evidencing their intrinsic relationship. The 2022 campaign was subject to higher evaporative demands (Table 2), resulting in lower Ψ_{stem} values and vine water status. The minimal Ψ_{stem} values in 2022 were 54.5 and 18.8% lower than in 2021 for the 9:00 and 12:00 measures, respectively. It is also remarkable that the lowest value at 9:00 in 2022 (−1.7 MPa) was even smaller than the lowest 12:00 value in 2021 (−1.6 MPa).

The results from both campaigns indicate that the Ψ_{stem} decreases throughout the day due to transpiration, obtaining lower values in almost all cases in the midday measure for the same treatment (Table 4). This tendency has been widely observed and studied [9]. In our study, the midday value was never higher than the morning one, only equal in one event (15 July 2022, pruning treatment). The midday Ψ_{stem} reading was 58.3% and 12.9% lower than at 9:00 in 2021 and 2022, respectively. Considering both seasons, the average Ψ_{stem} values significantly differed between the measuring times: −0.92 and −1.32 MPa at 9:00 and 12:00, respectively. This indicates that the Ψ_{stem} reduction during the day is less notable under more stressed conditions, as with the VPD (Table 2). The average VPD values were 2.7 and 4.5 kPa at 9:00 and 12:00, respectively. This is also evidenced by the fact that in both campaigns, the treatment with the lowest average Ψ_{stem} values presented the most negligible perceptual differences between the measuring times (no pruning in 2021, pruning in 2022).

The effect of pruning severity on canopy development (Table 3) did reveal some interesting differences that were not statistically significant in most cases. Ref. [42] concluded that less intensive pruning approaches resulted in higher nodes and shoots per vine but shorter shoot lengths. This can explain our results: the pruning treatment vines grew longer shoots, which developed more laterally, resulting in a higher percentage of soil coverage, although it has not been reflected in the rest of the parameters of the geometry of the vine, which did not show significant differences. These results suggest that remote-sensing tools may be more accurate in determining the breathable surface of plants than field measurements. Remote sensing has used airborne imagery to map the relative differences in vine canopies, which are used to characterise the grapevine canopy shape and vegetative expression throughout a vineyard [32].

Concerning the effect of water stress on vine production and berry quality (Table 6), our results are in agreement with other evaluated studies. Like [6], we found that more heavily water-stressed vines (no pruning treatment in 2021; pruning treatment in 2022) reported a lower yield per plant and smaller berries in both study seasons. The same results were observed by [26], who also reported that irrigated vines maintained at Ψ_{stem} values above -1.0 MPa until harvest produced higher berry flesh mass and had a lower skin to pulp relation. Regarding the fact that the no pruning treatment had a higher yield in 2022 compared to the pruning treatment, it could be explained by a higher precipitation in the spring of 2022 compared to 2021 (Figure 1). This could have influenced the sprouting of buds, greater number of bunches and then the bunch weight [1]

In our study, the Ψ_{stem} variation was not caused by irrigation management but by pruning management; however, the physiological response of the vines regarding production was the same.

Many studies have addressed the relationship between the VIs and healthy vegetative growth of crops. Our study observed that the VIs exhibited higher values between the two pruning treatments (Figure 4). However, when extreme climatic conditions occur, as in the 2022 season, plant stress increases. Such conditions affected the behaviour of the VIs and not all of them could detect the water stress experienced by the vines [27]. This stress, however, was detected by the Ψ_{stem} measurement.

The 2022 conditions limited the performance in the VIs. Severe water deficits decrease the leaf area and intercepted light [28] but also induce stomatal closure, which limits photosynthesis [25]. In addition, in grapevines, changes in the leaf pigment composition have been associated with water stress [43–45]. This situation makes it difficult to detect differences between treatments. However, another factor to consider in the evolution of the indexes is the angle of illumination from the sun, as it is not the same at the beginning of summer as at the end.

The central part of our study, the prediction of the vine Ψ_{stem} using high-resolution (12 cm/pixel) multispectral sensors mounted on UAVs, proved to be a successful tool in vineyard water status variability management. However, in most cases, the prediction of the Ψ_{stem} was stronger at solar noon than at 9:00 (Table 8), possibly due to the better quality spectral data derived from a more vertical sun position.

These values show that the NDVI is the one most strongly correlated with vine water status. However, indexes such as the OSAVI and RDVI, which have shown significant differences between the treatments in conditions where the Ψ_{stem} did not exhibit differences, suggest that they are related to more structural aspects than physiological ones of the crop. This would explain why both the OSAVI and RDVI were not influenced by the year (Table 7). Indexes integrating the red and NIR bands are often more associated with the structural characteristics of plants, such as the vigour, biomass, leaf area, etc. [32,46]. The NDVI includes these bands, although its use has also been linked to canopy vigour [32]; in our study, it was also able to correlate with water status. Other authors have found that the NDVI can be a good indicator of the plant water status [21,47].

The NDVI may be more related to the amount of intercepted light than to the leaf area or biomass. Ref. [43] observed that the Ψp vs. NDVI relationship was consistent across their study, vegetative development was strongly determined by water availability, and the vegetation index NDVI effectively characterised the effects of water availability on vine canopy growth.

Other authors did not find a relationship between the NDVI and water status under severe stress conditions. Reflectance indexes such as the NDVI or the simple ratio (SR) are useful for characterising the canopy structure and pigment concentration, and, thus, the potential photosynthetic activity [43], but they have proven less useful for monitoring photosynthetic functioning under stress conditions [48,49]. Thus, these indexes have proven less useful for monitoring plant physiological status (e.g., photosynthesis and/or water status) under stress conditions [43].

Different regression model approaches yielded different results (Table 8). The single-season and single-time simple Ψ_{stem} regressions were strongly associated with the calculated Vis. For this model type, the correlations were quite variable (R^2 range: 0.12–0.73), and the OSAVI index achieved the best fit (R^2 = 0.73) for the average 9:00 measure in 2021.

Ref. [24] estimated the Ψ_{stem} for a single day with a high-resolution multispectral sensor (10 cm/pixel) multiple camera array (MCA-6, Tetracam, Inc., California, USA) on board a UAV but obtained an R^2 value of only 0.35 when using the OSAVI. However, our results were similar to theirs with respect to the capacity of the NDVI and TCARI for estimating the Ψ_{stem}. They obtained a significant fit for both indexes, with R^2 values of 0.68 and 0.45 (NDVI and TCARI, respectively). In our case, the TCARI expressed the highest correlation in the 9:00 multi-seasonal model ($R^2 = 0.56$, $n = 88$) and the NDVI at 12:00 for the same type of model ($R^2 = 0.58$, $n = 88$). When combining data from both flying times, the NDVI was once again the best index for predicting the Ψ_{stem} ($R^2 = 0.43$, $n = 176$).

Additionally, ref. [5] could relate the NDVI and stomatal conductance on a single day with a correlation of r = 0.56, revealing that VIs can predict different vine water status indicators. Proximal-sensing methodologies have also proven useful in correlating VIs with the Ψ_{stem}. For example, ref. [50] reported a significant linear relationship between the NDVI and Ψ_{stem} ($R^2 = 0.69$) when using a hand-held spectrometer at canopy level, which is fundamentally similar to a UAV-mounted sensor.

Contrarily to other studies [13,14,51], we did not obtain satisfactory findings from the chlorophyll concentration measurements (Table 5), nor were we able to correlate it with the VIs. This could be because their studies used more precise destructive methods for determining the leaf chlorophyll concentration levels.

Normally, there were more works relating the Ψ_{stem} with several VIs than with spectral bands. Recently, ref. [52] showed that using spectral bands as Ψ_{stem} predictors provided better results than VIs in olive orchards, using machine-learning techniques. Our results revealed some interesting features of this approach. In some cases (2022, 9:00 and 12:00; 2021 and 2022, 9:00 and 12:00) (Table 8), the R^2 values were higher when using individual bands than VIs. The red band correlated strongly with the $\Psi_{stem,}$ obtaining an R^2 as high as 0.71. On the other hand, a downside of using individual bands is that the variation in the R^2 values was higher than with the VIs: the NDVI ranged from 0.5 to 0.65, while the red band ranged from 0.02 to 0.71.

To our mind, the most significant achievement of this work has been the development of a robust multi-seasonal linear model based on the red and red-edge bands, which predicts the Ψ_{stem} better than any of the tested VIs (Table 9). Further validation is needed to ensure its applicability using linear and non-linear methods. However, in the first instance, it does seem like a very valid method for detecting spatial and temporal water status variability in vineyards.

5. Conclusions

Using high-resolution (12 cm/pixel) multispectral imagery acquired by UAV-mounted sensors has proven useful in predicting the vine water status (Ψ_{stem}) in a semi-arid commercial vineyard in central Spain. Other authors have already found it useful in vineyards [19,20]. Our findings have revealed that simple or multiple-regression models using individual bands reflectance values and vegetation indexes yield significant correlations with the Ψstem independently of canopy development. The multispectral image acquisition time is an important factor to consider, given that the solar midday flights obtained better fittings than at 9:00. Nonetheless, by including the time as a categorical factor in our multivariable model, its effect was reduced, and a robust model was achieved. This multivariable model ($R^2 = 0.74$) with the red (620–750 nm) and red-edge (670–760 nm) bands could only be used to predict the Ψstem. Although other authors have found that the NDVI can be a good indicator of the plant water status [21,50]. In our study, the relation of the Ψstem with the NDVI is lower ($R^2 = 0.65$). This results in us being able to use cheaper custom-made sensors with just the two necessary bands.

The most significant achievement of this work has been the development of a robust multi-seasonal linear model based on the red and red-edge bands, which predicts the Ψstem better than any of the tested VIs. This research helps farmers to determine different irrigation management zones.

6. Future Research

This work opens up the possibility for future research focused on different aspects. As a future prospective, this study should be applied to other vineyards with different cultivars to validate the model. Exploration of the integration of AI and machine-learning methods should focus on automated data analysis, potentially enhancing the accuracy and enabling real-time application in diverse settings.

Author Contributions: Conceptualisation, J.C.N., L.K.A.-P., M.G.-d.-C. and A.M.T.; data curation, J.C.N.; formal analysis, J.C.N., L.K.A.-P., M.G.-d.-C. and A.M.T.; investigation, J.C.N.; methodology, J.C.N., M.G.-d.-C. and A.M.T.; resources, J.C.N.; software, J.C.N.; supervision, J.C.N., L.K.A.-P., M.G.-d.-C. and A.M.T.; validation, M.G.-d.-C. and A.M.T.; visualisation, J.C.N.; writing—original draft, J.C.N.; writing—review and editing, L.K.A.-P., M.G.-d.-C. and A.M.T. All authors have read and agreed to the published version of the manuscript.

Funding: This research was funded by Comunidad de Madrid through calls for grants for the completion of Doctorado Industrial IND2020/AMB-17341.

Data Availability Statement: Data are contained within the article.

Acknowledgments: The authors thank Bodegas y Viñedos Casa del Valle for allowing us to work in their vineyards, and the company UTW for supplying technical support and drone images.

Conflicts of Interest: Author Luz K. Atencia-Payares was employed by the company Unmanned Technical Works (UTW). The remaining authors declare that the research was conducted in the absence of any commercial or financial relationships that could be construed as a potential conflict of interest.

Appendix A

Table A1. Different vegetation indexes calculated from a high-resolution (12 cm/pixel) multispectral sensor mounted on board a UAV at two solar times (09:00 and 12:00), for two campaigns (2021 and 2022), under mechanical pruning (P) and no pruning (NP), in a commercial vineyard in central Spain (Yepes, Toledo).

		9:00 Solar Time									
Date		NDVI		RDVI		TCARI		OSAVI		NDRE	
25 June 2021	P	0.79	ns	0.69	*	0.05	ns	0.76	*	0.22	ns
	NP	0.75		0.60		0.19		0.69		0.22	
6 July 2021	P	0.82	ns	0.72	*	−0.05	*	0.79	*	0.28	ns
	NP	0.77		0.62		0.16		0.62		0.27	
20 July 2021	P	0.79	*	0.57	*	−0.02	*	0.70	*	0.21	*
	NP	0.69		0.44		0.16		0.58		0.19	
30 July 2021	P	0.78	*	0.60	*	0.02	*	0.71	*	0.21	ns
	NP	0.65		0.44		0.20		0.56		0.19	
19 August 2021	P	0.72	*	0.45	*	0.16	ns	0.59	*	0.23	*
	NP	0.52		0.29		0.18		0.40		0.18	
Average 2021	P	0.78	*	0.61	*	0.03	*	0.71	*	0.23	*
	NP	0.67		0.47		0.17		0.58		0.21	
30 June 2022	P	0.75	ns	0.64	ns	0.14	ns	0.71	ns	0.19	*
	NP	0.77		0.64		0.08		0.73		0.20	

Table A1. *Cont.*

Date		NDVI		RDVI		TCARI		OSAVI		NDRE	
		9:00 Solar Time									
15 July 2022	P	0.68	*	0.55	ns	0.29	*	0.64	ns	0.18	*
	NP	0.73		0.58		0.21		0.67		0.19	
5 August 2022	P	0.65	ns	0.56	ns	0.34	ns	0.62	ns	0.17	*
	NP	0.66		0.53		0.31		0.62		0.19	
12 August 2022	P	0.64	ns	0.52	ns	0.33	ns	0.60	ns	0.18	ns
	NP	0.64		0.51		0.32		0.59		0.18	
Average 2022	P	0.67	ns	0.56	ns	0.29	ns	0.63	ns	0.18	*
	NP	0.70		0.56		0.23		0.65		0.19	

ns: non-significant and * significant at $p < 0.05$.

Table A2. Different vegetation indexes calculated from a high-resolution (12 cm/pixel) multispectral sensor mounted on board a UAV at 12:00 solar time for two campaigns in a commercial vineyard under two pruning treatments (mechanical and no pruning) in central Spain (Yepes, Toledo).

Date		NDVI		RDVI		TCARI		OSAVI		NDRE	
		12:00 Solar Time									
25 June 2021	P	0.84	*	0.63	*	−0.26	*	0.76	*	0.22	ns
	NP	0.80		0.54		0.002		0.69		0.24	
6 July 2021	P	0.74	ns	0.58	*	0.18	ns	0.68	*	0.24	ns
	NP	0.71		0.51		0.19		0.63		0.24	
20 July 2021	P	0.71	*	0.53	*	0.15	ns	0.64	*	0.18	*
	NP	0.66		0.45		0.18		0.57		0.16	
30 July 2021	P	0.78	*	0.58	*	0.02	*	0.70	*	0.21	*
	NP	0.70		0.49		0.18		0.61		0.20	
19 August 2021	P	0.66	*	0.45	*	0.19	ns	0.57	*	0.17	ns
	NP	0.56		0.36		0.21		0.46		0.15	
Average 2021	P	0.74	*	0.55	*	0.06	*	0.67	*	0.2	ns
	NP	0.69		0.47		0.15		0.59		0.2	
30 June 2022	P	0.78	ns	0.57	ns	0.03	ns	0.69	ns	0.20	ns
	NP	0.79		0.57		−0.01		0.70		0.21	
15 July 2022	P	0.70	*	0.51	ns	0.18	*	0.62	ns	0.16	*
	NP	0.74		0.51		0.10		0.64		0.17	
5 August 2022	P	0.65	ns	0.48	ns	0.28	ns	0.59	ns	0.25	ns
	NP	0.66		0.48		0.27		0.59		0.25	
12 August 2022	P	0.65	ns	0.49	ns	0.24	ns	0.58	ns	0.17	ns
	NP	0.66		0.49		0.23		0.59		0.17	
Average 2022	P	0.68	ns	0.51	ns	0.18	ns	0.61	ns	0.19	ns
	NP	0.71		0.51		0.15		0.63		0.20	

ns: non-significant and * significant at $p < 0.05$.

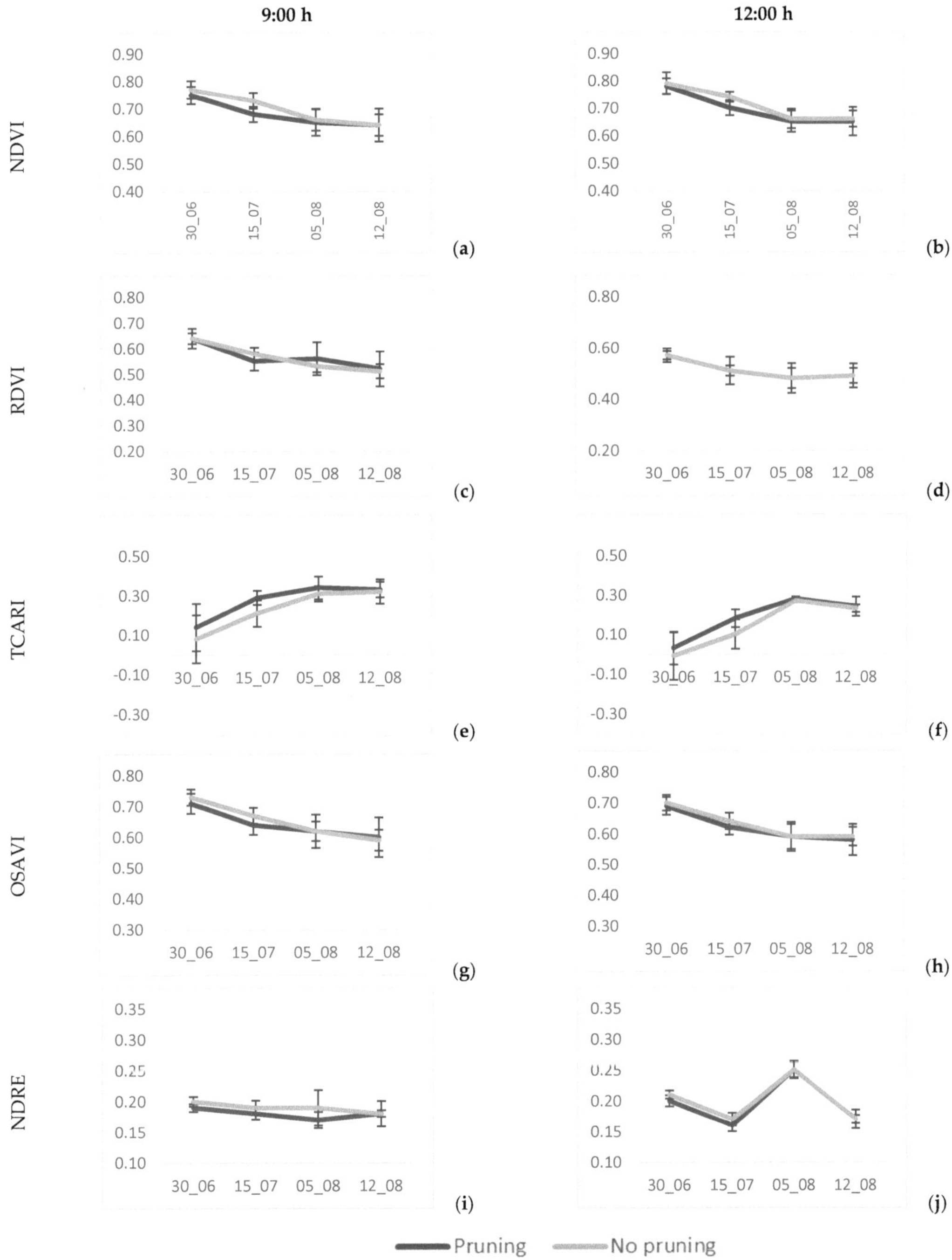

Figure A1. Evolution along the season of different vegetation indexes calculated from a high-resolution (12 cm/pixel) multispectral sensor mounted on board a UAV at 09:00 and 12:00 solar time in 2022 on a commercial vineyard under two pruning treatments (mechanical and no pruning) in central Spain (Yepes, Toledo). (**a**,**c**,**e**,**g**,**i**) in the morning, (**b**,**d**,**f**,**h**,**j**) at noon.

References

1. Ortega-Farias, S.; Mejías, R.S.; Simunovic, Y.M. Efecto de Distintos Niveles de Poda y Reposición Hídrica Sobre El Crecimiento Vegetativo, Rendimiento y Composición de Bayas En Vides Cv. Cabernet Sauvignon. *Agric. Tec.* **2007**, *67*, 401–413. [CrossRef]
2. Striegler, R.K.; Lake, C.B.; Morris, J.R.; Threlfall, R.T.; Main, G.L.; Graves, S.G. Minimal Input Production Systems Affect Yield and Juice Quality of "Sunbelt" Grapes in California's San Joaquin Valley. *HortScience* **2002**, *37*, 867–870. [CrossRef]
3. Sabbatini, P.; Wierba, K.; Clearwater, L.; Howell, G.S. Impact of Training System and Pruning Severity on Yield, Fruit Composition, and Vegetative Growth of 'Niagara' Grapevines in Michigan. *Int. J. Fruit Sci.* **2015**, *15*, 237–250. [CrossRef]
4. Munitz, S.; Schwartz, A.; Netzer, Y. Water Consumption, Crop Coefficient and Leaf Area Relations of a Vitis Vinifera Cv. "Cabernet Sauvignon" Vineyard. *Agric. Water Manag.* **2019**, *219*, 86–94. [CrossRef]
5. Espinoza, C.Z.; Khot, L.R.; Sankaran, S.; Jacoby, P.W. High Resolution Multispectral and Thermal Remote Sensing-Based Water Stress Assessment in Subsurface Irrigated Grapevines. *Remote Sens.* **2017**, *9*, 961. [CrossRef]
6. Wenter, A.; Zanotelli, D.; Montagnani, L.; Tagliavini, M.; Andreotti, C. Effect of Different Timings and Intensities of Water Stress on Yield and Berry Composition of Grapevine (Cv. Sauvignon Blanc) in a Mountain Environment. *Sci. Hortic.* **2018**, *236*, 137–145. [CrossRef]
7. Van Leeuwen, C.; Destrac-Irvine, A.; Dubernet, M.; Duchêne, E.; Gowdy, M.; Marguerit, E.; Pieri, P.; Parker, A.; De Rességuier, L.; Ollat, N. An Update on the Impact of Climate Change in Viticulture and Potential Adaptations. *Agronomy* **2019**, *9*, 514. [CrossRef]
8. Kliewer, W.M.; Freeman, B.M.; Hosssom, C. Effect of Irrigation, Crop Level and Potassium Fertilisation on Carignane Vines. I. Degree of Water Stress and Effect on Growth and Yield. *Am. J. Enol. Vitic.* **1983**, *34*, 186–196. [CrossRef]
9. Cole, J.; Pagay, V. Usefulness of Early Morning Stem Water Potential as a Sensitive Indicator of Water Status of Deficit-Irrigated Grapevines (*Vitis vinifera* L.). *Sci. Hortic.* **2015**, *191*, 10–14. [CrossRef]
10. Deloire, A.; Carbonneau, A.; Wang, Z.; Ojeda, H. Vine and Water a Short Review. *J. Int. des Sci. la Vigne du Vin* **2004**, *38*, 1–13. [CrossRef]
11. Medrano, H.; Bota, J.; Cifre, J.; Flexas, J.; Ribas-Carbó, M.; Gulias, J. Eficiencia En El Uso Del Agua Por Las Plantas. Available online: https://www.redalyc.org/pdf/176/17604304.pdf (accessed on 20 February 2024).
12. Knipling, E.B. Physical and Physiological Basis for the Reflectance of Visible and Near-Infrared Radiation from Vegetation. *Remote Sens. Environ.* **1970**, *1*, 155–159. [CrossRef]
13. Caruso, G.; Tozzini, L.; Rallo, G.; Primicerio, J.; Moriondo, M.; Palai, G.; Gucci, R. Estimating Biophysical and Geometrical Parameters of Grapevine Canopies ('Sangiovese') by an Unmanned Aerial Vehicle (UAV) and VIS-NIR Cameras. *Vitis—J. Grapevine Res.* **2017**, *56*, 63–70. [CrossRef]
14. Zarco-Tejada, P.J.; Berjón, A.; López-Lozano, R.; Miller, J.R.; Martín, P.; Cachorro, V.; González, M.R.; De Frutos, A. Assessing Vineyard Condition with Hyperspectral Indices: Leaf and Canopy Reflectance Simulation in a Row-Structured Discontinuous Canopy. *Remote Sens. Environ.* **2005**, *99*, 271–287. [CrossRef]
15. Matese, A.; Capraro, F.; Primicerio, J.; Gualato, G.; Gennaro, S.F.D.; Agati, G. Mapping of Vine Vigor by UAV and Anthocyanin Content by a Non-Destructive Fluorescence Technique. In *Precision Agriculture'13*; Wageningen Academic: Wageningen, The Netherlands, 2013. [CrossRef]
16. Lamb, D.W.; Weedon, M.M.; Bramley, R.G.V. Using Remote Sensing to Predict Grape Phenolics and Colour at Harvest in a Cabernet Sauvignon Vineyard: Timing Observations against Vine Phenology and Optimising Image Resolution. *Aust. J. Grape Wine Res.* **2004**, *10*, 46–54. [CrossRef]
17. Hall, A.; Louis, J.P.; Lamb, D.W. Low-Resolution Remotely Sensed Images of Winegrape Vineyards Map Spatial Variability in Planimetric Canopy Area Instead of Leaf Area Index. *Aust. J. Grape Wine Res.* **2008**, *14*, 9–17. [CrossRef]
18. Mathews, A.J.; Jensen, J.L.R. Visualizing and Quantifying Vineyard Canopy LAI Using an Unmanned Aerial Vehicle (UAV) Collected High Density Structure from Motion Point Cloud. *Remote Sens.* **2013**, *5*, 2164–2183. [CrossRef]
19. Berry, A.; Vivier, M.A.; Poblete-Echeverría, C. Evaluation of Canopy Fraction-Based Vegetation Indices, Derived from Multispectral UAV Imagery, to Map Water Status Variability in a Commercial Vineyard. *Irrig. Sci.* **2024**. [CrossRef]
20. López-García, P.; Intrigliolo, D.S.; Moreno, M.A.; Martínez-Moreno, A.; Ortega, J.F.; Pérez-álvarez, E.P.; Ballesteros, R. Assessment of Vineyard Water Status by Multispectral and Rgb Imagery Obtained from an Unmanned Aerial Vehicle. *Am. J. Enol. Vitic.* **2021**, *72*, 285–297. [CrossRef]
21. Acevedo-Opazo, C.; Tisseyre, B.; Guillaume, S.; Ojeda, H. The Potential of High Spatial Resolution Information to Define Within-Vineyard Zones Related to Vine Water Status. *Precis. Agric.* **2008**, *9*, 285–302. [CrossRef]
22. Giovos, R.; Tassopoulos, D.; Kalivas, D.; Lougkos, N.; Priovolou, A. Remote Sensing Vegetation Indices in Viticulture: A Critical Review. *Agric.* **2021**, *11*, 457. [CrossRef]
23. Arab, S.T.; Noguchi, R.; Matsushita, S.; Ahamed, T. Prediction of Grape Yields from Time-Series Vegetation Indices Using Satellite Remote Sensing and a Machine-Learning Approach. *Remote Sens. Appl. Soc. Environ.* **2021**, *22*, 100485. [CrossRef]
24. Baluja, J.; Diago, M.P.; Balda, P.; Zorer, R.; Meggio, F.; Morales, F.; Tardaguila, J. Assessment of Vineyard Water Status Variability by Thermal and Multispectral Imagery Using an Unmanned Aerial Vehicle (UAV). *Irrig. Sci.* **2012**, *30*, 511–522. [CrossRef]
25. Flexas, J.; Bota, J.; Escalona, J.M.; Sampol, B.; Medrano, H. Effects of Drought on Photosynthesis in Grapevines under Field Conditions: An Evaluation of Stomatal and Mesophyll Limitations. *Funct. Plant Biol.* **2002**, *29*, 461. [CrossRef]

26. Triolo, R.; Roby, J.P.; Pisciotta, A.; Di Lorenzo, R.; van Leeuwen, C. Impact of Vine Water Status on Berry Mass and Berry Tissue Development of Cabernet Franc (*Vitis vinifera* L.), Assessed at Berry Level. *J. Sci. Food Agric.* **2019**, *99*, 5711–5719. [CrossRef] [PubMed]
27. Pôças, I.; Rodrigues, A.; Gonçalves, S.; Costa, P.M.; Gonçalves, I.; Pereira, L.S.; Cunha, M. Predicting Grapevine Water Status Based on Hyperspectral Reflectance Vegetation Indices. *Remote Sens.* **2015**, *7*, 16460–16479. [CrossRef]
28. Pellegrino, A.; Lebon, E.; Voltz, M.; Wery, J. Relationships between Plant and Soil Water Status in Vine (*Vitis vinifera* L.). *Plant Soil* **2005**, *266*, 129–142. [CrossRef]
29. Ramón Rodríguez-Pérez, J.; Riaño, D.; Ustin, S.L. Evaluation of Hyperspectral Reflectance Indexes to Detect Grapevine Water Status in Vineyards Modelling-Rainfall-Evapotranspiration View Project Nitrous Oxide (N2O) Emissions in Croplands View Project. *Am. J. Enol. Vitic.* **2006**, *58*, 302–317. [CrossRef]
30. Borgogno-Mondino, E.; Lessio, A.; Tarricone, L.; Novello, V.; de Palma, L. A Comparison between Multispectral Aerial and Satellite Imagery in Precision Viticulture. *Precis. Agric.* **2018**, *19*, 195–217. [CrossRef]
31. Khaliq, A.; Comba, L.; Biglia, A.; Ricauda Aimonino, D.; Chiaberge, M.; Gay, P. Comparison of Satellite and UAV-Based Multispectral Imagery for Vineyard Variability Assessment. *Remote Sens.* **2019**, *11*, 436. [CrossRef]
32. Hall, A.; Lamb, D.W.; Holzapfel, B.; Louis, J. Optical Remote Sensing Applications in Viticulture—A Review. *Aust. J. Grape Wine Res.* **2002**, *8*, 36–47. [CrossRef]
33. Helman, D.; Bahat, I.; Netzer, Y.; Ben-Gal, A.; Alchanatis, V.; Peeters, A.; Cohen, Y. Using Time Series of High-Resolution Planet Satellite Images to Monitor Grapevine Stem Water Potential in Commercial Vineyards. *Remote Sens.* **2018**, *10*, 1615. [CrossRef]
34. Atencia Payares, L.K.; Tarquis, A.M.; Hermoso Peralo, R.; Cano, J.; Cámara, J.; Nowack, J.; Gómez del Campo, M. Multispectral and Thermal Sensors Onboard UAVs for Heterogeneity in Merlot Vineyard Detection: Contribution to Zoning Maps. *Remote Sens.* **2023**, *15*, 4024. [CrossRef]
35. Rouse, J.W.; Haas, R.H.; Schell, J.A.; Deering, D.W.; Harlan, J.C. *Monitoring the Vernal Advancements and Retrogradation of Natural Vegetation*; NASA/GSFC, Final Report; NASA: Greenbelt, MD, USA, 1974; pp. 1–137.
36. Roujean, J.L.; Breon, F.M. Estimating PAR Absorbed by Vegetation from Bidirectional Reflectance Measurements. *Remote Sens. Environ.* **1995**, *51*, 375–384. [CrossRef]
37. Haboudane, D.; Miller, J.R.; Tremblay, N.; Zarco-Tejada, P.J.; Dextraze, L. Integrated Narrow-Band Vegetation Indices for Prediction of Crop Chlorophyll Content for Application to Precision Agriculture. *Remote Sens. Environ.* **2002**, *81*, 416–426. [CrossRef]
38. Rondeaux, G.; Steven, M.; Baret, F. Optimization of Soil-Adjusted Vegetation Indices. *Remote Sens. Environ.* **1996**, *55*, 95–107. [CrossRef]
39. Barnes, E.M.; Clarke, T.R.; Richards, S.E.; Colaizzi, P.D.; Haberland, J.; Kostrzewski, M.; Waller, P.; Choi, C.; Riley, E.; Thompson, T.; et al. Coincident Detection of Crop Water Stress, Nitrogen Status and Canopy Density Using Ground Based Multispectral Data. In Proceedings of the Fifth International Conference on Precision Agriculture, Bloomington, MN, USA, 16–19 July 2000.
40. Medrano, H.; Flexas, J.; Pou, A.; Tomàs, M.; Martorell, S.; Gulias, J. Agua y Vino: Mejorar La Eficiencia Del Uso Del Agua Aumenta La Calidad de La Cosecha. *ACTAS Hortic.* **2010**, *60*, 599–602.
41. Williams, L.E.; Baeza, P. Relationships among Ambient Temperature and Vapor Pressure Deficit and Leaf and Stem Water Potentials of Fully Irrigated, Field-Grown Grapevines. *Am. J. Enol. Vitic.* **2007**, *58*, 173–181. [CrossRef]
42. Intrieri, C.; Poni, S.; Lia, G.; Del Campo, M.G. Vine Performance and Leaf Physiology of Conventionally and Minimally Pruned Sangiovese Grapevines. *Vitis* **2001**, *40*, 123–130.
43. Serrano, L.; González-Flor, C.; Gorchs, G. Assessing Vineyard Water Status Using the Reflectance Based Water Index. *Agric. Ecosyst. Environ.* **2010**, *139*, 490–499. [CrossRef]
44. Escalona, J.M.; Flexas, J.; Medrano, H. Stomatal and Non-Stomatal Limitations of Photosynthesis under Water Stress in Field-Grown Grapevines. *Funct. Plant Biol.* **2000**, *27*, 87. [CrossRef]
45. Schultz, H.R. Water relations and photosynthetic responses of two grapevine cultivars of different geographical origin during water stress. *Acta Hortic.* **1996**, *427*, 251–266. [CrossRef]
46. Rossini, M.; Fava, F.; Cogliati, S.; Meroni, M.; Marchesi, A.; Panigada, C.; Giardino, C.; Busetto, L.; Migliavacca, M.; Amaducci, S.; et al. Assessing Canopy PRI from Airborne Imagery to Map Water Stress in Maize. *ISPRS J. Photogramm. Remote Sens.* **2013**, *86*, 168–177. [CrossRef]
47. Bianchi, D.; Brancadoro, L. Water Use Efficiency and Nutritional Status of a New Grapevine Rootstock Selection. *Horticulturae* **2021**, *7*, 503. [CrossRef]
48. Gamon, J.A.; Field, C.B.; Goulden, M.L.; Griffin, K.L.; Hartley, A.E.; Joel, G.; Penuelas, J.; Valentini, R. Relationships Between NDVI, Canopy Structure, and Photosynthesis in Three Californian Vegetation Types. *Ecol. Appl.* **1995**, *5*, 28–41. [CrossRef]
49. Gamon, J.A.; Serrano, L.; Surfus, J.S. The Photochemical Reflectance Index: An Optical Indicator of Photosynthetic Radiation Use Efficiency across Species, Functional Types, and Nutrient Levels. *Oecologia* **1997**, *112*, 492–501. [CrossRef] [PubMed]
50. Rodríguez-Pérez, J.R.; Riaño, D.; Carlisle, E.; Ustin, S.; Smart, D.R. Evaluation of Hyperspectral Reflectance Indexes to Detect Grapevine Water Status in Vineyards. *Am. J. Enol. Vitic.* **2007**, *58*, 302–317. [CrossRef]

51. Zarco-Tejada, P.J.; Guillén-Climent, M.L.; Hernández-Clemente, R.; Catalina, A.; González, M.R.; Martín, P. Estimating Leaf Carotenoid Content in Vineyards Using High Resolution Hyperspectral Imagery Acquired from an Unmanned Aerial Vehicle (UAV). *Agric. For. Meteorol.* **2013**, *171*, 281–294. [CrossRef]
52. Garofalo, S.P.; Giannico, V.; Costanza, L.; Alhajj Ali, S.; Camposeo, S.; Lopriore, G.; Pedrero Salcedo, F.; Vivaldi, G.A. Prediction of Stem Water Potential in Olive Orchards Using High-Resolution Planet Satellite Images and Machine Learning Techniques. *Agronomy* **2024**, *14*, 1. [CrossRef]

plants

MDPI

Article

Optimizing the Mulching Pattern and Nitrogen Application Rate to Improve Maize Photosynthetic Capacity, Yield, and Nitrogen Fertilizer Utilization Efficiency

Hengjia Zhang [1,*,†], Tao Chen [1,†], Shouchao Yu [1], Chenli Zhou [1], Anguo Teng [2], Lian Lei [2] and Fuqiang Li [3]

1 College of Agronomy and Agricultural Engineering, Liaocheng University, Liaocheng 252059, China; gsauct@163.com (T.C.); ysc@lcu.edu.cn (S.Y.); zhouchenli2021@126.com (C.Z.)
2 Yimin Irrigation Experimental Station, Zhangye 734500, China; gsmltag@163.com (A.T.); ll4426072@163.com (L.L.)
3 College of Water Conservancy and Hydropower Engineering, Gansu Agricultural University, Lanzhou 730070, China; lifuq@gsau.edu.cn
* Correspondence: zhanghengjia@lcu.edu.cn
† These authors contributed equally to this work.

Abstract: Residual film pollution and excessive nitrogen fertilizer have become limiting factors for agricultural development. To investigate the feasibility of replacing conventional plastic film with biodegradable plastic film in cold and arid environments under nitrogen application conditions, field experiments were conducted from 2021 to 2022 with plastic film covering (including degradable plastic film (D) and ordinary plastic film (P)) combined with nitrogen fertilizer 0 (N0), 160 (N1), 320 (N2), and 480 (N3) kg·ha^{-1}. The results showed no significant difference ($p > 0.05$) in dry matter accumulation, photosynthetic gas exchange parameters, soil enzyme activity, or yield of spring maize under degradable plastic film cover compared to ordinary plastic film cover. Nitrogen fertilizer is the main factor limiting the growth of spring maize. The above-ground and root biomass showed a trend of increasing and then decreasing with the increase in nitrogen application level. Increasing nitrogen fertilizer can also improve the photosynthetic gas exchange parameters of leaves, maintain soil enzyme activity, and reduce soil pH. Under the nitrogen application level of N2, the yield of degradable plastic film and ordinary plastic film coverage increased by 3.74~42.50% and 2.05~40.02%, respectively. At the same time, it can also improve water use efficiency and irrigation water use efficiency, but it will reduce nitrogen fertilizer partial productivity and nitrogen fertilizer agronomic use efficiency. Using multiple indicators to evaluate the effect of plastic film mulching combined with nitrogen fertilizer on the comprehensive growth of spring maize, it was found that the DN2 treatment had the best complete growth of maize, which was the best model for achieving stable yield and income increase and green development of spring maize in cold and cool irrigation areas.

Keywords: film mulching; nitrogen; maize; yield; nitrogen use efficiency; soil quality

Citation: Zhang, H.; Chen, T.; Yu, S.; Zhou, C.; Teng, A.; Lei, L.; Li, F. Optimizing the Mulching Pattern and Nitrogen Application Rate to Improve Maize Photosynthetic Capacity, Yield, and Nitrogen Fertilizer Utilization Efficiency. *Plants* **2024**, *13*, 1241. https://doi.org/10.3390/plants13091241

Academic Editor: Youzhen Xiang

Received: 5 March 2024
Revised: 11 April 2024
Accepted: 23 April 2024
Published: 30 April 2024

1. Introduction

As one of the C4 crops with the most extensive planting area in the world, maize is a crop for food and feed and an essential source of industrial raw materials [1]. China has become the second largest producer of maize, with the planting area accounting for more than 30% of the national grain crop planting area, reaching over 40 million hm^2 [2]. (https://www.stats.gov.cn/sj/. accessed on 25 March 2024). Thus, increasing maize yield is essential to ensuring food security, achieving self-sufficiency in food supply, and stabilizing economic development. However, the frequent occurrence of extreme drought and the shrinking of arable land area has brought enormous pressure on agricultural production and even caused decreases in food production [3,4]. How to alleviate the pressure of reduced grain production and implement a food security strategy is a major challenge currently faced.

Plastic film mulching is a key technology to improve crop yield [5] and change the agricultural production mode in areas with water shortages [6]. Plastic film mulching may improve soil moisture and heat status, promote the decomposition and transformation of soil organic matter, improve soil nutrient content, enzyme activity, and microbial richness, inhibit weed growth, reduce nutrient competition, and create a good soil environment for crop growth [7]. Plastic film covering not only increases crop yield but also hinders gas exchange between soil and atmosphere, enhances crop root respiration intensity, and strengthens soil nitrification and denitrification processes to increase N_2O and CH_4 gas emission pressure [8]. Ordinary film used in mulching is mainly made of polyethylene, which is degraded very slowly in soil [9,10]. However, long-term plastic film mulching and a lack of effective recovery measures make the residue of farmland mulching plastic film a critical environmental problem. Film residue may destroy the continuity of soil pores [11], change the composition of the soil microbial community [12], affect water [13] and nutrient migration [14], hinder seed germination [15] and crop root development, and ultimately reduce crop yield [16], even threatening food security. To solve the adverse effects caused by continuous plastic film covering for many years, it is essential to develop and apply new covering materials, such as degradable plastic film.

As a new plastic film, the degradable type could be degraded into CO_2 and H_2O with the aid of the soil and natural environments [17], which could alleviate the pressure on the agricultural ecological environment and was considered an effective way to solve the problem of farmland residual film concentration. Therefore, the degradable plastic film has been applied in many countries, such as China [18], Italy [19], Thailand [20], etc. In addition, the soil water and heat preservation effects of degradable film are equivalent to those of ordinary film [21], which can effectively improve the soil moisture and heat status [22]. In the middle and late stages of crop growth, the soil moisture and heat of the degradable plastic film were lower than those of the ordinary due to the expansion of the degradation area [23]. The difference in crop yield and water use efficiency was slight between the degradable film mulching and the ordinary [1]. Also, soil nitrate nitrogen accumulation could even be reduced under degradable film mulching [24]. At present, nitrogen fertilizer application is widespread in agricultural production to maintain high yields. However, excessive application of nitrogen fertilizer does not significantly improve crop yield and may result in yield reduction; furthermore, surplus nitrogen might be discharged into the atmosphere in gaseous form, causing environmental pollution [25]. In addition, the low nitrogen use efficiency and decline in recovery rate would lead to more soil residual nitrogen or nitrogen leaching [26], resulting in soil salinization and groundwater pollution [27,28]. Therefore, optimizing nitrogen application rate and improving nitrogen utilization are of great significance for improving grain quality, efficiency, and environmental protection.

Film mulching combined with nitrogen fertilizer application is an essential measure in agricultural production, which can significantly improve crop yield and water use efficiency, increase soil microbial nitrogen content and particulate organic nitrogen, and improve soil fertility, conducive to sustainable development of the agricultural system. Therefore, the objectives of this study were to determine: (1) effects of ordinary and degradable plastic film on dry matter accumulation and physiological aspects of spring maize under different nitrogen application gradients; (2) performance in crop yield, water and nitrogen use efficiency, and soil quality subjected to nitrogen application; and (3) the possibility of degradable plastic film replacing the ordinary by multiple indicators.

2. Materials and Methods

2.1. Description of the Study Site

The trial was conducted at the Yimin irrigation experimental station in Minle County, Gansu Province, China, from April 2021 to October 2022. The area is located at 100°43′ east longitude, 38°39′ north latitude, and 1970 m above sea level, belonging to a temperate continental climate (Figure 1). The average annual precipitation is about 200 mm, with the evaporation of 1680–2270 mm; the average sunshine duration is about 2592–2997 h;

the average yearly temperature is 3.4–5.6 °C; and the frost-free period is about 78–188 d. The tested soil was light loam, with a maximum field water capacity of 24% and a soil bulk density of 1.46 g·cm^{-3} in topsoil. The soil pH is 7.2 and soil fertility is medium within 0–20 cm soil layer, with organic matter of 12.6 g·kg^{-1} and the available phosphorus, potassium, and alkali hydrolyzed nitrogen of 15.8 mg·g^{-1}, 192.1 mg·kg^{-1} and 57.5 mg·kg^{-1}, respectively. The rainfall in 2021 and 2022 was 244.7 mm and 237.4 mm, respectively (Figure 2).

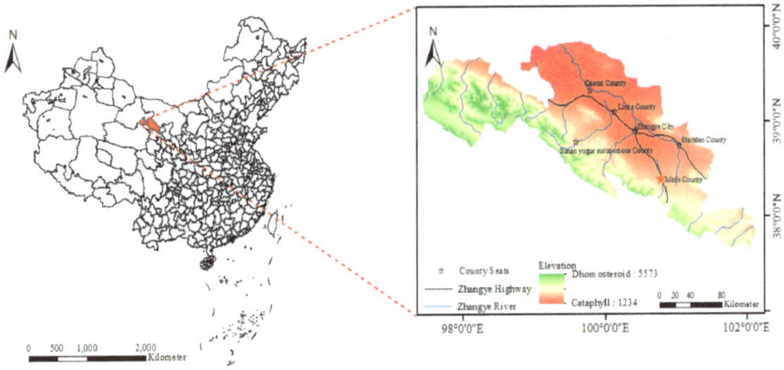

Figure 1. Location of the experimental site. Red star represents the city where the experimental site is located.

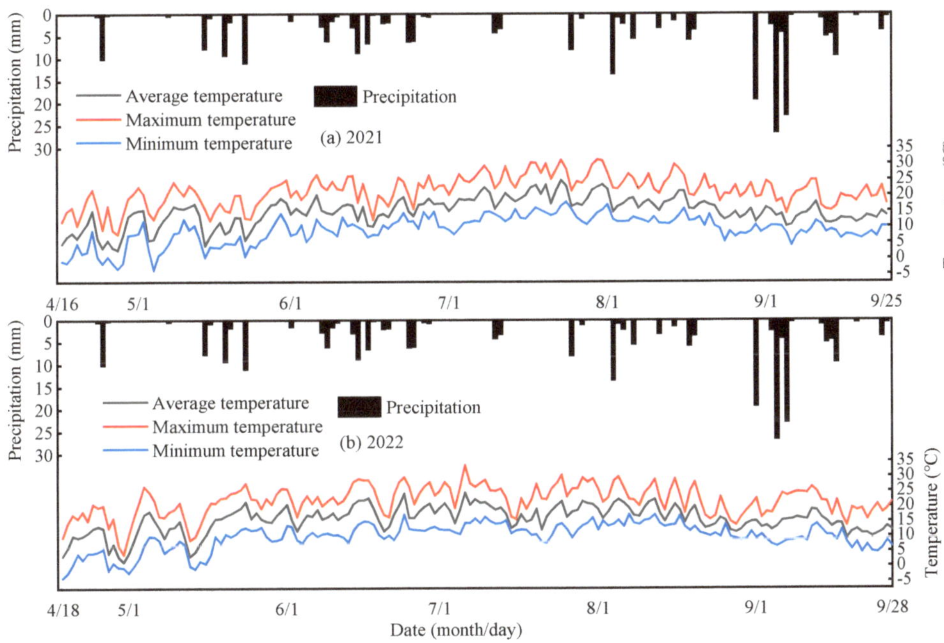

Figure 2. Precipitation and temperature during the spring maize growth period in 2021 (**a**) and 2022 (**b**).

2.2. Experimental Design and Field Management

The spring maize crop was film-mulched and fertilized with nitrogen. There were two kinds of mulching film: the ordinary mulching film and the egradable mulching film (produced by Shandong Tianzhuang environmental protection Co., Ltd. with a thickness of 0.008 mm and a width of 70 cm, Jinan, China), respectively recorded as P, D. There were four nitrogen application levels: 0, 160, 320, and 480 kg·ha^{-1}, respectively, recorded as N0, N1, N2, and N3. There were 8 treatments in total, with 3 replications. There were 18 plots with each area of 28 m^2 (2 m × 14 m), a 0.2 m interval between communities. The field plots were arranged with random blocks. The spring maize was planted with a row spacing of 40 and plant spacing of 35cm, and a planting density of 74,000 plants per hectare, sown on 16 April and harvested on 25 September 2021 or sown on 18 April and harvested on 28 September 2022, respectively. The fields were rotary tilled and leveled before sowing. The application amounts of phosphorus fertilizer (P_2O_5) and potassium fertilizer (K_2O) were the same at 120 kg·ha^{-1} and 80 kg·ha^{-1}, respectively, and all fertilizers were supplied as base fertilizer, which was applied to the soil when the soil was turned. Nitrogen fertilizer was applied four times at different growth stages, namely, 20% as base fertilizer, 30% at jointing, 30% at tasseling, and 20% at grain filling. The crop was watered using plastic film-mulched drip irrigation with the same irrigation amount according to 100%ETc (ETc = Kc × ET0, ET0 is calculated based on Penman–Monteith equation recommended by FAO, while Kc refers to the standard of the China Meteorological Administration; the Kc values for April, May, June, July, August, and September are 0.3, 0.4, 0.88, 1.26, 1.25, and 0.73, respectively) [29] (https://hbba.sacinfo.org.cn/, accessed on 10 January 2024). The meteorological parameters were provided by the micro-meteorological instrument system in the experimental station. The effective rainfall in the 2021 and 2022 crop growing seasons was 138.86 mm and 123.66 mm, respectively (Figure 3), and the irrigation amount in the above two growing seasons was 627 mm and 609 mm, respectively.

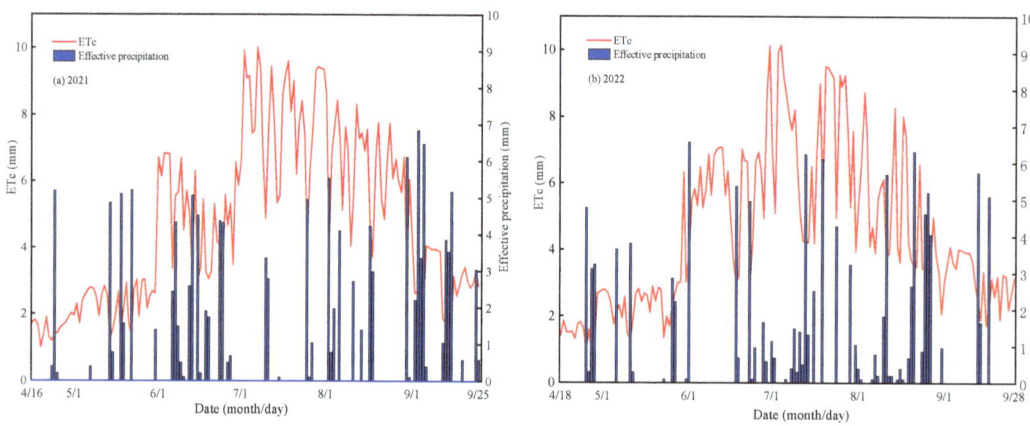

Figure 3. ETc and the effective precipitation during the spring maize growth period in 2021 (**a**) and 2022 (**b**).

2.3. *Measurements and Calculations*

2.3.1. Above-Ground and Underground Biomass

Three plants were randomly selected with uniform jointing, tasseling, and grain filling of spring maize. The root sampling area was 15 × 15 cm around the plant, and the sampling depth was determined according to the depth of spring maize roots. The plants were decomposed into different organs, then killed at 105 °C for 30 min, and finally dried at 80 °C to a constant weight. The dry weight of each organ was weighed, and the root/shoot ratio was calculated according to equation R/S(%) = root biomass/above-ground biomass.

2.3.2. Photosynthetic Gas Exchange Characteristics

Photosynthetic gas exchange parameters at the third leaf of the spring maize ear with three repetitions were measured at 9:00–11:00 a.m. on sunny days during spring maize jointing, tasseling, and grain filling using a LI-6400 portable photosynthesis instrument, including photosynthetic rate, stomatal conductance, and transpiration rate.

2.3.3. Soil Quality

During the maize harvest period, 0–20 cm of soil was taken in the middle of two corn plants, and 3 points were randomly taken from each treatment as mixed samples to measure soil enzyme activity. The urease was measured by sodium phenol and sodium hypochlorite colorimetry. The enzyme activity was expressed by the milligrams of NH^3-N produced by 1 g of soil after incubation at 37 °C for 24 h under the action of urease. The sucrase was measured using the 3,5-dinitrosalicylic acid colorimetric method, and enzyme activity was expressed as the milligrams of glucose produced in 1 g of soil after being incubated at 37 °C for 24 h under the action of sucrase. The soil pH was measured by a pH meter (PHS-3C), with a soil mass extract of 2.5:1.

2.3.4. Grain Yield and Its Components

Ten spring maize plants were randomly selected in each plot at spring maize ripening to determine the grain yield after measuring the yield components, including grain number per ear, row number per ear, ear longitudinal diameter, and ear diameter.

2.3.5. Water and Nitrogen Use Efficiency

The crop evapotranspiration (ET, mm) was calculated using the following equation [30]:

$$ET = P + I + U - D - S + \Delta W \quad (1)$$

where P is the effective precipitation (mm); I is the amount of irrigation (mm); U is the amount of groundwater recharge (mm). The depth of groundwater is below 20 m, so groundwater recharge can be ignored. D is the amount of deep leakage (mm) (the tested area is flat, thus there is no surface runoff, therefore D = 0). ΔW is the soil water storage change between plant sowing and harvest (mm).

The water use efficiency (WUE, kg·m^{-3}) was calculated according to following formula:

$$WUE = Y/ET \quad (2)$$

where Y is spring maize grain yield (kg·ha^{-1}).

The irrigation water use efficiency (IWUE, kg·m^{-3}) was calculated using the following formula:

$$IWUE = Y/I \quad (3)$$

The nitrogen fertilizer partial productivity (NPF, kg·kg^{-1}) was calculated according to the following formula:

$$NPF = Y_N/N \quad (4)$$

where Y_N is the spring maize yield in nitrogen application area (kg·ha^{-1}), and N is the amount of nitrogen fertilizer input (kg·ha^{-1}).

The nitrogen fertilizer agronomic use efficiency (NFA, kg·kg^{-1}) was calculated using the following formula:

$$NFA = (Y_N - Y_0)Y_N/N \quad (5)$$

where Y_0 is the spring maize yield in the area without nitrogen application (kg·ha^{-1}).

2.4. Statistical Analysis

The SPSS 22.0 software was used to analyze the difference in the measured data ($p < 0.05$), and the Origin 2021 software was used for plotting. The Yaaph software (http://www.jeffzhang.cn/, accessed on 25 March 2024) was used to draw the comprehensive analysis hierarchy model of spring maize and the weight analysis of each index. The Matlab software (https://ww2.mathworks.cn/products/matlab.html, accessed on 25 March 2024) was used to calculate the weight of the combination based on the game theory and the comprehensive score of TOPSIS.

3. Result

3.1. Root and Shoot Growth

Nitrogen fertilizer is the main factor affecting spring maize root and shoot growth. The degradable plastic film was gradually degraded with spring maize growth, and the effects of different film types on spring maize growth were quite different, showing significant ($p < 0.05$) effects on spring maize growth at tasseling and grain filling and significant ($p < 0.01$) effects on spring maize root and shoot growth (Table 1).

Table 1. Significance test on spring maize root and shoot growth at different growth stages; ns means no significant difference ($p > 0.05$); * means significant at $p < 0.05$ level; ** means significant at $p < 0.01$ level.

Year	F Fest	Jointing		Tasseling		Grain Filling	
		Above-Ground Dry Matter	Root Dry Matter	Above-Ground Dry Matter	Root Dry Matter	Above-Ground Dry Matter	Root Dry Matter
2021	F	1.16 ns	3.83 ns	4.35 ns	5.12 *	5.39 *	8.16 *
	N	12.44 **	29.36 **	43.94 **	102.72 **	45.27 **	139.07 **
	F×N	0.04 ns	0.16 ns	0.16 ns	0.27 ns	0.15 ns	0.19 ns
2022	F	4.09 ns	4.31 ns	2.38 ns	5.52 *	7.01 *	10.18 **
	N	29.27 **	27.05 **	18.33 **	58.71 **	63.02 **	157.72 **
	F×N	0.02 ns	0.06 ns	0.05 ns	0.26 ns	0.20 ns	0.04 ns
	Y	34.57 **	9.79 **	3.48 ns	12.39 **	16.10 **	8.94 **
	F	4.73 *	8.05 **	5.87 *	10.55 **	12.31 *	18.20 **
	N	39.39 **	54.90 **	51.40 **	151.20 **	106.92 **	295.61 **
	Y×F	0.38 ns	0.21 ns	0.03 ns	0.19 ns	0.03 ns	0.01 ns
	Y×N	1.52 ns	0.90 ns	0.03 ns	0.25 ns	0.66 ns	0.10 ns
	F×N	0.02 ns	0.01 ns	0.16 ns	0.52 ns	0.28 ns	0.16 ns
	Y×F×N	0.05 ns	0.19 ns	0.00 ns	0.01 ns	0.07 ns	0.08 ns

3.1.1. Above-Ground Dry Matter

The spring maize above-ground dry matter accumulation showed an increasing trend with plant growth (Figure 4). At jointing, there was no significant difference ($p > 0.05$) between the degradable plastic film mulching and the ordinary above-ground dry matter accumulation. The above-ground dry matter was significantly improved under nitrogen application, and that in N2 treatment marked the maximum with 7.76~31.43% increase under the degradable plastic film mulching and 6.50~28.85% increase under the ordinary mulching. At tasseling, nitrogen fertilizer was the main factor affecting spring maize above-ground dry matter. Compared with N0, N1, and N3, N2 treatment, the above-ground dry matter of spring maize was increased by 49.89%, 22.36%, and 7.12% under the degradable plastic film mulching and 40.39%, 20.45%, and 5.27% under the ordinary mulching, respectively. At the grain-filling stage, nitrogen fertilizer had a more significant effect on increasing above-ground dry matter accumulation. Compared with N0 and N1, N2 significantly increased by 61.45%, 28.66% under degradable plastic film mulching and 52.87%, 28.74% under ordinary mulching, and the effect of nitrogen fertilizer under degradable plastic film mulching was better than that of ordinary plastic film mulching. It can be seen that a reasonable amount of nitrogen fertilizer can promote the growth and development of spring maize and improve the dry matter quality of above-ground parts. When the nitrogen fertilizer level exceeds N2, it will inhibit the growth of spring maize and affect the accumulation of dry matter. There was no significant difference in the development of spring maize under ordinary plastic film mulching and degradable plastic film mulching ($p > 0.05$).

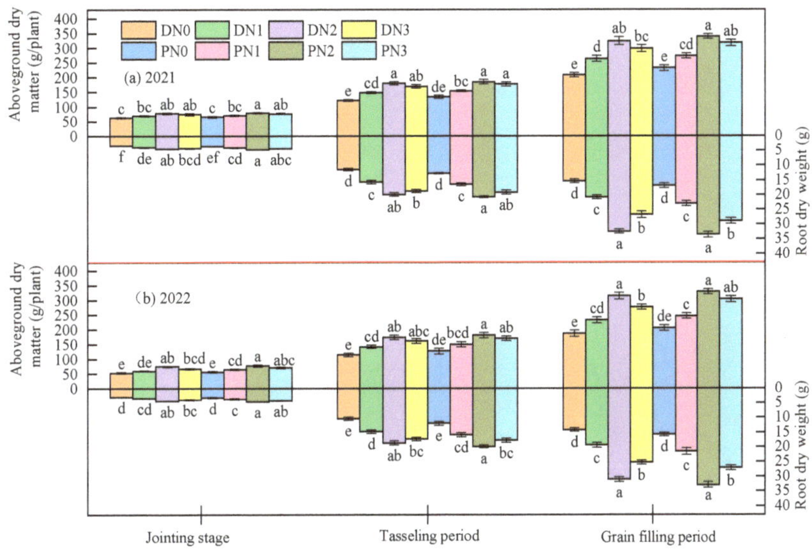

Figure 4. Effects of different mulching and nitrogen application on root and shoot growth of spring maize in 2021 (**a**) and 2022 (**b**). D represents degradable plastic film, P represents ordinary plastic film, and N0, N1, N2, and N3 represent 0, 160, 320, and 480 kg·ha^{-1} nitrogen fertilizer. The letters above the histogram indicate that there are significant differences among different treatments ($p < 0.05$). The data in the figure are the average of multiple repeated sets (n = 3). The bar above the bar graph represents the standard error.

3.1.2. Root Dry Matter

The root biomass of spring maize reached its maximum, with the growth stage advancing to the filling phase (Figure 4). Nitrogen fertilizer can promote the growth of the spring maize root system and improve its quality. At the jointing stage, N2 treatment increased the root system by 40.57%, 20.58%, and 8.07% under the degradable plastic film mulching and 37.97%, 20.50%, and 7.66% under the ordinary mulching, respectively, compared with N0, N1, and N3, indicating that N3 nitrogen application can inhibit the growth of the spring maize root system. At the tasseling stage, under the cover of degradable plastic film and ordinary plastic film, the growth rate from N0 to N1 increased by 40.39% and 31.06%, while from N1 to N2 it increased by 26.38% and 25.43%. It can be seen that the effect of root mass growth gradually decreased with the increase in nitrogen application level, and even the N3 treatment of degradable plastic film and ordinary plastic film decreased by 6.58% and 8.92%, respectively. At the grain filling stage, compared with N0, N1, and N3, N2 treatment significantly increased 115.73%, 58.17%, and 22.02% under the degradable plastic film mulching and 103.56%, 49.18%, and 18.53% under the ordinary mulching ($p < 0.05$). From the jointing stage to the grain filling period, the degradable mulching film improved root quality more than the ordinary one.

3.1.3. Root Shoot Ratio

At the jointing stage, film mulching type, nitrogen application level, and their interaction had no significant effect on root shoot ratio ($p > 0.05$). From the tasseling stage to the grain filling stage, the impact of nitrogen fertilizer on the root shoot ratio reached $p < 0.01$ level, and film mulching and its interaction had no significant effect on the root shoot ratio ($p > 0.05$) (Table 2).

Table 2. Significance test of spring maize root shoot ratio at different growth stages; ns means no significant difference ($p > 0.05$); * means significant at $p < 0.05$ level; ** means significant at $p < 0.01$ level.

Year	F Fest	Jointing Stage	Tasseling Period	Grain Filling Period
2021	F	0.41 ns	0.03 ns	0.22 ns
	N	1.77 ns	19.24 **	17.72 **
	F×N	0.12 ns	0.38 ns	0.32 ns
2022	F	0.11 ns	0.27 ns	0.13 ns
	N	0.60 ns	6.23 **	10.43 **
	F×N	0.07 ns	0.25 ns	0.17 ns
	Y	4.73 *	1.41 ns	0.86 ns
	F	0.45 ns	0.28 ns	0.33 ns
	N	2.03 ns	19.40 **	27.09 **
	Y×F	0.02 ns	0.13 ns	0.00 ns
	Y×N	0.13 ns	0.10 ns	0.25 ns
	F×N	0.01 ns	0.56 ns	0.35 ns
	Y×F×N	0.17 ns	0.01 ns	0.12 ns

The root shoot ratio increased first and then decreased with the growth period (Figure 5). At a jointing stage, film mulching type and nitrogen application level had no significant effect on the root shoot ratio ($p > 0.05$). At the tasseling stage, nitrogen fertilizer could significantly improve the root shoot ratio of spring maize. The root shoot ratio of spring maize under degradable plastic film mulching increased with the increase in nitrogen application level; from N0 to N1 increased by 14.59%, from N2 to N2 increased by 3.35%, and from N2 to N3 increased by 0.15%; Under ordinary plastic film mulching, the root shoot ratio of the N3 treatment was 4.37% lower than that of N2, indicating that degradable mulching was more conducive to the growth of the spring maize root shoot and coordinated the root shoot ratio. There was no significant difference in the root shoot ratio of the nitrogen application treatment ($p > 0.05$). At the grain-filling stage, the nitrogen application level of N2 was significantly higher than that of N0 and N1 by 33.35%, 22.56% under the degradable plastic film mulching, and 32.54%, 15.69% under the ordinary mulching ($p < 0.05$), respectively. At the same time, there was no significant difference between N3 and N2 ($p > 0.05$).

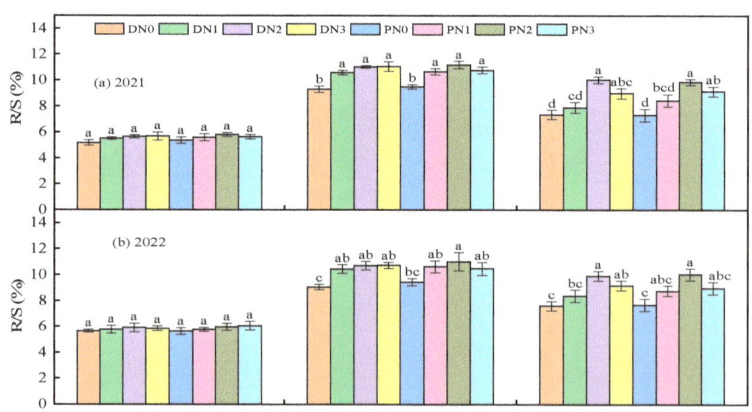

Figure 5. Effect of different film mulching and nitrogen application on spring maize root shoot ratio in 2021 (**a**) and 2022 (**b**). R/S represents root-to-shoot ratio, D represents degradable plastic film, P represents ordinary plastic film, and N0, N1, N2, and N3 represent 0, 160, 320, and 480 kg·ha^{-1} nitrogen fertilizer. The letters above the histogram indicate that there are significant differences among different treatments ($p < 0.05$). The data are the figure is the average of multiple repeated sets (n = 3). The bar above the bar graph represents the standard error.

3.2. Photosynthetic Gas Exchange Characteristics

Nitrogen fertilizer was the main factor affecting the net photosynthetic rate, transpiration rate, and stomatal conductance of spring maize, reaching a level of $p < 0.01$. Film mulching and its interaction at the jointing stage did not significantly affect the net photosynthetic rate, transpiration rate, or stomatal conductance. Nitrogen fertilizer from the tasseling location to the grain filling stage had significant ($p < 0.05$) and highly significant ($p < 0.01$) effects on net photosynthetic rate, transpiration rate, and stomatal conductance, and the interaction between film mulching and nitrogen fertilizer had no significant impact ($p > 0.05$) (Table 3).

Table 3. Significance test of photosynthetic gas exchange parameters of spring maize at different growth stages. Pn represents the net photosynthetic rate, Tr represents the transpiration rate, and Gs represents stomatal conductance. ns means no significant difference ($p > 0.05$); * means significant at $p < 0.05$ level; ** means significant at $p < 0.01$ level.

Year	F Fest	Jointing Stage			Tasseling Period			Grain filling Period		
		Pn	Tr	Gs	Pn	Tr	Gs	Pn	Tr	Gs
2021	F	0.84 ns	3.02 ns	4.00 ns	2.10 ns	3.73 ns	3.88 ns	5.74 *	8.71 **	21.70 **
	N	15.14 **	24.54 **	33.63 **	17.82 **	26.54 **	21.23 **	38.33 **	81.79 **	86.84 **
	F×N	0.12 ns	0.89 ns	0.44 ns	0.13 ns	0.08 ns	0.04 ns	0.12 ns	1.19 ns	4.31 *
2022	F	3.16 ns	4.22 ns	2.24 ns	3.39 ns	5.83 *	8.07 *	3.42 ns	4.57 *	8.84 **
	N	19.89 **	34.39 **	8.29 **	25.51 **	34.77 **	56.66 **	23.08 **	21.97 **	136.09 **
	F×N	0.27 ns	0.61 ns	0.23 ns	0.50 ns	0.22 ns	0.11 ns	0.10 ns	0.92 ns	0.04 ns
	Y	7.13 *	1.15 ns	71.52 **	35.52 **	2.26 ns	21.89 **	7.42 *	1.15 ns	0.72 ns
	F	3.97 ns	7.14 *	5.56 *	5.42 *	9.36 **	10.16 **	8.85 **	12.50 **	28.98 **
	N	34.23 **	58.08 **	31.60 **	43.00 **	60.68 **	62.68 **	59.28 **	88.49 **	213.84 **
	Y×F	0.97 ns	0.02 ns	0.02 ns	0.08 ns	0.06 ns	0.02 ns	0.04 ns	0.07 ns	1.28 ns
	Y×N	2.68 ns	0.21 ns	0.88 ns	0.39 ns	0.07 ns	0.23 ns	0.28 ns	5.04 **	10.15 **
	F×N	0.38 ns	1.47 ns	0.44 ns	0.55 ns	0.26 ns	0.11 ns	0.21 ns	0.77 ns	2.03 ns
	Y×F×N	0.07 ns	0.05 ns	0.16 ns	0.09 ns	0.03 ns	0.01 ns	0.01 ns	1.30 ns	2.23 ns

3.2.1. Net Photosynthetic Rate

With the advance of the spring maize growth period, the net photosynthetic rate reached its maximum at the tasseling stage and slightly decreased at the grain filling stage. The net photosynthetic rate increased with the increase in nitrogen application level (Figure 6). At jointing, there was no significant difference in net photosynthetic rate between N2 and N3 treatments under degradable plastic film mulching ($p > 0.05$), which was significantly increased by 19.40%, 8.95% and 24.25%, 13.37% compared with N0 and N1, respectively. Under ordinary plastic film mulching, N3 was increased by 18.90%, 10.54%, and 3.88% compared with N2, N1, and N0, respectively. At tasseling, the net photosynthetic rate of degradable plastic film and ordinary plastic film mulching increased by 16.57% and 11.45% from N0 to N1, increased by 9.24% and 7.78% from N1 to N2, and increased by 2.87% and 2.17% from N2 to N3, respectively. It can be seen that the effect of nitrogen fertilizer gradually weakened with the increase in nitrogen application level. At the grain filling stage, the nitrogen application level of N3 was significantly higher than that of N0, N1, and N2 ($p < 0.05$), increasing by 45.71%, 26.64%, and 15.10% under the degradable plastic film mulching, and 43.44%, 29.41%, and 13.32% under the ordinary mulching ($p < 0.05$), respectively. From the jointing to the grain filling stage, there was no significant difference in the net photosynthetic rate between degradable plastic film and ordinary plastic film under the same nitrogen application level ($p > 0.05$), and the increase in the net photosynthetic rate of degradable plastic film combined with nitrogen fertilizer was higher than that of ordinary plastic film.

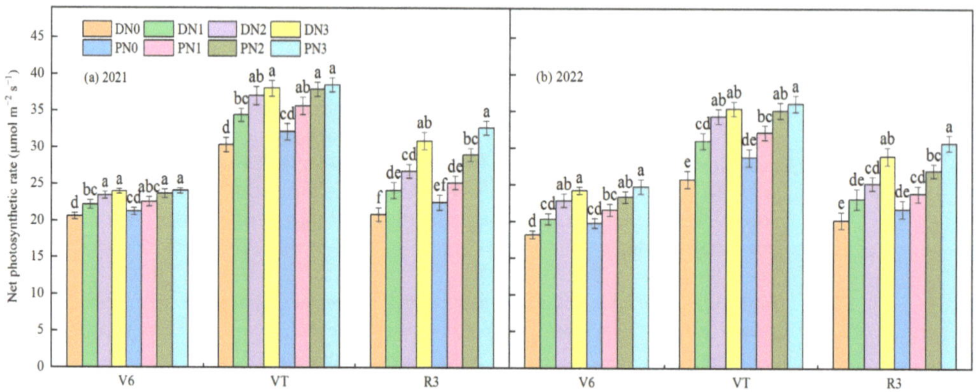

Figure 6. Effect of different film mulching combined with nitrogen fertilizer on the net photosynthetic rate of spring maize (2021 (**a**) and 2022 (**b**)). D represents degradable plastic film, P represents ordinary plastic film, and N0, N1, N2, and N3 represent 0, 160, 320, and 480 kg·ha^{-1} nitrogen fertilizer. The letters above the histogram indicate that there are significant differences among different treatments ($p < 0.05$). The data in the figure are the average of multiple repeated sets (n = 3). The bar above the bar graph represents the standard error.

3.2.2. Transpiration Rate

The transpiration rate of spring maize increased first, then decreased with the advance of the growth period, and increased with the increase in nitrogen application level (Figure 7). At a jointing stage, the degradable plastic film and common plastic film increased by 23.20% and 10.96%, respectively, from N0 to N1, increased by 14.29% and 12.63%, respectively, from N1 to N2, and increased by 5.23% and 4.62%, respectively, from N2 to N3. This showed that nitrogen application significantly increased the transpiration rate. At the tasseling stage, the transpiration rate of nitrogen application treatment was significantly higher than that of no nitrogen application treatment ($p < 0.05$). N1, N2, and N3 under degradable plastic film cover increased by 29.50%, 54.07%, and 68.84% compared to N0, respectively. N1, N2, and N3 under ordinary plastic film cover increased by 22.81%, 40.40%, and 52.66% compared to N0. Moreover, there was no significant difference in transpiration rate between degradable plastic film cover and ordinary plastic film cover under the same nitrogen application level ($p > 0.05$). At the grain filling stage, the nitrogen application level treatment of N3 was 69.97%, 21.08%, and 5.06% under the degradable plastic film mulching, and 55.42%, 15.13%, and 8.15% under the ordinary mulching, higher than that of N0, N1, and N2, respectively. From the jointing to the tasseling stage, the transpiration rate of common plastic film mulching was higher than that of degradable plastic film mulching, but there was no significant difference ($p > 0.05$).

3.2.3. Stomatal Conductance

As the growth period progressed, the stomatal conductance of spring maize reached its maximum at the tasseling stage and slightly decreased during the filling phase (Figure 8). During the jointing stage, the stomatal conductance of nitrogen application treatments was significantly higher than that of non-nitrogen application treatments ($p < 0.05$). N3, N2, and N1 increased by 11.87%, 23.50%, and 32.00% under the degradable plastic film mulching, and 11.57%, 16.67%, and 24.07% under the ordinary mulching, respectively, compared to N0. Moreover, the stomatal conductance of ordinary plastic film was higher than that of degradable plastic film, but there was no significant difference ($p > 0.05$). During the tasseling period, N3 treatment under degradable plastic film coverage was significantly higher than N2, N1, and N0 ($p < 0.05$), with increases of 48.95%, 23.48%, and 12.70%, respectively. There was no significant difference in nitrogen application levels between N3 and N2 under ordinary plastic film coverage ($p > 0.05$), both of which were significantly higher than N1 and N0 ($p < 0.05$). During the grain-filling period, under the cover of degradable and ordinary plastic film, the increase from N0 to N1 was 12.83% and 16.74%, respectively. The increase from N1 to N2 was 29.02% and 35.48%, and the increase from N2 to N3 was 17.63% and 9.79%, respectively. Under the same nitrogen application level, the stomatal conductance of ordinary plastic film was higher than that of degradable plastic film, and when the nitrogen application level was lower than N2, the amplification effect of average plastic film was better than that of degradable plastic film. Under the nitrogen application level of N3, the amplification effect of degradable plastic film was better than that of ordinary plastic film.

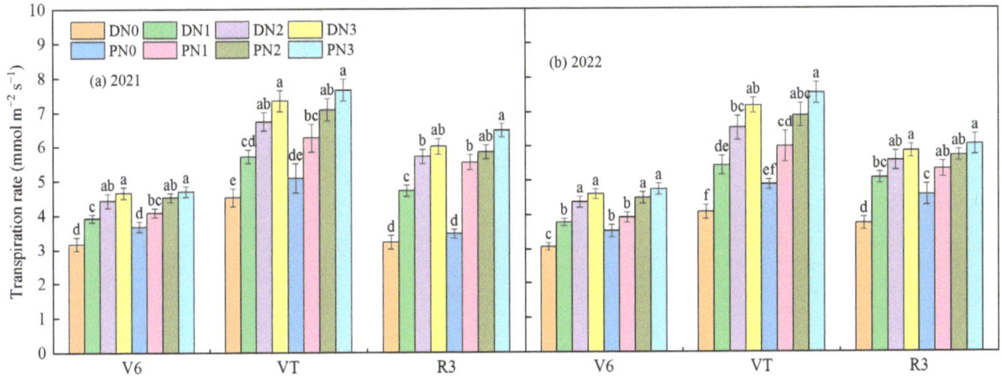

Figure 7. The effect of different coverage and nitrogen fertilizer applications on the spring maize transpiration rate (2021 (**a**) and 2022 (**b**)). D represents degradable plastic film, P represents ordinary plastic film, and N0, N1, N2, and N3 represent 0, 160, 320, and 480 kg·ha^{-1} nitrogen fertilizer. The letters above the histogram indicate that there are significant differences among different treatments ($p < 0.05$). The data in the figure are the average of multiple repeated sets (n = 3). The bar above the bar graph represents the standard error.

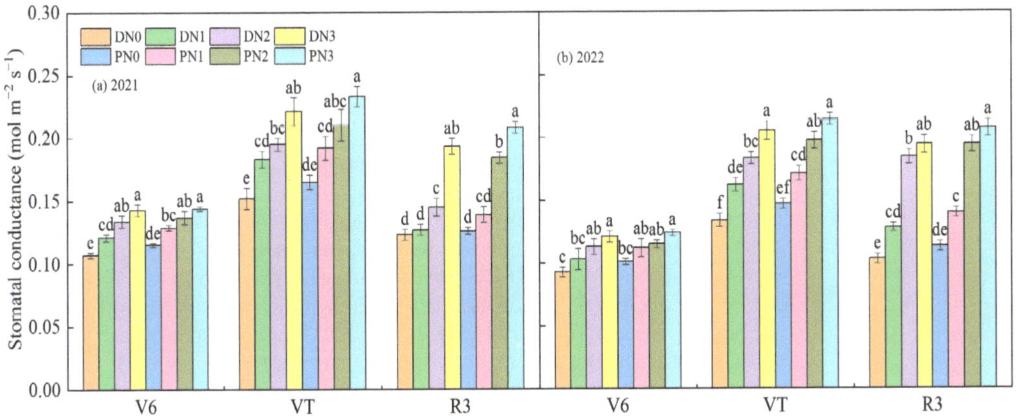

Figure 8. The effect of different coverage and nitrogen fertilizer application on the stomatal conductance of spring maize (2021 (**a**) and 2022 (**b**)). D represents degradable plastic film, P represents ordinary plastic film, and N0, N1, N2, and N3 represent 0, 160, 320, and 480 kg·ha^{-1} nitrogen fertilizer. The letters above the histogram indicate that there are significant differences among different treatments ($p < 0.05$). The data in the figure are the average of multiple repeated sets (n = 3). The bar above the bar graph represents the standard error.

3.3. Yield and Water and Nitrogen Use Efficiency

Nitrogen fertilizer was the main factor affecting spring maize yield, water consumption, and water and nitrogen utilization efficiency, reaching $p < 0.05$ and $p < 0.01$. Film mulching has a significant ($p < 0.05$) and highly effective ($p < 0.01$) impact on irrigation water use efficiency (IWUE) and nitrogen fertilizer agronomic utilization efficiency. The interaction between film mulching and nitrogen fertilizer had a significant ($p < 0.01$) impact on nitrogen fertilizer agronomic utilization efficiency (Table 4).

Table 4. Effects of different mulching and nitrogen application on spring maize yield and water and nitrogen use efficiency. The data in the table is the mean ± standard deviation, n = 3. Different letters after the same column of numbers indicate significant differences ($p < 0.05$); ns means no significant difference ($p > 0.05$); * means significant at $p < 0.05$ level; ** means significant at $p < 0.01$ level.

Year	Treatments	Yield (kg·ha^{-1})	ET (mm)	WUE (kg·m^{-3})	IWUE (kg·m^{-3})	NPF (kg·kg^{-1})	NFA (kg·kg^{-1})
2021	DN0	10,056.06 ± 723.38 d	786.20 ± 16.74 ab	1.28 ± 0.07 d	1.60 ± 0.12 d	--	--
	DN1	12,275.78 ± 497.49 bc	818.83 ± 14.01 ab	1.50 ± 0.05 bcd	1.96 ± 0.08 bc	76.72 ± 3.11 a	28.83 ± 3.03 a
	DN2	14,016.28 ± 542.33 ab	839.86 ± 15.91 a	1.67 ± 0.10 a	2.24 ± 0.11 ab	43.80 ± 2.04 b	12.38 ± 0.40 b
	DN3	13,680.4 ± 653.61 ab	844.57 ± 20.03 a	1.62 ± 0.03 abc	2.18 ± 0.08 ab	28.50 ± 1.13 c	7.55 ± 0.83 c
	PN0	11,014.06 ± 622.97 cd	770.25 ± 21.94 b	1.43 ± 0.10 cd	1.76 ± 0.10 cd	--	--
	PN1	12,973.96 ± 514.01 ab	793.97 ± 16.97 ab	1.63 ± 0.05 abc	2.07 ± 0.08 ab	81.09 ± 3.21 a	12.25 ± 1.28 b
	PN2	14,895.24 ± 586.73 a	827.76 ± 20.65 ab	1.80 ± 0.03 a	2.38 ± 0.10 a	46.55 ± 1.83 b	12.13 ± 0.23 b
	PN3	14,621.44 ± 607.53 a	835.34 ± 14.52 a	1.75 ± 0.09 a	2.33 ± 0.10 a	30.46 ± 1.27 c	7.52 ± 0.24 c
F fest	F	4.23 ns	1.52 ns	6.38 *	4.25 ns	2.71 ns	24.17 **
	N	18.00 **	5.05 *	9.52 **	17.95 **	251.76 **	44.21 **
	F×N	0.02 ns	0.07 ns	0.01 ns	0.02 ns	0.15 ns	22.97 **
2022	DN0	9446.10 ± 446.82 d	746.38 ± 11.50 cd	1.26 ± 0.04 d	1.55 ± 0.07 d	--	--
	DN1	11,207.88 ± 488.95 cd	765.96 ± 16.27 bcd	1.46 ± 0.07 cd	1.84 ± 0.08 cd	70.05 ± 3.06 a	25.21 ± 0.88 a
	DN2	13,773.96 ± 766.25 ab	801.97 ± 14.61 ab	1.72 ± 0.12 ab	2.26 ± 0.13 ab	43.04 ± 2.39 b	13.52 ± 1.31 bc
	DN3	13,107.30 ± 518.96 abc	813.98 ± 15.09 a	1.61 ± 0.03 abc	2.15 ± 0.09 abc	27.31 ± 1.08 c	7.63 ± 0.18 d
	PN0	10,028.44 ± 895.66 d	727.34 ± 10.15 d	1.38 ± 0.11 cd	1.65 ± 0.15 d	--	--
	PN1	12,017.44 ± 626.16 bc	758.35 ± 13.62 bcd	1.59 ± 0.10 bc	1.97 ± 0.10 bc	75.11 ± 3.91 a	12.43 ± 2.92 bcd
	PN2	14,567.82 ± 498.79 a	784.20 ± 11.10 abc	1.86 ± 0.05 a	2.39 ± 0.08 a	45.52 ± 1.56 b	14.19 ± 1.25 b
	PN3	14,248.88 ± 573.33 a	795.25 ± 16.05 ab	1.79 ± 0.07 ab	2.34 ± 0.09 a	29.69 ± 1.19 c	8.79 ± 1.16 cd
F fest	F	3.61 ns	2.65 ns	6.64 *	3.75 ns	2.78 ns	8.60 *
	N	21.63 **	9.98 **	14.52 **	21.65 **	169.04 **	24.24 **
	F×N	0.07 ns	0.08 ns	0.09 ns	0.07 ns	0.20 ns	13.45 **
	Y	4.45 *	25.85 **	0.00 ns	0.83 ns	4.09 ns	0.02 ns
	F	7.81 **	3.88 ns	13.29 **	7.97 **	5.48 *	11.74 **
	N	39.48 **	13.72 **	24.50 **	39.48 **	412.74 **	25.52 **
	Y×F	0.00 ns	0.00 ns	0.00 ns	0.00 ns	0.01 ns	0.53 ns
	Y×N	0.29 ns	0.06 ns	0.35 ns	0.38 ns	1.77 ns	0.54 ns
	F×N	0.05 ns	0.01 ns	0.03 ns	0.05 ns	0.34 ns	13.78 **
	Y×F×N	0.04 ns	0.14 ns	0.07 ns	0.05 ns	0.01 ns	0.12 ns

Nitrogen fertilizer significantly increased the spring maize yield. Under biodegradable plastic film and ordinary plastic film coverage, the yield increased from N0 to N1 by 20.42% and 18.77%, respectively, and from N1 to N2 by 18.34% and 17.89%, respectively. As the nitrogen application rate increased, the yield of spring maize gradually weakened and even decreased by 2.01% to 3.61% at the N3 nitrogen application level. Moreover, under the same nitrogen application level, the yield of biodegradable plastic film was not significantly different from that of ordinary plastic film ($p > 0.05$). Nitrogen fertilizer promoted spring maize water absorption and increased water consumption. N3 is 8.22%, 4.65%, and 1.02% higher under the degradable plastic film mulching than N2, N1, and N0, respectively, and 8.88%, 5.04%, and 1.16% under the ordinary mulching. Increasing nitrogen fertilizer application can improve spring maize WUE and IWUE. Under degradable plastic film coverage, N2 treatment increased WUE and IWUE by 33.55%, 14.53%, 42.55%, and 18.40% compared to N1 and N0, respectively. Under ordinary plastic film coverage, N2 treatment increased WUE and IWUE by 30.08%, 13.48%, 40.09%, and 17.94% compared to N1 and N0, respectively. However, it reduced nitrogen fertilizer productivity. Degradable plastic film treatment reduced N2 and N3 by 69.01% and 163.00%, respectively, while ordinary plastic film treatment reduced N2 and N3 by 69.65% and 159.69%, respectively, compared to N1.

3.4. Soil Enzyme Activity and pH

Nitrogen fertilizer significantly affected the soil urease and sucrase ($p < 0.01$), while film mulching had a critical ($p < 0.05$) and highly effective ($p < 0.01$) effect on urease and sucrase. Film mulching and nitrogen fertilizer had no significant effect on pH ($p > 0.05$), and the interaction had no considerable impact on urease, sucrase, and pH ($p > 0.05$) (Table 5).

Table 5. Significance test of soil enzyme activity and pH. ns means no significant difference ($p > 0.05$); * means significant at $p < 0.05$ level; ** means significant at $p < 0.01$ level.

Year	F Fest	Urease	Sucrase	pH
2021	F	9.66 **	4.33 ns	0.02 ns
	N	76.05 **	15.13 **	0.48 ns
	F×N	0.06 ns	0.12 ns	0.002 ns
2022	F	6.16 *	8.61 *	0.06 ns
	N	64.21 **	37.80 **	1.32 ns
	F×N	0.24 ns	0.20 ns	0.003 ns
	Y	1.21 ns	0.45 ns	0.08 ns
	F	15.49 **	11.18 **	0.06 ns
	N	139.26 **	43.66 **	1.57 ns
	Y×F	0.10 ns	0.02 ns	0.00 ns
	Y×N	0.24 ns	0.04 ns	0.04 ns
	F×N	0.14 ns	0.25 ns	0.01 ns
	Y×F×N	0.17 ns	0.04 ns	0.00 ns

The soil urease, sucrase, and pH under plastic film mulching were lower than those under ordinary plastic film mulching, but there was no significant difference ($p > 0.05$) (Figure 9). Under ordinary plastic film coverage, urease and sucrase increased by 34.24% and 13.67% from N0 to N1, 26.72% and 15.24% from N1 to N2, and 11.18% and 1.44% from N2 to N3, respectively. Under biodegradable plastic film coverage, urease and sucrase increased by 42.21% and 21.92% from N0 to N1, 32.88% and 14.48% from N1 to N2, and 12.71% and 2.33% from N2 to N3, respectively. It can be seen that increasing nitrogen fertilizer can significantly improve soil urease and sucrase activities, but the increase gradually decreases with increasing nitrogen application. Increasing nitrogen fertilizer application can reduce soil pH, but the treatments have no significant difference ($p > 0.05$).

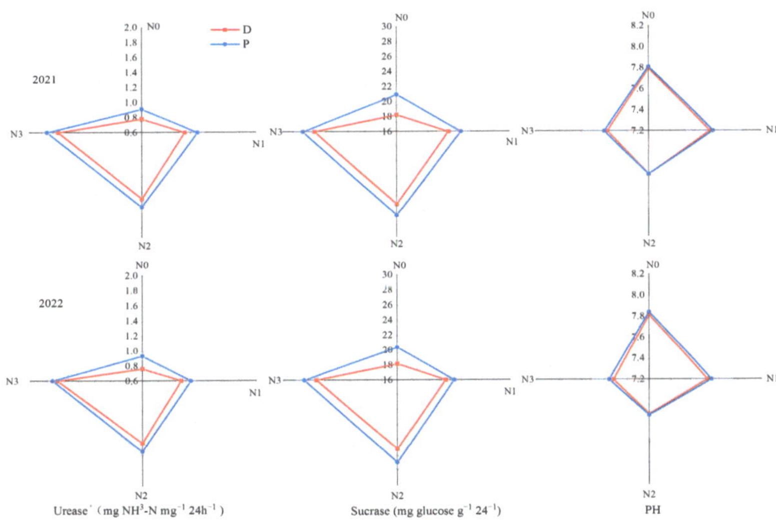

Figure 9. The effect of film mulching combined with nitrogen fertilizer on the soil quality of spring maize. D represents degradable plastic film, P represents ordinary plastic film, and N0, N1, N2, and N3 represent 0, 160, 320, and 480 kg·ha^{-1} nitrogen fertilizer.

3.5. Correlation Analysis between Various Indicators of Spring Maize

The yield of spring maize mainly comes from the photosynthetic products during the filling period [31], so the root and crown growth and photosynthetic gas exchange parameters during the grain filling period are selected as evaluation indicators. Figure 10 showed a significant positive correlation ($p < 0.05$) between yield and water consumption, transpiration rate, above-ground biomass, root biomass, root-to-shoot ratio, urease, and sucrase under degradable plastic film coverage. The yield under plastic film coverage was positively correlated with water consumption, above-ground biomass, root biomass, root-to-shoot ratio, urease, and sucrase ($p < 0.05$) and negatively correlated with pH.

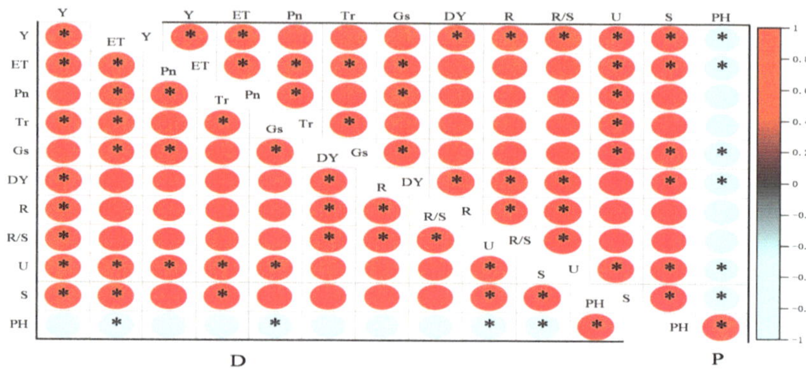

Figure 10. Correlation analysis between various indicators of spring maize under different treatments, * Significant difference at $p < 0.05$ level. D represents degradable plastic film, P represents ordinary plastic film, Y represents yield, ET represents crop evapotranspiration, Pn represents net photosynthetic rate, Tr represents transpiration rate, Gs represents stomatal conductance, DY represents above-ground biomass, R represents root biomass, R/S represents root to shoot ratio, U represents urease, and S represents sucrase.

3.6. Construction of a Comprehensive Growth Evaluation Model for Spring Maize

3.6.1. Comprehensive Evaluation Hierarchy Model

They were using Yaaph software to establish a hierarchical model for the comprehensive evaluation of spring maize (Figure 11). The total growth index (C) target layer includes four criteria layers: yield and water use index (C1), photosynthetic index (C2), root and crown growth index (C3), and soil index (C4). The yield indicators include two indicator layers: yield (C11) and water consumption (C12). The photosynthetic indicators include three indicator layers: net photosynthetic rate (C21), transpiration rate (C22), and stomatal conductance (C23). The root cap growth indicators include three indicator layers: above-ground dry matter mass (C31), root mass (C32), and root cap ratio (C33). Soil quality indicators include three indicator layers: urease (C41), sucrase (C42), and pH (C43).

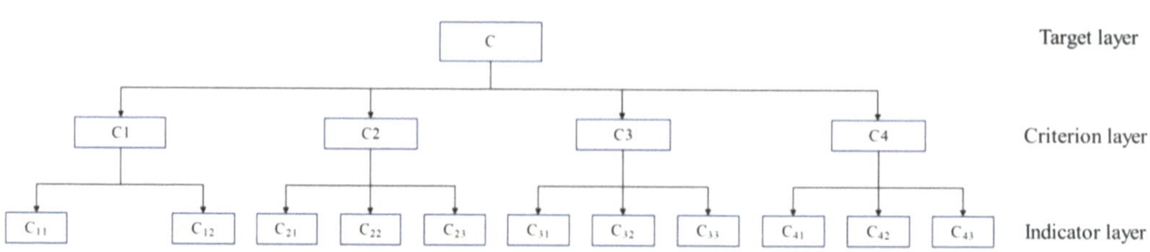

Figure 11. Comprehensive growth evaluation model diagram of spring maize.

3.6.2. Indicator Weights

AHP Method for Determining Indicator Weights

After establishing the hierarchical model, a judgment matrix is specified using a scale of 1–9. According to Figure 10, values are assigned to each indicator, and the consistency of the judgment matrix is checked. The judgment matrices for the comprehensive growth indicator (C), yield indicator (C1), root and crown growth indicator (C3), and soil indicator (C4) are as follows:

Degradable plastic film:

$$C = \begin{bmatrix} 1.0000 & 2.0000 & 2.5000 & 3.0000 \\ 0.5000 & 1.0000 & 0.5000 & 0.5000 \\ 0.4000 & 2.0000 & 1.0000 & 2.0000 \\ 0.3333 & 2.0000 & 0.5000 & 1.000 \end{bmatrix} C1 = \begin{bmatrix} 1.0000 & 2.0000 \\ 0.5000 & 1.0000 \end{bmatrix}$$

$$C2 = \begin{bmatrix} 1.0000 & 0.5000 & 1.0000 \\ 2.0000 & 1.0000 & 2.0000 \\ 1.0000 & 0.5000 & 1.0000 \end{bmatrix} C3 = \begin{bmatrix} 1.0000 & 2.0000 & 1.5000 \\ 0.5000 & 1.0000 & 1.1000 \\ 0.6667 & 0.9091 & 1.0000 \end{bmatrix}$$

$$C4 = \begin{bmatrix} 1.0000 & 0.5000 & 1.5000 \\ 2.0000 & 1.0000 & 1.5000 \\ 0.6667 & 0.6667 & 1.0000 \end{bmatrix}$$

Ordinary plastic film:

$$C = \begin{bmatrix} 1.0000 & 2.0000 & 2.5000 & 2.0000 \\ 0.5000 & 1.0000 & 0.5000 & 1.5000 \\ 0.4000 & 2.0000 & 1.0000 & 2.0000 \\ 0.5000 & 0.6667 & 0.5000 & 1.000 \end{bmatrix} C1 = \begin{bmatrix} 1.0000 & 1.5000 \\ 0.6667 & 1.0000 \end{bmatrix}$$

$$C2 = \begin{bmatrix} 1.0000 & 2.0000 & 2.5000 \\ 0.5000 & 1.0000 & 0.5000 \\ 0.4000 & 2.0000 & 1.0000 \end{bmatrix} C3 = \begin{bmatrix} 1.0000 & 2.5000 & 1.5000 \\ 0.4000 & 1.0000 & 0.5000 \\ 0.6667 & 0.9091 & 1.0000 \end{bmatrix}$$

$$C4 = \begin{bmatrix} 1.0000 & 0.3333 & 0.5000 \\ 3.0000 & 1.0000 & 2.0000 \\ 2.0000 & 0.5000 & 1.0000 \end{bmatrix}$$

The consistency test coefficients CR of the comprehensive growth index (C), yield index (C1), root and shoot growth index (C3), and soil index (C4) of the two plastic film mulchings were all less than 0.1, indicating that the consistency test results were good. The established judgment matrix was reliable and reasonable (Table 6, λmax is the maximum eigenvalue). The results showed that the weight of each index under degradable plastic film mulching was in the order of yield, water consumption, above-ground dry matter quality, sucrase, transpiration rate, root-shoot ratio, root quality, urease, pH, net photosynthetic rate, and stomatal conductance. The weight of each index under ordinary plastic film mulching was in the order of yield, water consumption, above-ground dry matter quality, net photosynthetic rate, root-shoot ratio, sucrase, stomatal conductance, root quality, pH, transpiration rate, and urease.

Table 6. Weight calculation results of AHP Analytic Hierarchy Process.

	Degradable Plastic Film			Ordinary Plastic Film		
	Local Weights	**Final Weight**	**Consistency Check Parameters**	**Local Weights**	**Final Weight**	**Consistency Check Parameters**
Target layer C	0.4435 0.1360 0.2493 0.1713	0.4435 0.1360 0.2493 0.1713	$C_R = 0.0664 < 0.1$ λmax = 4.1774	0.4115 0.1781 0.2604 0.1460	0.4115 0.1781 0.2604 0.1460	$C_R = 0.0477 < 0.1$ λmax = 4.1274

Table 6. *Cont.*

	Degradable Plastic Film			Ordinary Plastic Film		
	Local Weights	Final Weight	Consistency Check Parameters	Local Weights	Final Weight	Consistency Check Parameters
Criterion layer C1	0.6667 0.3333	0.2957 0.1478	$C_R = 0.0000 < 0.1$ $\lambda max = 2.0000$	0.6000 0.4000	0.2469 0.1646	$C_R = 0.0000 < 0.1$ $\lambda max = 2.0000$
Criterion layer C2	0.2500 0.5000 0.2500	0.0340 0.0680 0.0340	$C_R = 0.0000 < 0.1$ $\lambda max = 3.0000$	0.5232 0.1928 0.2840	0.0932 0.0343 0.0506	$C_R = 0.0904 < 0.1$ $\lambda max = 3.0940$
Criterion layer C3	0.4641 0.2636 0.2723	0.1157 0.0657 0.0679	$C_R = 0.0157 < 0.1$ $\lambda max = 3.0163$	0.4797 0.1805 0.3398	0.1249 0.0470 0.0885	$C_R = 0.0036 < 0.1$ $\lambda max = 3.0037$
Criterion layer C4	0.2918 0.4632 0.2451	0.0500 0.0793 0.0420	$C_R = 0.0516 < 0.1$ $\lambda max = 3.0536$	0.1634 0.5396 0.2970	0.0239 0.0788 0.0434	$C_R = 0.0088 < 0.1$ $\lambda max = 3.0092$

Entropy Weight Method for Determining Indicator Weights

The weights of various indicators of spring maize were calculated using Matlab programming, as shown in Table 7. According to the table, the consequences of multiple indicators under degradable plastic film cover, in descending order, were: pH, stomatal conductance, root-to-shoot ratio, root mass, net photosynthetic rate, urease, water consumption, above-ground dry matter mass, yield, sucrase, and transpiration rate. Under ordinary plastic film cover, the weights of various indicators were in descending order: pH, stomatal conductance, net photosynthetic rate, root mass, above-ground dry matter mass, urease, sucrase, and water consumption root-to-shoot ratio, yield, and transpiration rate.

Table 7. Single index weights of spring maize calculated based on Entropy Weight Method.

Treatments	Index	C_{11}	C_{12}	C_{21}	C_{22}	C_{23}	C_{31}	C_{32}	C_{33}	C_{41}	C_{42}	C_{43}
D P	Weight	0.0797 0.0787	0.0837 0.0817	0.0922 0.0980	0.0754 0.0717	0.1102 0.0999	0.0834 0.0858	0.0923 0.0862	0.0983 0.0802	0.0864 0.0834	0.0785 0.0822	0.1199 0.1522

Combination Weight Determination Based on the Game Theory

To avoid the influence of subjective factors on evaluation, an essential weight set formula was constructed based on two weighting values obtained from the AHP method and the entropy weighting method:

$$w = \sum_{k=1}^{1} \alpha_k \times w_k^T (\alpha_k > 0)$$

where α_k, w_k are the weights obtained from the AHP method and the entropy weight method.

Calculate the weight set model based on game theory and derive the formula for the game model: $Min = \| \sum_{j=1}^{i} a_j \times u_i^T - u_i^T \| (i = 1, 2)$. The normalized combination coefficients of the formula can be obtained using Matlab: $a_1 = 0.8507$, $a_2 = 0.1493$ (D); $a_1 = 0.7881$, $a_2 = 0.2119$ (P). Thus, the combined weight vector was obtained, and the final result is shown in Table 8. As shown in the table, the weights of various indicators under degradable plastic film cover in descending order were yield, water consumption, above-ground dry matter mass, sucrase, root-to-shoot ratio, root mass, transpiration rate, urease, pH, stomatal conductance, and net photosynthetic rate. Under ordinary plastic film cover, the weights of various indicators in descending order were yield, water consumption, above-ground dry matter mass, net photosynthetic rate, root-to-shoot ratio, sucrase, pH, stomatal conductance, root quality, transpiration rate, and urease.

Table 8. Determination of Single Index Weights for spring maize Based on Game Theory through Combination Weighting.

Treatments	Index	C_{11}	C_{12}	C_{21}	C_{22}	C_{23}	C_{31}	C_{32}	C_{33}	C_{41}	C_{42}	C_{43}
D	Weight	0.2634	0.1382	0.0427	0.0691	0.0454	0.1109	0.0697	0.0724	0.0554	0.0792	0.0536
P		0.2113	0.1470	0.0942	0.0422	0.0610	0.1166	0.0553	0.0867	0.0365	0.0795	0.0665

3.6.3. Comprehensive Growth Evaluation of Spring Maize Based on TOPSIS Method

Established a TOPSIS comprehensive evaluation model with combined weighting, normalize the decision matrix, established a weighted matrix, and calculated the ideal solution and fit degree C_i of the evaluation index. The calculation results were shown in Table 9. As shown in the table, DN3 treatment had the highest comprehensive index of adhesion (0.8522) for spring maize, followed by PN2 treatment (0.8435), and DN0 treatment had the lowest bonding (0.0194), indicated that poor comprehensive performance of spring maize.

Table 9. Comprehensive indicators and ranking of spring maize based on TOPSIS method. S^+ represents the ideal solution, S^- represents the inverse perfect solution, D^+ represents the distance between each processing and the ideal solution, and D^- represents the distance between each processing and the inverse perfect solution.

Treatments	C_{11}	C_{12}	C_{21}	C_{22}	C_{23}	C_{31}	C_{32}	C_{33}	C_{41}	C_{42}	C_{43}	D^+	D^-	C_i	Sorted
DN0	0.3963	0.4774	0.4066	0.3435	0.3685	0.3703	0.3066	0.4288	0.2998	0.3930	0.5091	0.2023	0.0040	0.0194	8
DN1	0.4772	0.4936	0.4678	0.4821	0.4175	0.4647	0.4181	0.4666	0.4263	0.4792	0.5048	0.1245	0.0825	0.3985	5
DN2	0.5647	0.5114	0.5147	0.5557	0.5381	0.5979	0.6614	0.5719	0.5664	0.5485	0.4941	0.0322	0.1855	0.8522	1
DN3	0.5443	0.5166	0.5924	0.5838	0.6327	0.5380	0.5420	0.5209	0.6384	0.5613	0.4918	0.0413	0.1755	0.8097	4
S^+	0.5647	0.5166	0.5924	0.5838	0.6327	0.5979	0.6614	0.5719	0.6384	0.5613	0.5091				
S^-	0.3963	0.4774	0.4066	0.3435	0.3685	0.3703	0.3066	0.4288	0.2998	0.3930	0.4918				
PN0	0.3999	0.4757	0.4121	0.3700	0.3570	0.3846	0.3169	0.4255	0.3284	0.4162	0.5091	0.1781	0.0043	0.0235	7
PN1	0.4749	0.4931	0.4568	0.4995	0.4165	0.4567	0.4325	0.4875	0.4408	0.4731	0.5055	0.1130	0.0689	0.3789	6
PN2	0.5599	0.5121	0.5216	0.5318	0.5623	0.5880	0.6452	0.5640	0.5586	0.5452	0.4928	0.0300	0.1616	0.8435	2
PN3	0.5487	0.5180	0.5911	0.5751	0.6188	0.5457	0.5443	0.5131	0.6211	0.5530	0.4925	0.0323	0.1584	0.8308	3
S^+	0.5599	0.5180	0.5911	0.5751	0.6188	0.5880	0.6452	0.5640	0.6211	0.5530	0.5091				
S^-	0.3999	0.4757	0.4121	0.3700	0.3570	0.3846	0.3169	0.4255	0.3284	0.4162	0.4925				

4. Discussion

4.1. Effect of Film Mulching Combined with Nitrogen Fertilizer Application on Root and Shoot Growth

Crop root and shoot growth is more sensitive to nitrogen fertilizer. Increasing nitrogen fertilizer application can accelerate crop growth, root and shoot growth and development, and increase nitrogen uptake. However, excessive or insufficient nitrogen application can change crop growth morphology, affecting dry matter distribution and accumulation [32,33]. In the early stage of maize growth, degradable plastic film and ordinary plastic film coverage can form a "diaphragm effect" to significantly promote maize growth. In the later growth stage, degradable plastic film coverage degrades, which is beneficial for rainfall infiltration. In addition, the same amount of irrigation provides a good water and fertilizer environment for maize growth, with little impact on crop reproductive growth. Therefore, the effect of ordinary plastic film coverage and the application of nitrogen fertilizer with plastic film coverage on crop root and crown growth is consistent; under the same nitrogen application level, there was no significant difference in the root and crown growth of maize covered with degradable plastic film and ordinary plastic film, which was similar to the research conclusions of Huang et al. [34] and Wang et al. [21]. The root system is the main organ for crops to absorb nutrients. Increasing nitrogen fertilizer application can promote the growth of maize roots and increase root biomass, and the relationship between root biomass and nitrogen application is non-linear. When nitrogen application exceeds 320 $kg \cdot ha^{-1}$, it will inhibit root growth and development and reduce root biomass. This was consistent with the research conclusion of Qi et al. [35], which indicated that reasonable nitrogen fertilizer management measures can contribute to the formation of maize root morphology and increase root quality. The results of this study also indicated that there was a non-linear relationship between the accumulation of above-ground dry matter in maize and nitrogen application; that is, nitrogen application exceeding 320 kg·ha^{-1} will affect maize growth and reduce above-ground biomass, which was consistent with the research results of Li et al. [36]. Appropriate nitrogen fertilizer management measures can promote the development of maize roots, benefit the accumulation of above-ground biomass, form a reasonable root cap ratio, and lay the foundation for high crop yield.

4.2. Effect of Film Mulching Combined with Nitrogen Fertilizer Application on Photosynthetic Gas Exchange Characteristics

Photosynthesis is the process by which crops convert inorganic substances in the atmosphere, such as water and carbon dioxide, into organic matter and release oxygen. Crops automatically adapt to environmental changes and develop in a direction that is conducive to photosynthesis [37]. The future way to increase crop yield will mainly rely on the increase in photosynthetic conversion rate [38]. The results of this study indicated that increasing nitrogen fertilizer application can significantly enhance the photosynthetic capacity of maize leaves, and the photosynthetic gas exchange parameters (net photosynthetic rate, transpiration rate, and stomatal conductance of maize) showed an approximately linear relationship with increasing nitrogen application rate. At a nitrogen application level of 480 kg·ha^{-1}, net photosynthetic rate, transpiration rate, and stomatal conductance were the highest, rising by 2.87~45.71%, 5.06~69.97%, and 12.70~71.24% under the degradable plastic film mulching, and 2.17~43.44%, 4.62~55.42%, and 9.79~73.64% under the ordinary mulching, respectively. This is because nitrogen can enhance the activity of mesophyll cells; increasing the SPAD value of leaves can improve photosynthesis [39], which was similar to the research conclusion of Gao et al. [40], and indicated that nitrogen fertilizer can improve the photosynthetic capacity of maize leaves. However, the degree of improvement varies due to factors such as crop variety, nitrogen fertilizer management measures, and the experimental environment. At the same time, this study also found that there was no significant difference in the photosynthetic gas exchange parameters between degradable plastic film-covered leaves and ordinary plastic film at the same nitrogen application level from the jointing stage to the grain-filling phase, indicating that the "diaphragm effect" formed by degradable plastic film and ordinary plastic film is the same [41], which can replace average plastic film to some extent.

4.3. Effect of Film Mulching Combined with Nitrogen Fertilizer Application on Maize Yield and Water and Nitrogen Use Efficiency

Reasonable nitrogen fertilizer management measures can promote root nutrient absorption, enhance crop assimilation, and increase yield. This study showed that the yield changed with nitrogen application rate in a quadratic parabolic relationship, and the yield-increasing effect slowed down with an increase in nitrogen application rate, which was in line with the diminishing returns effect. Moreover, excessive nitrogen application will reduce yield because it will affect crop nitrogen absorption efficiency, reduce nitrogen transport rate, and even affect root water absorption, resulting in a decreased yield [42,43]. Increasing the application of nitrogen fertilizer can enhance the water absorption capacity of the root system [44]. The results of this study indicated that increasing the application of nitrogen fertilizer can improve the water use efficiency and irrigation water use efficiency of maize. However, with the increase in nitrogen application level, the agronomic use efficiency and partial productivity of nitrogen fertilizer tended to decrease, which was consistent with the research conclusions of Li et al. [1]. Therefore, the appropriate amount of nitrogen fertilizer application provided a good soil environment for root growth, which improved crop yield and water use efficiency.

4.4. Effect of Film Mulching Combined with Nitrogen Fertilizer Application on Soil Enzyme Activity and pH

Soil enzyme activity, as an essential component of soil microbial activity and soil fertility, plays a critical catalytic role in soil nutrient cycling and energy conversion and can reflect the impact of fertilization on soil fertility and quality [45]. The results of this study indicated that the coverage area affected urease and sucrase activities. Under the same nitrogen application level, soil urease and sucrase activities under ordinary plastic film cover were higher than those under degradable plastic film, but there was no significant difference. This was similar to the research conclusions of Yang et al. [46] and Chen et al. [47]. Still, the reduction amplitude varies due to factors such as experimental materials, the experimental area environment, and field management measures. The application of nitrogen fertilizer can significantly increase the activities of urease and sucrase, which were due to the promotion of microbial activity by nitrogen fertilizer, changes in microbial composition, and thus affected soil enzyme activity [48], which was consistent with the research findings of Li et al. [48]. This study also found that increasing nitrogen fertilizer application can control soil salinity, reduce soil pH, and avoid soil salinization, which was consistent with the findings of Fudjoe et al. [49]. Increasing the application of nitrogen fertilizer can promote root development, enhance soil microbial activity, improve soil fertility, and reduce soil salinity, which is conducive to the sustainable development of agriculture.

5. Conclusions

Plastic film mulching and nitrogen fertilizer application are essential in agricultural production. The results of this study indicated that although the root and shoot growth, photosynthesis, and grain yield of spring maize under degradable plastic film mulching were lower than those under ordinary film mulching, there was no significant difference found. Nitrogen fertilizer was the main factor affecting spring maize growth and grain yield formation. When the nitrogen application rate approached 320 kg·ha^{-1}, spring maize root growth and root biomass might be promoted, and there was no significant difference in net photosynthetic rate, transpiration rate, and stomatal conductance compared to the nitrogen application rate of 480 kg·ha^{-1}. Under the nitrogen application level of 320 kg·ha^{-1}, the yield of degradable plastic film and ordinary plastic film coverage increased by 3.74~42.50% and 2.05~40.02%, respectively, while spring maize had the highest water use efficiency and irrigation water use efficiency. However, nitrogen fertilizer's agronomic utilization efficiency and partial productivity showed a decreasing trend. At the same time, there was no significant difference in soil enzyme activity between the nitrogen application level and 480 kg·ha^{-1}. After conducting a comprehensive evaluation of the impact of plastic film mulching combined with nitrogen fertilizer on the growth of spring maize, using multiple indicators, it was found that the best overall growth of corn was achieved by using a nitrogen application rate of 320 kg·ha^{-1} with degradable plastic film mulching. Therefore, this strategy is optimal for plastic film mulching combined with nitrogen fertilizer application.

Author Contributions: Data curation, T.C. and C.Z.; formal analysis, T.C. and L.L.; funding acquisition, H.Z.; writing—original draft, T.C.; supervision, H.Z., S.Y. and A.T.; writing—review and editing, T.C., H.Z. and F.L. All authors have read and agreed to the published version of the manuscript.

Funding: This work was supported by the National Natural Science Foundation of China (No. 52269008, 51669001), the Industrial Support Plan Project of Gansu Provincial Department of Education (No. 2022CYZC–51), the Key Research and Planning Projects of Gansu Province (No. 18YF1NA073), and the Open Project of Liaocheng University Landscape Architecture Discipline (No. 31946221236).

Data Availability Statement: The datasets used and/or analyzed during the current study are available from the corresponding author upon reasonable request and the approval of the data owner.

Acknowledgments: We thank everyone who helped during the manuscript writing. We also thank the reviewers for their useful comments and suggestions.

Conflicts of Interest: The authors declare no conflicts of interest.

References

1. Li, Y.; Gu, X.; Li, Y.; Fang, H.; Chen, P. Ridge-furrow mulching combined with appropriate nitrogen rate for enhancing photosynthetic efficiency, yield and water use efficiency of summer maize in a semi-arid region of China. *Agric. Water Manag.* **2023**, *287*, 108450. [CrossRef]
2. Zhu, G.; Liu, Z.; Qiao, S.; Zhang, Z.; Huang, Q.; Su, Z. How could observed sowing dates contribute to maize potential yield under climate change in Northeast China based on APSIM model. *Eur. J. Agron.* **2022**, *136*, 126511. [CrossRef]
3. Dong, M.; Zhao, J.; Li, E.; Liu, Z.; Guo, S.; Zhang, Z. Effects of Changing Climate Extremes on Maize Grain Yield in Northeast China. *Agronomy* **2023**, *13*, 1050. [CrossRef]
4. Lee, C.; He, Z.; Luo, H. Spatio-temporal characteristics of land ecological security and analysis of influencing factors in cities of major grain-producing regions of China. *Environ. Impact Rev.* **2024**, *104*, 107344. [CrossRef]
5. Yang, B.; Yue, S.; Gao, N.; Wei, Y.; Shen, Y.; Ai, Z.; Li, S. Effects of the key environmental and management factors on the advantages of film mulching spring maize in northwest China: A meta-analysis. *Eur. J. Agron.* **2023**, *150*, 126947. [CrossRef]
6. Ahmad, S.; Saqlain, Z.M.; Haider, A.H.; Erinle, K.O.; Wani, S.H.; Iqbal, R.; Okone, O.; Raza, A.; Waqas, M.M.; Nawaz, M. Physiological and biochemical properties of wheat (*Triticum aestivum* L.) under different mulching and water management systems in the semi-arid region of Punjab, Pakistan. *Arid Land Res. Manag.* **2022**, *362*, 181–196. [CrossRef]
7. Abbate, C.; Scavo, A.; Pesce, G.R.; Fontanazza, S.; Restuccia, A.; Mauromicale, G. Soil Bioplastic Mulches for Agroecosystem Sustainability: A Comprehensive Review. *Agriculture* **2023**, *13*, 197. [CrossRef]
8. Shinde, Y.A.; Jagtap, M.P.; Patil, M.G.; Khatri, N. Experimental investigation on the effect of soil solarization incorporating black, silver, and transparent polythene, and straw as mulch, on the microbial population and weed growth. *Chemosphere* **2023**, *336*, 139263. [CrossRef] [PubMed]
9. Cuello, J.P.; Hwang, H.Y.; Gutierrez, J.; Kim, S.Y.; Kim, P.J. Impact of plastic film mulching on increasing greenhouse gas emissions in temperate upland soil during maize cultivation. *Appl. Soil Ecol.* **2015**, *91*, 48–57. [CrossRef]
10. Khalil, A.H.P.S.; Chong, E.W.N.; Owolabi, F.A.T.; Asniza, M.; Tye, Y.Y.; Tajarudin, H.A.; Paridah, M.T.; Rizal, S. Microbial-induced CaCO 3 filled seaweed-based film for green plasticulture application. *J. Cleaner Prod.* **2018**, *199*, 150–163. [CrossRef]
11. Sintim, H.Y.; Bandopadhyay, S.; English, M.E.; Bary, A.I.; DeBruyn, J.M.; Schaeffer, S.M.; Miles, C.A.; Reganold, J.P.; Flury, M. Impacts of biodegradable plastic mulches on soil health. *Agric. Ecosyst. Environ.* **2019**, *273*, 36–49. [CrossRef]
12. Wang, Z.; Li, X.; Shi, H.; Li, W.; Yang, W.; Qin, Y. Estimating the water characteristic curve for soil containing residual plastic film based on an improved pore-size distribution. *Geoderma* **2020**, *370*, 114341. [CrossRef]
13. Fu, F.; Long, B.; Huang, Q.; Li, J.; Yang, C.; Zhou, W. Integrated effects of residual plastic films on soil-rhizosphere microbe-plant ecosystem. *J. Hazard. Mater.* **2023**, *445*, 130420. [CrossRef] [PubMed]
14. Li, Y.; Zhao, C.; Yan, C.; Mao, L.; Liu, Q.; Li, Z.; He, W. Effects of agricultural plastic film residues on transportation and distribution of water and nitrate in soil. *Chemosphere* **2020**, *242*, 125131.
15. Zhu, J.; Wang, Z.; Li, W.; Jia, H. Effect of residual plastic film on soil water salt, nutrient and cotton growth under long-term mulched drip irrigation. *J. Arid. Res. Environ.* **2021**, *35*, 151–156.
16. Li, B.; Huang, S.; Wang, H.; Liu, M.; Xue, S.; Tang, D.; Cheng, W.; Fan, T.; Yang, X. Effects of plastic particles on germination and growth of soybean (*Glycine max*): A pot experiment under field condition. *Environ. Pollut.* **2021**, *272*, 116418. [CrossRef]
17. Chen, P.; Gu, X.; Li, Y.; Qiao, L.; Li, Y.; Fang, H.; Yin, M.; Zhou, C. Effects of residual film on maize root distribution, yield and water use efficiency in Northwest China. *Agric. Water Manag.* **2022**, *260*, 107289. [CrossRef]
18. Kasirajan, S.; Ngouajio, M. Polyethylene and biodegradable mulches for agricultural applications: A review. *Agron. Sustain. Dev.* **2012**, *32*, 501–529. [CrossRef]
19. Li, M.; Zhang, Q.; Wei, S.; Liu, Z.; Zong, R.; Jin, T.; Zhang, M.; Sun, C.; Li, Q. Biodegradable film mulching promotes better soil quality and increases summer maize grain yield in North China Plain. *Arch. Agron. Soil Sci.* **2023**, *69*, 2493–2509. [CrossRef]
20. Eugenio, C.; Ida, D.; Lucia, O.; Maurizio, B.; Milena, P.; Elvira, F.; Mori, M.; Morra, L. Assessing Yield and Quality of Melon (*Cucumis melo* L.) Improved by Biodegradable Mulching Film. *Plants* **2023**, *12*, 219. [CrossRef] [PubMed]
21. Rugchati, O.; Thanacharoenchanaphas, K. Application of Biodegradable Film from Yam (*Dioscorea alata*) Starch in Thailand for Agriculture Activity. *In. J. Environ. Rural. Dev.* **2020**, *6*, 28–33.
22. Wang, Z.; Li, M.; Flury, M.; Schaeffer, S.M.; Chang, Y.; Zhao, T.; Jia, Z.; Li, S.; Ding, F.; Wang, J. Agronomic performance of polyethylene and biodegradable plastic film mulches in a maize cropping system in a humid continental climate. *Sci. Total Environ.* **2021**, *786*, 147460. [CrossRef] [PubMed]
23. Dubenok, N.N.; Mayer, A.V. Long-term Studies of the Hydrothermal Regime of Agrocenoses and a Combined Irrigation System for Its Regulation. *Russ. Agric. Sci.* **2022**, *48*, 123–127. [CrossRef]

24. Zhang, L.; Xu, C.; Yan, W.; Tan, G.; Sun, N.; Zhao, H.; Li, F.; Yu, J.; Meng, X.; Bian, S. Effects of Surface Mulching on Soil Water Temperature, Physiological Characteristics and Yield of Maize. *Int. J. Plant Prod.* **2023**, *17*, 283–296. [CrossRef]
25. Wang, W.; Xie, Y.; Li, H.; Dong, H.; Li, B.; Guo, Y.; Wang, Y.; Guo, X.; Yin, T.; Liu, X.; et al. Responses of lettuce (*Lactuca sativa* L.) growth and soil properties to conventional non-biodegradable and new biodegradable microplastics. *Environ. Pollut.* **2023**, *341*, 122897. [CrossRef] [PubMed]
26. Li, S.; Shen, C. Research progress on regulation of nitrate metabolism and utilization in crops. *J. Nanjing Agric. Univ.* **2020**, *45*, 848–855.
27. Qiu, Z.; Liao, J.; Chen, J.; Li, A.; Lin, M.; Liu, H.; Huang, W.; Sun, B.; Liu, S.; Zheng, P. Comprehensive analysis of fresh tea (*Camellia sinensis* cv. *Lingtou dancong*) leaf quality under different nitrogen fertilization regimes. *Food Chem.* **2024**, *439*, 138127. [CrossRef]
28. Zou, Y.; Zhang, Y.; Cui, J.; Gao, J.; Guo, L.; Zhang, Q. Nitrogen fertilization application strategies improve yield of the rice cultivars with different yield types by regulating phytohormones. *Sci. Rep.* **2023**, *13*, 21803. [CrossRef]
29. Liu, X.; Fu, N.; Li, C.; Wang, L.; Yang, Q. Trends and Causes Analysis of Water Requirement for Main Grain Crops in Henan Province. *Trans. Chin. Soc. Agric. Mach.* **2015**, *46*, 188–197.
30. Zhang, H.; Wang, Z.; Yu, S.; Teng, A.; Zhang, C.; Lei, L.; Ba, Y.; Chen, X. Crop coefficient determination and evapotranspiration estimation of watermelon under water deficit in a cold and arid environment. *Front. Plant Sci.* **2023**, *14*, 1153835. [CrossRef]
31. Antonietta, M.; Fanello, D.D.; Acciaresi, H.A.; Guiamet, J.J. Senescence and yield responses to plant density in stay green and earlier-senescing maize hybrids from Argentina. *Field Crops Res.* **2014**, *155*, 111–119. [CrossRef]
32. Zhao, B.; Ata, U.; Duan, A.; Gao, Y.; Lou, H.; Liu, Z.; Qin, A.; Ning, D.; Ma, S.; Liu, Z. Estimating the Impacts of Plant Internal Nitrogen Deficit at Key Top Dressing Stages on maize Productivity and Intercepted Photosynthetic Active Radiation. *Front. Plant Sci.* **2022**, *13*, 864258. [CrossRef] [PubMed]
33. Zheng, C.; Li, C.; Tian, L.; Shen, Z.; Feng, G.; Hou, W.; Liu, F.; Gao, Q.; Wang, Y. Mixture of controlled-release and normal urea to improve maize root development, post-silking plant growth, and grain filling. *Eur. J. Agron.* **2023**, *151*, 126994. [CrossRef]
34. Huang, F.; Liu, Z.; Li, Z.; Wang, B.; Zhang, P.; Jia, Z. Is biodegradable film an alternative to polyethylene plastic film for improving maize productivity in rainfed agricultural areas?—Evidence from field experiments. *Agric. Water Manag.* **2022**, *272*, 107868. [CrossRef]
35. Qi, D.; Chen, P. Responses of shoot biomass accumulation, distribution, and nitrogen use efficiency of maize to nitrogen application rates under waterlogging. *Agric. Water Manag.* **2022**, *261*, 107352. [CrossRef]
36. Li, G.; Cheng, G.; Lu, W.; Lu, D. Differences of yield and nitrogen use efficiency under different applications of slow release fertilizer in spring maize. *J. Integr. Agric.* **2021**, *20*, 554–564. [CrossRef]
37. Wang, Z.; Zhu, Y.; Zhang, J.; Li, W.; Bian, Q. Effects of Water and Nitrogen Fertilization on Physiological Characteristics and Yield of Cotton under Drip Irrigation in Mildly Salinized Soil. *Trans. Chin. Soc. Agric. Mach.* **2018**, *49*, 296–308.
38. Zhu, X.; Stephen, P.L.; Donald, R.O. Improving Photosynthetic Efficiency for Greater Yield. *Annu. Rev. Plant Biol.* **2010**, *61*, 235–261. [CrossRef]
39. Tian, G.; Qi, D.; Zhu, J.; Xu, Y. Effects of nitrogen fertilizer rates and waterlogging on leaf physiological characteristics and grain yield of maize. *Arch. Agron. Soil Sci.* **2020**, *67*, 863–875. [CrossRef]
40. Gao, F.; Khan, R.; Yang, L.; Chi, Y.; Wang, Y.; Zhou, X. Uncovering the potentials of long-term straw return and nitrogen supply on subtropical maize (*Zea mays* L.) photosynthesis and grain yield. *Field Crops Res.* **2023**, *302*, 109062. [CrossRef]
41. Li, H.; Kuang, N.; Gou, Q.; Ma, Y.; Li, Q. Effects of different film mulches on photosynthetic characteristics and yield of summer maize (*Zea mays* L.) in the North China Plain. *Arch. Agron. Soil Sci.* **2020**, *67*, 179–190. [CrossRef]
42. Duan, F.; Wei, Z.; Soualiou, S.; Zhou, W. Nitrogen partitioning in maize organs and underlined mechanisms from different plant density levels and N application rate in China. *Field Crops Res.* **2023**, *294*, 1088. [CrossRef]
43. Shao, G.; Cheng, H.; Dai, H.; Zhang, H.; Ai, J.; Liu, K.; Li, Z.; Zamanian, K.; Qian, X. Nitrogen uptake and utilization of two maize hybrids with contrasting nitrogen use efficiencies depending on fertilization amount. *Arch. Agron. Soil Sci.* **2023**, *69*, 2202–2217. [CrossRef]
44. Gao, Y.; Chen, J.; Wang, G.; Liu, Z.; Sun, W.; Zhang, Y.; Zhang, X. Different Responses in Root Water Uptake of Summer Maize to Planting Density and Nitrogen Fertilization. *Front. Plant Sci.* **2022**, *13*, 918043. [CrossRef]
45. Li, P.; Jia, L.; Chen, Q.; Zhang, H.; Deng, J.; Lu, J.; Xu, L.; Li, H.; Hu, F.; Jiao, J. Adaptive evaluation for agricultural sustainability of different fertilizer management options for a green manure-maize rotation system: Impacts on crop yield, soil biochemical properties and organic carbon fractions. *Sci. Total Environ.* **2023**, *908*, 168170. [CrossRef]
46. Yang, W.; Yan, N.; Zhang, J.; Yan, J.; Ma, D.; Wang, S.; Yin, L. The applicability of water-permeable plastic film and biodegradable film as alternatives to polyethylene film in crops on the Loess Plateau. *Agric. Water Manag.* **2022**, *274*, 107952. [CrossRef]
47. Chen, N.; Li, X.; Shi, H.; Hu, Q.; Zhang, Y.; Leng, X. Effect of biodegradable film mulching on crop yield, soil microbial and enzymatic activities, and optimal levels of irrigation and nitrogen fertilizer for the Zea mays crops in arid region. *Sci. Total Environ.* **2021**, *776*, 145970. [CrossRef]

48. Li, Y.; Ihsan, M.; Chi, Y.; Wang, D.; Zhou, X. Straw Return and Nitrogen Fertilization to Maize Regulate Soil Properties, Microbial Community, and Enzyme Activities Under a Dual Cropping System. *Front. Microbiol.* **2022**, *13*, 823963.
49. Fudjoe, S.; Li, L.; Anwar, S.; Shi, S.; Xie, J.; Wang, L.; Xie, L.; Zhou, Y. Nitrogen fertilization promoted microbial growth and N_2O emissions by increasing the abundance of nirS and nosZ denitrifiers in semiarid maize field. *Front. Microbiol.* **2023**, *14*, 1265562. [CrossRef]

plants

Article

Monitoring Plant Height and Spatial Distribution of Biometrics with a Low-Cost Proximal Platform

Giovanni Bitella [1], Rocco Bochicchio [1], Donato Castronuovo [1], Stella Lovelli [1], Giuseppe Mercurio [1], Anna Rita Rivelli [1], Leonardo Rosati [1], Paola D'Antonio [1], Pierluigi Casiero [2], Gaetano Laghetti [3], Mariana Amato [1,*] and Roberta Rossi [4]

1 School of Agriculture, Forestry, Food and Environmental Sciences, University of Basilicata, 85100 Potenza, Italy; gianfranco.bitella@gmail.com (G.B.); bochicchiorocco@gmail.com (R.B.); donato.castronuovo@unibas.it (D.C.); stella.lovelli@unibas.it (S.L.); giuseppe.mercurio@unibas.it (G.M.); annarita.rivelli@unibas.it (A.R.R.); leonardo.rosati@unibas.it (L.R.); paola.dantonio@unibas.it (P.D.)
2 Department of European and Mediterranean Cultures, Environment, and Cultural Heritage, University of Basilicata, 85100 Potenza, Italy; pierluigi.casiero@unibas.it
3 Istituto di Bioscienze e BioRisorse (CNR-IBBR), Consiglio Nazionale delle Ricerche, 80055 Bari, Italy; gaetano.laghetti@ibbr.cnr.it
4 Research Centre for Animal Production and Aquaculture (CREA-ZA), Council for Agricultural Research and Economics, 81100 Bella Muro, Italy; roberta.rossi@crea.gov.it
* Correspondence: mariana.amato@unibas.it

Abstract: Measuring canopy height is important for phenotyping as it has been identified as the most relevant parameter for the fast determination of plant mass and carbon stock, as well as crop responses and their spatial variability. In this work, we develop a low-cost tool for measuring plant height proximally based on an ultrasound sensor for flexible use in static or on-the-go mode. The tool was lab-tested and field-tested on crop systems of different geometry and spacings: in a static setting on faba bean (*Vicia faba* L.) and in an on-the-go setting on chia (*Salvia hispanica* L.), alfalfa (*Medicago sativa* L.), and wheat (*Triticum durum* Desf.). Cross-correlation (CC) or a dynamic time-warping algorithm (DTW) was used to analyze and correct shifts between manual and sensor data in chia. Sensor data were able to reproduce with minor shifts in canopy profile and plant status indicators in the field when plant heights varied gradually in narrow-spaced chia (R^2 = 0.98), faba bean (R^2 = 0.96), and wheat (R^2 = up to 0.99). Abrupt height changes resulted in systematic errors in height estimation, and short-scale variations were not well reproduced (e.g., R^2 in widely spaced chia was 0.57 to 0.66 after shifting based on CC or DTW, respectively)). In alfalfa, ultrasound data were a better predictor than NDVI (Normalized Difference Vegetation Index) for Leaf Area Index and biomass (R^2 from 0.81 to 0.84). Maps of ultrasound-determined height showed that clusters were useful for spatial management. The good performance of the tool both in a static setting and in the on-the-go setting provides flexibility for the determination of plant height and spatial variation of plant responses in different conditions from natural to managed systems.

Keywords: plant spatial variation; canopy height; stress response monitoring

Citation: Bitella, G.; Bochicchio, R.; Castronuovo, D.; Lovelli, S.; Mercurio, G.; Rivelli, A.R.; Rosati, L.; D'Antonio, P.; Casiero, P.; Laghetti, G.; et al. Monitoring Plant Height and Spatial Distribution of Biometrics with a Low-Cost Proximal Platform. *Plants* **2024**, *13*, 1085. https://doi.org/10.3390/plants13081085

Academic Editor: Youzhen Xiang

Received: 26 February 2024
Revised: 4 April 2024
Accepted: 6 April 2024
Published: 12 April 2024

1. Introduction

Plant height is the result of many genetic, environmental, and management factors and is one of the most important traits in plant ecology [1]. It has been discussed in terms of strategy linked to latitude and correlations with environmental variables. It affects the ability of species and individuals to compete for light and, therefore, to acquire carbon through photosynthesis and is a determinant of water evapotranspiration and seed dispersion [2,3]. A general relationship between height and biomass of vegetation across a range of ecosystems has been proposed [4] as a way to assess aboveground carbon stocks in natural systems. The height of plants is a very useful parameter in agronomy; it is a commonly recognized indicator of crop growing status. It is related to crop yield [5] and is the most relevant parameter for the fast determination of crop response to management and crop variability alone or in conjunction with other plant characteristics [6]. Phenotyping for the selection of high-yielding or stress-resistant genotypes uses plant height as one of the fast indicators of plant response. In wheat breeding programs, the importance of plant height has long been recognized [7]. Height variability at early stages of development could be used as a proxy of crop establishment and productive performances [8]. Monitoring plant growth at a high spatial density is essential in precision agriculture [9].

However, measuring plant height repeatedly during the season and at high resolution is a long and expensive task if performed manually. Technologies currently used to provide this parameter are as follows: stereo vision, laser image detection and ranging (LiDAR), and ultrasonic sensors. Stereo vision is a method of distance measurement based on the different perspectives on the same scene [10,11]. Its use is in expansion, related to the increasing diffusion of drones, but requires expensive computation [12]. The LiDAR yields a distance measure based on the time of flight of a laser beam emitted by the sensor and captured back by a photodetector [13]. It has been used to estimate features of herbaceous vegetation like crop density [14] and weed infestation [15]. Terrestrial LiDAR offers high resolution and cover, but its cost is still high. Like the LiDAR, an ultrasonic sensor is capable of detecting the distance from an obstacle indirectly by measuring the time of flight (TOF) of an ultrasonic pulse emitted by the sensor and echoed back by an obstacle. Compared to LiDAR, ultrasonic sensors are less expensive and do not involve possible harm to the naked eye. LiDAR, on the other hand, can make measurements at a larger distance and features narrow point detection; this can be a pro as it gives a higher accuracy but also a con as given objects may be missed in a prospection. Ultrasonic sensors have been used in agriculture to provide cheap and repeatable distance information related to crop biomass [16], canopy density, to optimize pesticide applications [17], and for weed detection [18].

Ultrasonic devices are still underutilized in agriculture and can offer, at low cost, very useful information to researchers and farmers. Integrating crop sensors through open-source hardware projects might help fill this gap. Open hardware platforms are easy to realize and use; their low cost, coupled with the increasing amount of information sharing, supports the rapid development of new-generation, inexpensive, and very flexible scientific instrumentations [19,20]. Ultrasound-based high-throughput sensors or platforms have been proposed for use in different crops [21–27], mostly in the on-the-go mode.

Montazeaud and co-authors [28] recently proposed a low-throughput sensor intended for manual use for small-scale applications such as plot experiments where measurements of the height of individual plants are needed. They tested it on single-plant measurements on sorghum. The device is easy to use although limited to static measurements.

We aimed to devise a low-cost system for measuring plant height which could be used in a flexible way, both on-the-go and in static mode, and to prove its usefulness in different applications from single plants to plots and whole fields. We therefore aimed to test the device across a range of vegetation types, from crop row transects to whole-field mapping.

2. Results

2.1. Lab Tests

Figure 1 reports the results of laboratory testing. Ultrasound-derived height (hus) followed the contour of manually measured height (h) with a significant correlation ($r = 0.86$, p-value < 0.05). However, a slight shift in data was observed and the sensor overestimated the height of shortest targets by 0.89 cm on average (Figure 1).

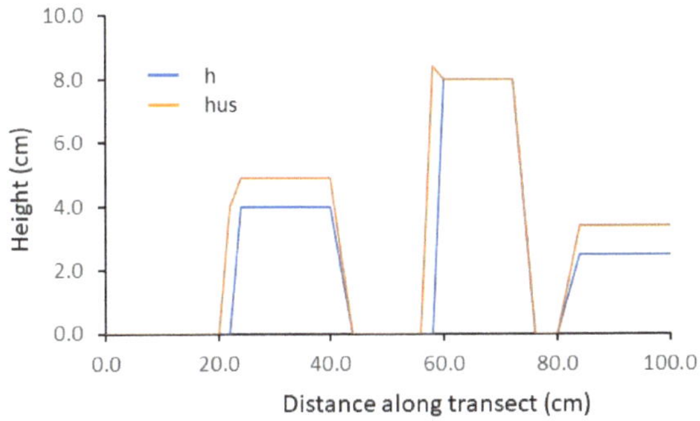

Figure 1. Comparison of sensor-determined and measured height of parallelepipeds in the laboratory. ultrasonic data (hus = orange line) overlaid by ground truth data (h = blue line).

The sample cross-correlation shows that the highest correlation (r = 0.88) occurs at lag 1. Hence, a one-lag shift between the two series slightly improved the overall correlation. The Dynamic Time Warping Normalized Distance (DTW) between sensor data and manually measured height was 0.39.

2.2. Faba Bean (*Vicia faba L.)*

Figure 2 reports ultrasound-measured heights as a function of ground truth data for *Vicia faba* L. Plant heights were significantly ($p < 0.05$) related to sensor-measured height data with high R^2 values (0.96 for the regression of pooled data), which were not significantly improved by dividing var. *major* from var. *minor* data. Values for single varieties were substantially aligned (Figure 3b,c).

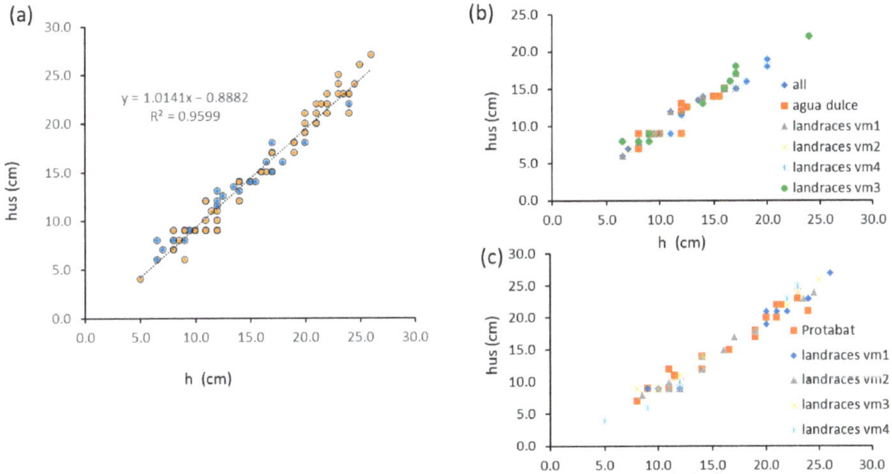

Figure 2. Height measurements in a *Vicia faba* L. field setting. (**a**) bivariate plot of ultrasound-measured heights as a function of ground truth data for *Vicia faba* L. var. *major* (blue dots) and *Vicia faba* L. var. minor (orange dots); (**b**) bivariate plot of ultrasound-measured heights as a function of ground truth data for *Vicia faba* L. var. *major* by variety; (**c**) bivariate plot of ultrasound-measured heights as a function of ground truth data for *Vicia faba* L. var. *minor* by variety.

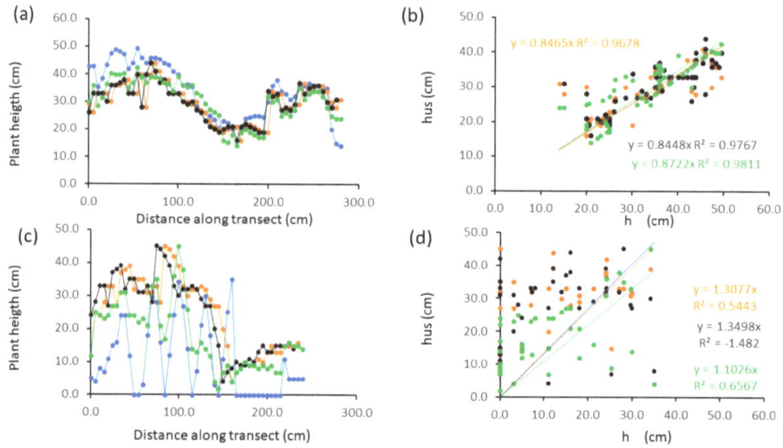

Figure 3. Height measurements in *Salvia hispanica* L. field setting. (**a**) row with plants spaced 5 cm: bivariate plot of ground truth data (blue dots), ultrasound-measured heights (orange dots) and ultrasound-measured heights shifted of 1 lag distance (black dots), data from dynamic time-warping (green dots); (**b**) row with plants spaced 5 cm: regression between measured height and ultrasound-measured height (orange dots and regression equations), ultrasound-measured height shifted of 1 lag distance (black dots and regression equations), and data from dynamic time-warping (green dots and regression equation); (**c**) row with plants spaced 30 cm: bivariate plot of ground truth data (blue dots) ultrasound-measured heights (orange dots), ultrasound-measured height shifted of 2 lag distance (black dots), data from dynamic time-warping (green dots); (**d**) row with plants spaced 30 cm: regression between measured height and ultrasound-measured height (orange dots) and ultrasound-measured height shifted of 2 lag distance (black dots and regression equations); data from dynamic time-warping (green dots and regression equations).

2.3. Chia (*Salvia hispanica* L.)

In *Salvia hispanica* field transect on rows with 5-cm plant spacing (Figure 3a,b) the crop height pattern obtained with manual measurements (Figure 3a blue dots) was closely reproduced by ultrasound-derived height (Figure 3a orange dots) in the zone between 40 and 260 cm of distance along the transect, but underestimated plant height of about 4.4 cm. At the transect edges (About 40 cm from each end of the row) sensor data failed to reproduce the height profile and differences were higher. The sample cross-correlation was maximum at lag 1 therefore a second series of height data derived from shifting ultrasound data by one lag was created (Figure 3a black dots). The use of a dynamic time warping algorithm yielded a DTW distance of 3.38, and data corrected for DTW are shown in Figure 3a as black dots.

Across the whole transect, including edges, regression between hus and ground truth (h) values (Figure 3b) yielded a significant relationship ($p < 0.05$) both at 0 (raw data) and 1 lag (respectively, orange and black dots and regression lines). The non-significant intercept was removed, and the regression line was forced to pass through the origin yielding high values of R^2 even for raw data (almost 0.97 in Figure 3b, orange dots), and slightly improved (0.98) for data shifted by 1 lag (Figure 3b black dots) and DTW-warped (Figure 3b green dots).

In the case of plant spacing of 30 cm (Figure 3c,d), hus data (orange dots) follow the overall crop profile but fail to reconstruct short-scale variability in plant height, and especially bare-soil points (e.g., at 50, 85 and 115 cm along the transect) and a whole bare-soil stretch between cm 165 and 215 along the transect. The maximum correlation is found at lag 2 but even data shifted in 2 lags (black dots in Figure 3c,d) do not reproduce short-scale variability and relevant very low height or zero values. The use of a dynamic time warping algorithm yielded a DTW distance of 7.94, and a better reproduction of the pattern of manual data is found with a data series calculated with dynamic time-warping (Figure 3c green dots). However, very low and zero values are not reproduced. The correlation between manual and sensor data is lower for plants spaced at 30 cm than for those at 5 cm but is still significant (p-value = 0.04). The univariate regression model still explains a considerable amount of the total variability if the line is forced to pass through the origin (R^2 = 0.57 for data after 2-lag shifting and 0.66 for data after DTW correction). Sensor data overestimate height by about 14.0 cm on average in this transect after 2-lag shifting (black dots) and 7.3 cm after correction with DTW (green dots).

2.4. Alfalfa (Medicago sativa L.)

Figure 4 shows maps of ultrasound-measured canopy height at three plant heights (Figure 4) in an alfalfa (*Medicago sativa* L.) field.

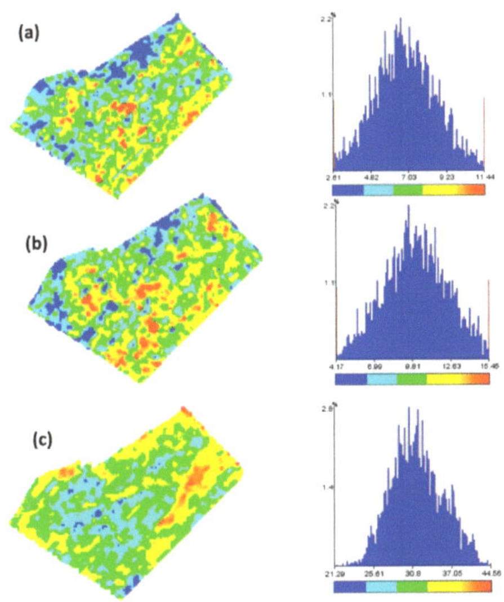

Figure 4. Maps of ultrasound-measured canopy height in a *Medicago sativa* L. field at (**a**) plant maximum height 12.07 cm; (**b**) maximum plant height 16.75 cm; (**c**) maximum plant height 57.56 cm. Left: plant height (hus) maps. Right: frequency distributions.

Values of ultrasound-measured plant height ranged between 2.51 and 12.07 cm on the first date, between 3.97 and 16.75 am on the second, and between 8.02 and 57.76 cm on the third. Plant height showed an unimodal frequency distribution at all three dates (Figure 4 right), but values were not spatially distributed at random in the field: maps show that clusters of low values (blue areas) were found in the top and left field zones, and high values (red areas) in the center of the field in the first two dates, whereas at the third date high values (red areas) were found at the top edge and in a strip inside the field, and low values (blue areas) in the center, and frequencies were slightly skewed to the left: height values lower than the mode were less frequent.

Figures 5–7 report data collected in the ground-truth areas and summary statistics for plant biometrics and sensor indices are reported in Table 1.

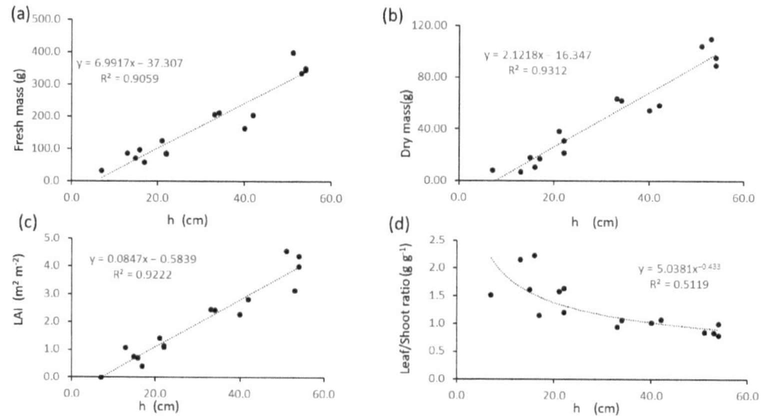

Figure 5. Bivariate plots of alfalfa biometrics as a function of plant height: (**a**) fresh biomass; (**b**) dry biomass; (**c**) Leaf Area Index; (**d**) leaf/stem mass ratio.

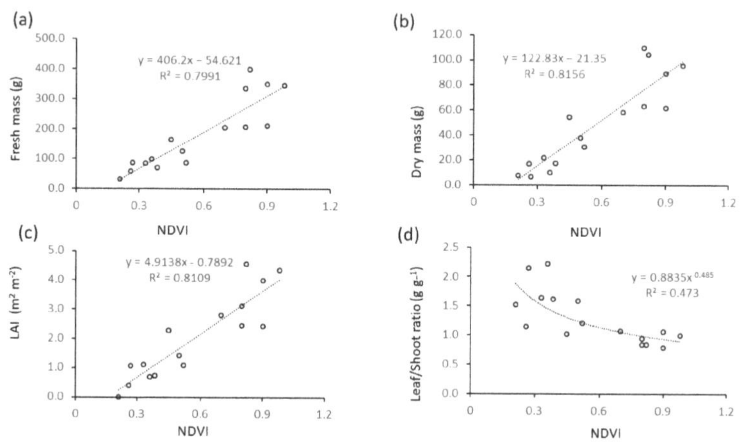

Figure 6. Bivariate plots of alfalfa biometrics as a function of NDVI: (**a**) fresh biomass; (**b**) dry biomass; (**c**) Leaf Area Index; (**d**) leaf/stem mass ratio.

Table 1. Plant biometrics and sensor indices in the ground-truth areas for *Medicago sativa* L. St dev = standard deviation; CV% = coefficient of variation; h = plant height; hus = ultrasound-measured plant height.

	LAI of Alfalfa	Fresh Biomass	Dry Biomass	NDVI	Leaf/Total Mass Ratio	h	hus
	(m^2 m^{-2})	(g m^{-2})	(g m^{-2})		(g g^{-1})	(cm)	(cm)
Min	0	128.00	26.40	0.21	0.44	13.02	6.11
Max	4.56	1592.00	440.00	0.98	0.69	54.28	45.07
Mean	2.03	714.25	196.65	0.57	0.55	32.47	27.14
St dev	1.44	479.58	143.55	0.26	0.08	15.56	14.84
CV%	47.91	67.14	73.00	45.92	14.44	47.91	54.66

Values showed a high variability: the coefficient of variation ranged between 45.82% of NDVI and 73.00% of plant dry mass, with the exception of the leaf/total mass ratio where values were lower than 15%.

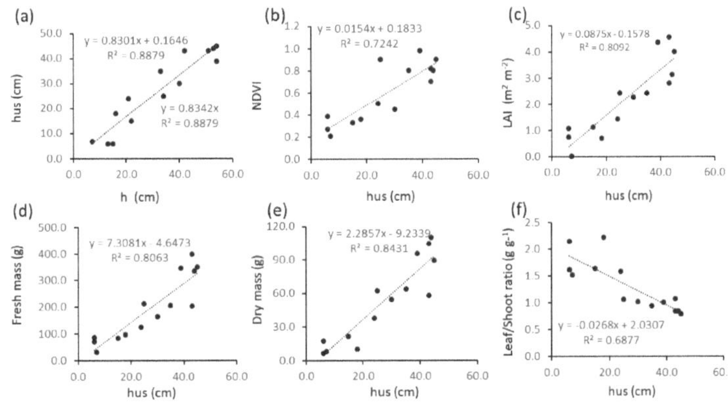

Figure 7. Relationships between alfalfa biometrics and ultrasound-measured plant height (hus): (**a**) hus as a function of manually measured plant height (h); (**b**) NDVI as a function of hus; (**c**) Leaf Area Index as a function of hus; (**d**) fresh mass as a function of hus; (**e**) dry mass as a function of hus; (**f**) leaf to stem mass ratio as a function of hus.

Figure 5 shows the relationships between manually measured plant height (h) and other biometrics. Significant ($p < 0.05$) linear regressions with R^2 values higher than 0.9 were found with fresh and dry mass and LAI (Figure 5a–c), and a power regression with leaf/stem ratio (Figure 5d).

The same regression models were fitted for relationships between normalized difference vegetation index (NDVI) and other biometrics (Figure 6), which were all significant ($p < 0.05$) and linear for fresh and dry mass and LAI (Figure 6a–c), and power regression for leaf/stem ratio (Figure 6d). Values of R^2 were lower than those found for regressions with plant height (Figure 5).

Figure 7 shows the relationships between manually (h) and ultrasound-measured (hus) plant height (Figure 7a) and between hus and other biometrics (Figure 7b–f). For all tested models, the highest R^2 values were found for linear regressions, and they were all significant ($p < 0.05$). Sensor-derived height (hus) was significantly related to ground-truth plant height (h) (Figure 7a), and the model explained 89 to 98% of the variability if the intercept is kept or removed, respectively. The hus variable was also significantly related to NDVI with a 0.72 R^2 value (Figure 7b). Regressions of LAI and biomass with hus (Figure 7c–e) were characterized by R^2 values higher than those of the same variables with NDVI (Figure 6) and lower than those with h (Figure 5). The regression of leaf/stem ratio with hus (Figure 7f) was linear and had a higher R^2 than those with h (Figure 5d) and NDVI (Figure 6d).

2.5. Wheat (*Triticum durum Desf.*)

Results from the wheat experimental field are reported in Figures 8 and 9. Maps of sensor values across the experimental field at the end of tillering are reported in Figure 8. A number of 481 to 493 values were acquired in each plot and the maps of hus (Figure 8a) and NDVI (Figure 8d) show differences between plots of the C and W treatments. In the C plot, the blue color dominates, corresponding to low values (color-scales for hus and NDVI at Figure 8b and e respectively), whereas in the W plots, the dominant color is red, corresponding to high values. Maps also show within-plot variability with different shades of blue in the C plots and colors from green to black in the W plots. Frequency distributions of values for the whole field (Figure 8b,e) were bimodal, corresponding to the different modes of the treatments, respectively for hus 4.14 cm in C and 12.98 cm in W, and for NDVI values of 0.11 in C and 0.27 in W. Average values of the W plots were more than twice those of the C plots for ìboth hus and NDVI. Data from single transects at the same phenological stage, when maps were made (end of tillering), are summarized across the whole field in Table 2.

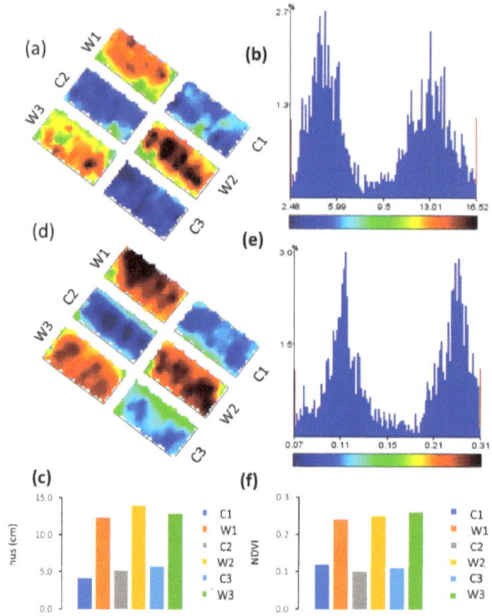

Figure 8. Maps of the experimental field of wheat; (**a**) hus values across the field; (**b**) hus values frequency distribution; (**c**) hus in each plot: colored bars = average values. Line bars: standard deviation; (**d**) NDVI values across the field; (**e**) NDVI values frequency distribution; (**f**) NDVI in each plot: colored bars = average values. Line bars: standard deviation. C1, C2, and C3 = three replicate plots of treatment C = wheat cut at stage 30 of the Zadoks scale; W1, W2, W3 = three replicate plots of treatment W: uncut wheat plants.

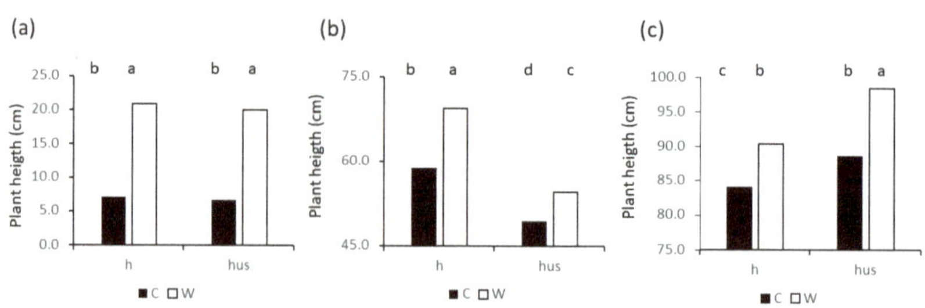

Figure 9. Height of wheat plants from manual measurements (h) or derived from an ultrasound sensor (hs). Values represent averages of treatments C = wheat cut at stage 30 of the Zadoks scale and W: uncut wheat plants. (**a**): the end of tillering; (**b**) booting; (**c**) grain filling. Different letters indicate significant ($p < 0.05$) differences in Tukey's *post-hoc* test.

Figure 9 reports averages of plant heights from cut (C) and uncut (W) treatments. Values were significantly ($p< 0.05$) different throughout the growth cycle for both h and hus.

At the end of tillering (Figure 9a) sensor data (hus) were not significantly different from manually measured heights for both cut and uncut wheat, whereas hus was significantly lower than h at the booting stage (Figure 9b) and significantly higher within cutting treatment at grain filling (Figure 9c). Nevertheless, the overall linear regression between h and hus on data from the three phenologoical stages was significant ($p < 0.05$) and explained 96% of the variability (hus = 1.0031 h + 1.68 $R^2 = 0.96$).

Table 2. Plant biometrics and sensor indices in *Triticum durum* Desf. transects. hus = ultrasound-measured plant height; NDVI = normalized difference vegetation index. LAI = Leaf area Index. St dev = standard deviation; CV% = coefficient of variation. Summary statistics are calculated on all data across the experimental field.

	hus	**h**	**NDVI**	**LAI**
	(cm)	**(cm)**		**(m^2 m^{-2})**
Min	5.53	6.02	0.10	0.32
Max	24.57	24.92	0.26	2.02
Mean	13.68	14.37	0.18	1.02
St dev	7.02	7.24	0.08	0.61
CV%	51.27	50.41	42.89	60.19

The variability of measurements across the whole experimental field, regardless of treatments, was quite high, with coefficients of variation ranging from 42.89% for NDVI to 60.19% for LAI. Significant ($p < 0.05$) linear regressions are found between h and hus (hus = 0.9638 h − 0.1637 R^2 = 0.99), and between hus and the leaf area index (LAI = 0.0731 hus + 0.034 R^2 = 0.79) and the normalized difference vegetation index (NDVI = 0.0103 hus + 0.042 R^2 = 0.98).

3. Discussion

In our data the bivariate relationship between ultrasound-measured height and manual data was always significant, therefore our sensor can be considered a useful tool, and hus can be taken as a proxy of plant height across a range of canopy types, from rows of spaced plants (e.g., *S. hispanica*) to canopies spread over whole fields (e.g., *M. sativa*), and this may be extended to different conditions of crops or natural vegetation.

This agrees with data in the literature where ultrasound measurements provide reliable plant height data in a range of crops, and in static or on-the-go modes [21–28].

Our sensor performed with different accuracy in the different modes we tested: the regression models explained up to 99% of the variability in wheat at the end of tillering, but as little as 54 to 66% of the variability when used on-the-go on wide-spaced chia plants (Figure 3c,d). Sharp variation of the target's contour are not completely caught as in lab setting (Figure 1) or at the edge of chia rows (Figure 3) or in case of wide-spaced chia plants (Figure 3c,d) where the sensor was much less accurate in reconstructing short scale variability in plant height and did not pick up narrow canopy voids. This is due to interference of neighboring features within the sensor's field of view and has been documented in the literature (e.g., [15]). More gradual height changes were reproduced by the sensor with higher accuracy, as in the chia transect at 5-cm plant spacing where changes in height corresponded to gradual variations in chia plant tops, or to the presence of shorter *Amaranthus retroflexus* L. plants which represented the main segetal species in the field.

No systematic error was recorded in some cases for both static or on-the go measurements (e.g., in faba bean, Figure 2 or alfalfa, Figure 7a or wheat during tillering, Figure 9a). Nevertheless they emerged in other cases: under-or over-estimation and minor shifts were found in laboratory measurements (Figure 1) and in chia (Figure 3a,d). A measure of misalignment was obtained by cross correlation or using a dynamic time warping algorithm. After quantifying misalignments, we used cross-correlation or dynamic time-warping distance (DTW) to shift hus measurements one of one or few lags and obtained a better correspondence of sensor and measured data (e.g., Figure 3c,d for chia plants spaced at 30 cm). Cross correlation is a feature-based measure of similarity between data series [29] and was used in this work to shift sensor data with respect to manual data in order to improve data matching. The need to shift data series can be ascribed to a lag time in data collection or ground-positioning system (gps) data recording linked to the speed of the moving sensor. The dynamic time warping algorithm is a shape-based procedure to measure similarity between data sequences, through quantifying the distance between similar elements in different series of data [29]. It originates from time-series analysis [30] but can be applied to shape-matching, and in particular to find out if similarity or matching in shape can be found between two data sequences which are out of phase. The Dynamic Time Warping (DTW) distance is therefore a measure of the level of dissimilarity between data series. In our data the DTW-normalized distance was 0.39 in lab setting, 3.38 in the chia row at 5-cm spacing and 7.94 in the chia row with 30-cm spacing, this showing an increasing level of dissimilarity between sensor and manually determined height.

Even where alignment between sensor and manual data series is improved through shifts based on cross-correlation or analyzed through time-warping, the problem of over- or under- estimation persists. In our data for instance the chia transect with plants spaced 30 cm (Figure 3c,d) showed overestimation of 7.3 to 14.4 on average, respectively for data after shifting based on DTW or cross-correlation. This was due to inability to detect very low or zero values when interspersed with tall plants. In *Salvia hispanica* (Figure 3) hus values closely reproduced plant height except for at the edges of measured transect. In this experiment the edges of transects corresponded to plot edges, therefore to regions where bare soil and/or segetal species were found, with height different from chia. We hypothesize that in this case specific edge effects coupled with shifts and systematic errors in sensor measurements we recorded in this dataset may be due to an imperfect perpendicularity of the sensor, which may have picked up reflections from bare soil at the beginning of the transect, thus underestimating plant height, and reflections from the crop at the transect end, thus overestimating plant height.

Ultrasound measurements underestimated wheat height at booting (Figure 9b), and overestimated it at grain filling (Figure 9c) and this may be ascribed to general factors and specific issues linked to the method of height measurement. For the booting stage the flag leaf possibly was not a consistent enough target for the ultrasound beam to pick it up completely. This is consistent to the type of underestimation explained by Sui and Thomasson [31] as occurring when the sensor is not centered on the top or top-leaf of the plant (And this may happen frequently for on-the go measurements): in such cases the closest leaf that echoes back the signal would not be the highest. At the wheat booting stage (Figure 3c) a lower h is explained by the fact that it did not include awns as described in the Materials and Methods section.

In general, biases between manual and sensor measures could have been caused by several reasons ranging from sensor misalignment, to soil micro-topography and to the influence of temperature, which we didn't consider in our analysis. Several factors can reduce ultrasonic accuracy: the presence of systematic features like ridges and furrows, the influence of air temperature and the inherent sensor transducer accuracy. Canopies are porous, hierarchical structures, therefore the echoing of the signal depends on leaf morphology, angle and canopy architecture. Also inconsistencies might be due to the multiple reflected signals within the signal field of view [31]. All these issues lead to a cumulative error, and must be taken into account if quantitative predictions are needed. Our data confirm that a crop- and even a crop-stage specific calibration would always be required [32,33]. Having experimented our sensor in a range of modes from static (faba bean) to on-the-go at the row (chia, wheat), plot (wheat, alfalfa) and field (alfalfa) scale and having found different merits and drawbacks of each model, we can add that sensors should be tested in different conditions and modes.

Nevertheless in our data even overall regression models encompassing different growth stages or crop varieties were significant and explained a large part of the variability (e.g., 96% in wheat across growth stages and in faba bean across botanical and varieties, accessions and commercial varieties) Also, even where the sensor systematically under-estimated or overestimated canopy height, areas where canopy height changed could be well mapped (Figures 2, 3a,b and 4–9), except for very narrow variations as in chia spaced plants (Figure 3c,d), and differences between experimental treatments were well quantified (Figures 8 and 9).

In our measurement settings we aimed at different conditions: we tested it on broadleaf (e.g., faba bean) and narrowleaf (wheat), plants with large (e.g., chia) or small (e.g., alfalfa) laves and on a wide range of plant heights over which the sensor was tested, not only between but also within experiments, as quantified by high values of coefficients of variation. We had satisfactory to excellent agreement with ground-truth data, and this confirms the sensor is a good tool for different herbaceous vegetation types and can pick up field variability and possibly help in spatial applications. Applications in ecology include the spatial distribution of primary production and of plant light interception. A general relationship between height and biomass of vegetation across a range of ecosystems has been proposed [4] as a way to assess aboveground carbon stocks in natural systems, and our sensor may provide a flexible tool for height measurements over large areas or along transects.

Monitoring plant growth at a high spatial density is essential in precision agriculture [8] therefore height maps as those we show for alfalfa (Figure 4) or wheat (Figure 8) can be of assistance in precision farming operations.

Important crop biometrics other than manually measured height are well related to ultrasound measurements in our data, such as the leaf area index (R^2 = 0.79 in wheat and 0.81 in alfalfa) and the NDVI (R^2 = 0.98 in wheat and 0.72 in alfalfa). Plant height proved to be a better predictor of biomass than NDVI in alfalfa, where both h (Figure 5) and hus (Figure 7) showed higher R^2 values in the bivariate regression models with fresh (Figures 5a and 7d) and dry (Figures 5b and 7e) mass than than NDVI (Figure 6a,b). As vegetative biomass is the commercial product of alfalfa, a map of biomass corresponds to a map of forage yield in this crop. Yield maps are crucial tools for driving precision farming operations [34], and given the high predictive value of hus over biomass (R^2 = 0.81 for fresh and 0.84 for dry mass, Figure 7a,e), maps of ultrasound-determined height can be considered proxies of yield maps and used for driving delineation of uniform management zones in spatially-aware farming [35].

A relationship between ultrasound-measured height and biomass in forage crops has also been found by Fricke and co-authors [15] but with lower R^2 than in our case. Other measurements on forage crops [36,37] were less accurate than in our case with R^2 values between 0.7 and 0.8, but over a wide range of conditions, and results nare also discussed in terms of vehicle speed. Authors, though, stress higher accuracy compared to other non-desctructive methods for predictiong forage biomass like the rising-plate meter [37] and airborne or satellite-based radiometric methods constrained by equipment cost, expertise and/or meteorological conditions.

In our data hus was significantly ($p < 0.05$) and strongly related to LAI (R^2 = 0.79 for wheat to 0.81 for alfalfa), therefore ultrasound-measured height may also be used as a fast measurement for agronomic decisions related to leaf area as Leaf Area Index is a crucial parameter for crop modeling and input management, especially irrigation (e.g., [38]. In the case of alfalfa we also found an inverse relationship of height with the leaf to shoot ratio, which is a parameter of forage quality, and this widens the range of agronomic decisions our sensor may assist with.

The board used for our sensor is especially amenable for flexible uses and therefore to address error sources linked to different settings (e.g., [39]) and for use in mono-sensors or for multi-sensor platform (e.g., [40]) Internet of Things applications given the different connectivity modes from Wi-Fi to Bluetooth Classic and Bluetooth Low Energy (BLE), Sim card or Ethernet support and additional interfaces, including UART, SPI, I2C, and ADC, enabling connection to a wide range of peripheral devices. Flexibility is also given by a high number of General-Purpose Input/Output pins and a large flash memory for programming and storage, and compatibility with other boards and development platforms.

4. Materials and Methods

4.1. Ultrasound Sensor Platform

The platform used in this study was composed of:

(1) An ESP32 board (Espressif Systems, Singapore) with a dual-core microcontroller. Tensilica Xtensa 32-bit LX6 microprocessor with wireless connectivity Wi-Fi: 802.11 b/g/n/e/i (802.11n @ 2.4 GHz up to 150 Mbit/s) and Bluetooth: v4.2 BR/EDR and Bluetooth Low Energy (BLE). The current cost of an ESP32 board ranges from 1.75 to 8 Euros depending on the source.

(2) An ultrasound sensor was an HC-SR04 (Picaxe, Revolution Education Ltd, Bathh, UK) transmitting at 40 KHz frequency, and operating between 3 and 400 cm of distance with accuracy of 3 mm with a cone of 45 degrees from the sensor. The HC-SR04 rapidly generates a series of ultrasound pulses which propagate in a straight line in front of the sensor. The ultrasounds hit an object in front of the sensor and are reflected back towards the sensor, which detects the time taken for the ultrasound pulses to travel from their source to the object and back. The sensor uses the elapsed time to calculate the distance between itself and the object as:

$$\text{Distance} = (\text{Elapsed time} \times \text{Speed of sound})/2$$

The current cost of an HC-SR04 ranges from 2 to 10 Euros depending on the source,

(3) A Zs-040 module which sends data via Bluetooth. This was added in order to simplify hardware and make data easily available in real time thanks to transmission to a PC or smartphone. The current cost ranges from 0.3 to 10 m Euros depending on the source.

Connections of electronic circuits are depicted in Figure 10.

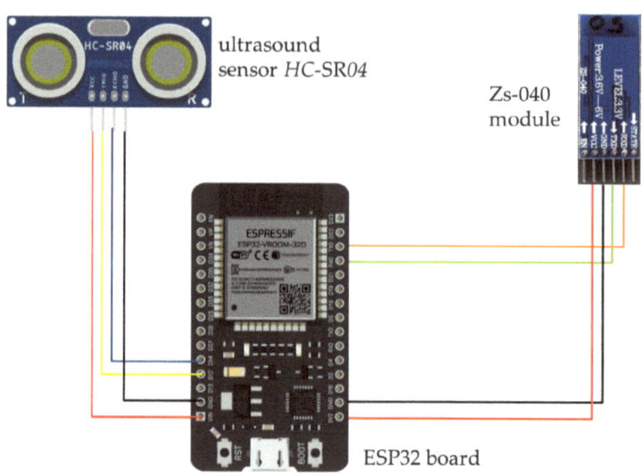

Figure 10. Diagram of connections between components of the platform.

All the electronic components were enclosed in a protective plastic case (current cost 5 Euros). The whole device can be easily powered through the USB-C port. We connected a 7860 mAh power bank, commonly used for charging smartphones, to the USB port (current cost 15 Euros), which allowed its use for several hours.

4.2. Data Collection

Ultrasound data were collected in both lab (static measures) and field setting (static or on-the-go).

Static measurements were made after mounting the sensor on a pole at a fixed distance from the ground (e.g., Figure 11a).

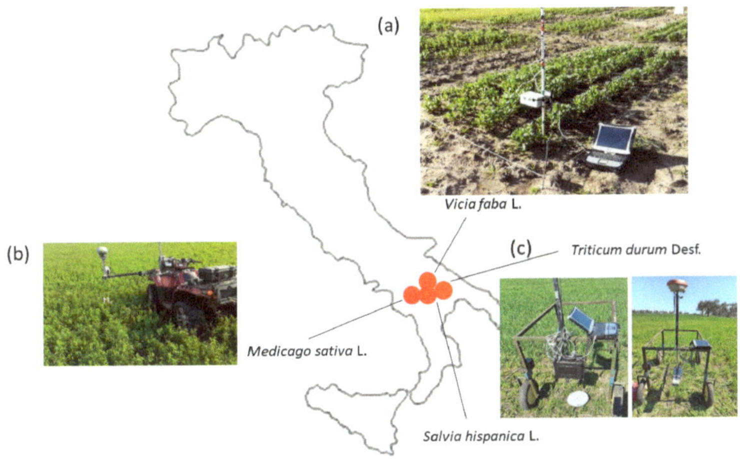

Figure 11. Sites of field measurements and sensor modes of data acquisition: (**a**) pole-mounted for static measurements in a faba bean filed; (**b**) mounted on a quad for on-the go or static measuremts in an alfalfa field; (**c**) mounted on a wheeled chassis for on-the-go measurements in wheat plots. The latter setting was used in a chia field.

On-the-go measurements were collected connecting a differential GPS and the sensor was mounted either on a quad (Figure 11b) or on straddle wheeled chassis (Figure 11c), kept at a fixed distance above the ground (sensor distance = hs) pointing vertically down to the row, and towed manually across the field.

4.3. Sensor Testing

To test the ultrasonic sensor accuracy data have been ground-calibrated taking manual measurements of target height.

In all tests the height of target objects or vegetation from ultrasound measurements (hus) was calculated as:

$$\text{hus} = \text{hs} - \text{usd (cm)}$$

where

hs = height of sensor from the ground

usd = ultrasound-measured distance between target objects or vegetation and sensor.

The sensor was tested in conditions of different complexity:

4.3.1. Lab Test

The sensor was mounted on a wheeled chassis at the distance of 30 cm from the floor where dark boxes of heights from 2.5 to 8 cm were aligned in a transect, and the chassis was shifted along the transect at an average speed of 0.1 m s^{-1}. The ultrasonic sensor triggered 5 measures per second, e.g., a measurement every 0.02 m.

Field tests were conducted at four sites in Southern Italy as indicated in Figure 11 on crops of different geometry and namely:

4.3.2. Faba Bean (*Vicia faba* L.)

The field was located at Lavello (Italy (Lat. N 41°07′28.49″, Long E 15°91′89.59″) and texture data were: 31% sand, 33% silt, 36% clay, and organic matter amounted to 16.4 g kg^{-1}. Measurements were made in static mode on row-planted single plants of faba bean (*Vicia faba* L.). For the botanical variety *Vicia faba* (L.) var. *maior* (Harz) Beck, we used the commercial variety Aguadulce and accessions sourced from the germplasm collection of the Institute of Biosciences and Bioresources (IBBR) of the Italian National Research Council (CNR) in Bari, Italy:

- Vma1 = Accession number 112906 from USA Vma1
- Vma2 = Accession number 103235 from Italy
- Vma3 = Accession number 107620 from Greece.
- Vma4 = Accession number 106374 from Algeria

For the botanical variety *Vicia faba* L. var. *minor* (Harz) Beck we used the commercial variety Prothabat and accessions sourced from the germplasm collection of the Institute of Biosciences and Bioresources (IBBR) of the Italian National Research Council (CNR) in Bari, Italy:

- Vmi1 = Accession number 113620 from Germany
- Vmi2 = Accession number 113620 from Germany
- Vmi3 = Accession number 109322 from Ethiopia
- Vmi4 = Accession number 118952 from Afghanistan

Measurements were made for *Vicia faba* (L.) var. *maior* at phenological stage 20 (no side shoots) for the BBCH scale and for *Vicia faba* (L.) var. *minor* at phenological stage 21 (Beginning of side shoot development: first side shoot detectable) for the BBCH scale. Acquisitions were made in static mode with the ultrasound sensor mounted on a pole and placed above single plants at 80 cm distance from the ground (Figure 10). For ground-truth height measurements (h) were made manually with a rigid measuring tape on the same plant where ultrasound data were acquired. The variable h = plant height was the distance between the ground and the top of the uppermost plant structure.

4.3.3. Chia (*Salvia hispanica* L.)

The field was located at Masserie Saraceno (Atella, Italy, Lat. N 40°51′37.59″, Long. E 15°38′49.43″) on loam soil with the following characteristics: sand 43.6%, silt 34.2%, clay 22.1%. The broad leaf crop *Salvia hispanica* L. was sown in rows with plant spacing of 5 cm or 30 cm on the row. One transect for each of the plant spacing treatments was chosen based on the presence of gradients of plant height and/or areas of bare soil along the row. Sensor measurements were acquired on-the-go with the ultrasound sensor mounted on straddle wheeled chassis and moved at an average speed of 0.25 m s^{-1}. The ultrasonic sensor triggered 5 measures per second, e.g., a measurement every 0.05 m. Ground-truth measurements were made manually with a rigid measuring tape every 0.05 m along the transect. This corresponded to plant tops or lateral leaves or bare soil, or to *Amaranthus retroflexus* L. plants which represented the main segetal species in the field and were found along chia rows or beyond the edges of the plot. Data points were geo-referenced with a RTK GPS k9t (Kolida Instrument Co., Ltd, Guangzhou, China) with an accuracy of ±2 cm. Sensor and manual measurements were paired after retrieving from the sensor dataset observations at the closest location to measured data.

4.3.4. Alfalfa (*Medicago sativa* L.)

Measurements were made in a 7-ha alfalfa (*Medicago sativa* L. cv. Altiva) stand in Palomonte (Italy, Lat N 40°61′39.52″ Long E 15°30′32.64″) at 210 m asl. The soil was classified as a Typic Eutrudept fine, mixed, thermic Calcaric Cambisols (Soil Survey Staff, 1999; IUSS Working Group WRB, 2006). The average soil texture within the first 0.5 m layer was 41.29% sand, 17.14% silt, 41.57% clay; the average soil organic matter content was 26 g kg^{-1}. The stand was planted at a seeding rate of 40 kg ha^{-1}.

Crop height was mapped on-the-go after mounting the ultrasound sensor on a quad at 0.6 m distance from the ground. Data points were geo-referenced with a RTK GPS k9t (Kolida Instrument Co., Ltd., Guangzhou, China) with an accuracy of ±2 cm. Maps were obtained at three times corresponding to three different heights of the alfalfa stand: a. maximum plant height 12.07 cm; b. maximum plant height 16.75 cm; c. maximum plant height 57.56 cm. On the latter date ground truth data for sensor testing were taken in 16 areas in the field which were chosen across the whole range of crop height values obtained from on-the-go maps with a surface-response-sampling method [41]. The following plant biometric and radiometric measurements were made:

NDVI

A radiometric Greenseeker® (Trimble, Sunnyvale, CA, USA) sensor was used to meaasure reflected radiation in the red (~660 nm) and near infrared (~770 nm) wavelengths for the calculation of the Normalized Differences Vegatation Index (NDVI):

$$NDVI = (NIR - VIS)/(NIR) + (VIS)$$

where

$$NIR = \text{reflectance in the infrared band } (\sim 770 \text{ nm})$$

$$VIS = \text{reflectance in the red band } (\sim 660 \text{ nm}).$$

Leaf Area Index

Leaf area index (LAI m^2 m^{-2}) was measured with a LI-COR 2200c (LI-COR, Lincoln, NE, USA) field leaf-area meter.

Vegetation Height

Vegetation height was measured with a rigid measuring tape as the distance from the ground of a 10-cm diameter disk mounted to a stick and placed on top of the canopy every 0.1 m on a 0.5 × 0.5 m area.

Biomass

Above-ground plant parts were harvested on 0.5 × 0.5 m areas and weighed fresh and after oven-drying at 65 °C until constant weight.

4.3.5. Wheat (*Triticum durum* Desf.)

Measurements were made in a hard wheat (*Triticum durum* Desf. cv. Tirex) stand in Genzano di Lucania (Italy Lat N 40°49′24.9″ Long E 16°05′33.8″. The average soil texture within the first 0.5 m layer was: sand 18.8%, silt 52.7%, clay 28.5%; organic matter 20.17 g kg^{-1}. The stand was planted in rows at a distance of 15 cm and at a seeding rate of 230 kg ha^{-1}.

Six plots of 2 × 3 m were set up to compare the height of wheat plants grown for grain (W = whole) with that of plants grown as a dual-purpose crop and, therefore, cut at the end of tillering at 0.07 m from the ground level (C = cut). Measurements were made at the end of tillering-beginning of stem elongation, booting, and grain filling stages (respectively stages 30, 41, 71) of the Zadoks scale [42]. Sensor measurements were acquired on the go along a row of each of the plots with the ultrasound sensor mounted on a straddle wheeled chassis. At the end of tillering and booting, plant height (h) was measured with a rigid measuring tape, which was the distance from the ground of a 10-cm diameter disk mounted to a stick and placed on top of the canopy. At grain filling, plant height (h) was measured manually with a rigid measuring tape as the distance from the ground to the top of the ear.

At the Zadoks stage, 30 ultrasound measurements were made along a row where LAI and NDVI were also measured with methods described in Section 4.3.3, and on the whole plots, where NDVI was also measured with methods described in Section 4.3.3.

4.4. Statistical Analysis

Data from the ultrasound sensor were analyzed in comparison with ground-truth height data and other plant biometrics with univariate regression models. On data from the wheat plot experiment, analysis of variance was conducted, and means were separated with the *post-hoc* test of Tukey. Where horizontal shifts of sensor data compared to manual data were found, we calculated feature-based and shape-based distance measures: respectively cross-correlation distance and Dynamic Time Warping normalized Distance (DTW) to account for possible misalignment between series [30]; DTW is based on the Euclidean distance computed after using dynamic programming to find the minimal path in a distance matrix between similar elements in two compared series [2]. All statistical analyses were performed within the R environment for statistical computing (version 3.1.2 [43]).

5. Conclusions

We devised and tested a low-cost platform for measuring plant height with an ultrasound sensor designed to be used in different modes, from static to on-the-go.

We generated point measurements with the static setting on faba bean, with a high regression coefficient where the ultrasound sensor proved able to reproduce height regardless of genotype and phenological stage, thus showing the potential of the platform for static single-plant measurements.

High regression coefficients were also shown in the on-the-go mode for settings with a quite continuous plant cover, such as narrow-spaced chia, wheat, and alfalfa, whereas sensor data were not able to closely reproduce crop height where sharp variations were found, such as in widely spaced chia.

We were able to quantify and correct some systematic errors, such as data misalignment with feature-based (cross-correlation) or shape-based (dynamic time-warping) measures of similarity between data series.

Ultrasound-measured plant height proved to be a better predictor than NDVI for plant biometrics relevant to water relations and yield behavior, such as Leaf Area Index and biomass in alfalfa.

Overall, ultrasound-measured height with our platform proved to be a fast and low-cost method of estimating crop parameters that is useful in ecological research or agriculture applications.

Important characteristics of our platform are that it is simple and flexible, given the possibility to be employed by users with different skills and inclinations for technology and in different settings, from mounted on simple poles for single-point static measurements to towed manually for the complete characterization of plots or rows in experiments. Further, the platform may be carried on vehicles for mapping large surfaces like open fields, prairies, and natural herbaceous vegetation sites, providing maps useful for spatial management and characterization of spatial variation of plant responses in different conditions from natural to managed systems.

One of the features of the platform's ESP32 board is wide connectivity. Therefore, future developments may include the design of a multi-sensor platform with the same flexibility of use in different settings.

Future work should also focus on the analysis and correction of errors linked to field settings and modes of platform use, such as data misalignment.

Author Contributions: Conceptualization, G.B., R.R., S.L., A.R.R., L.R., P.D., G.L. and M.A.; methodology, G.B., R.R., R.B. and P.D.; investigation, G.B., R.R., R.B., D.C., G.M., P.C. and L.R. data curation, R.R., G.B., R.B., G.M., S.L., A.R.R., P.D., D.C. and M.A.; writing—original draft preparation, G.B., R.R. and M.A.; writing—review and editing R.B., D.C., S.L., A.R.R., L.R., P.C. and G.L.; funding acquisition, M.A. All authors have read and agreed to the published version of the manuscript.

Funding: This research was partly funded by the project PROTEAGRI: Tecnologie per la Riduzione del Gap Proteico in Sistemi Agricoli Destinati all'Alimentazione Umana ed Animale. P.O. FESR 2014–2020—Azione 1B.1.2.1. "Avviso pubblico per il sostegno alla creazione e sviluppo di cluster tecnologici della Regione Basilicata. CUP G49J19001360006".

Data Availability Statement: Data are contained within the article.

Conflicts of Interest: The authors declare no conflicts of interest. The funders had no role in the design of the study; in the collection, analyses, or interpretation of data; in the writing of the manuscript, or in the decision to publish the results.

References

1. Moles, A.T.; Warton, D.I.; Warman, L.; Swenson, N.; Laffan, S.W.; Zanne, A.; Pitman, A.J.; Hemmings, F.; Leishman, M. Global patterns in plant height. *J. Ecol.* **2009**, *97*, 923–932. [CrossRef]
2. Heady, H.F. The Measurement and Value of Plant Height in the Study of Herbaceous Vegetation. *Ecology* **1957**, *38*, 313–320. [CrossRef]
3. Westoby, M.; Falster, D.S.; Moles, A.T.; Vesk, P.A.; Wright, I.J. Plant ecological strategies: Some leading dimensions of variation between species. *Ann. Rev. Ecol. Syst.* **2002**, *33*, 125–159. [CrossRef]
4. Proulx, R. On the general relationship between plant height and aboveground biomass of vegetation stands in contrasted ecosystems. *PLoS ONE* **2021**, *16*, e0252080. [CrossRef] [PubMed]
5. Yin, X.; McClure, M.A.; Jaja, N.; Tyler, D.D.; Hayes, R.M. In-season prediction of corn yield using plant height under major production systems. *Agron. J.* **2011**, *103*, 923–929. [CrossRef]
6. Mourtzinis, S.; Arriaga, F.J.; Balkcom, K.S.; Ortiz, B.V. Corn grain and stover yield prediction at R1 growth stage. *Agron. J.* **2013**, *105*, 1045–1050. [CrossRef]
7. Cui, F.; Li, J.; Ding, A.; Zhao, C.; Wang, L.; Wang, X.; Li, S.; Bao, Y.; Li, X.; Feng, D.; et al. Conditional QTL mapping for plant height with respect to the length of the spike and internode in two mapping populations of wheat. *Theor. Appl. Genet.* **2011**, *122*, 1517–1536. [CrossRef] [PubMed]
8. Boomsma, C.R.; Santini, J.B.; West, T.D.; Brewer, J.C.; McIntyre, L.M.; Vyn, T.J. Maize grain yield responses to plant height variability resulting from crop rotation and tillage system in a long-term experiment. *Soil Till. Res.* **2010**, *106*, 227–240. [CrossRef]
9. Machado, S.; Bynum, E.D.; Archer, T.L.; Lascano, R.J.; Wilson, L.T.; Bordovsky, J.; Segarra, E.; Bronson, K.; Nesmith, D.M.; Xu, W. Spatial and temporal variability of corn growth and grain yield. *Crop Sci.* **2002**, *42*, 1564–1576. [CrossRef]
10. Dandois, J.P.; Ellis, E.C. High spatial resolution three-dimensional mapping of vegetation spectral dynamics using computer vision. *Remote Sens. Environ.* **2013**, *136*, 259–276. [CrossRef]
11. Jin, J.; Tang, L. Corn plant sensing using real-time stereo vision. *J. Field Robot.* **2009**, *26*, 591–608. [CrossRef]
12. Baha, N. Real-Time Obstacle Detection Approach using Stereoscopic Images. *Int. J. Inf. Eng. Electron. Bus.* **2014**, *6*, 42–48. [CrossRef]
13. Omasa, K.; Hosoi, F.; Konishi, A. 3D lidar imaging for detecting and understanding plant responses and canopy structure. *J. Exp. Bot.* **2014**, *58*, 881–898. [CrossRef] [PubMed]
14. Saeys, W.; Lenaerts, B.; Craessaerts, G.; De Baerdemaeker, J. Estimation of the crop density of small grains using LiDAR sensors. *Biosyst. Eng.* **2009**, *102*, 22–30. [CrossRef]
15. Andújar, D.; Escolà, A.; Rosell-Polo, J.R.; Fernández-Quintanilla, C.; Dorado, J. Potential of a terrestrial LiDAR-based system to characterise weed vegetation in maize crops. *Comput. Electron. Agric.* **2013**, *92*, 11–15. [CrossRef]
16. Fricke, T.; Richter, F.; Wachendorf, M. Assessment of forage mass from grassland swards by height measurement using an ultrasonic sensor. *Comput. Electron. Agric.* **2011**, *79*, 142–152. [CrossRef]
17. Llorens, J.; Gil, E.; Llop, J.; Queraltó, M. Georeferenced LiDAR 3D vine plantation map generation. *Sensors* **2011**, *11*, 6237–6256. [CrossRef] [PubMed]
18. Andújar, D.; Weis, M.; Gerhards, R. An ultrasonic system for weed detection in cereal crops. *Sensors* **2012**, *12*, 17343–17357. [CrossRef] [PubMed]
19. Fisher, D.K.; Gould, P.J. Open-Source Hardware Is a Low-Cost Alternative for Scientific Instrumentation and Research. *Mod. Instrum.* **2012**, *1*, 8–20. [CrossRef]
20. Leeuw, T.; Boss, E.S.; Wright, D.L. In situ measurements of phytoplankton fluorescence using low cost electronics. *Sensors* **2013**, *13*, 7872–7883. [CrossRef]
21. Bronson, K.F.; French, A.N.; Conley, M.M.; Barnes, E.M. Use of an Ultrasonic Sensor for Plant Height Estimation in Irrigated Cotton. *Agron. J.* **2021**, *113*, 2175–2183. [CrossRef]
22. Scotford, I.M.; Miller, P.C.H. Combination of Spectral Reflectance and Ultrasonic Sensing to Monitor the Growth of Winter Wheat. *Biosyst. Eng.* **2004**, *87*, 27–38. [CrossRef]
23. Fisher, D.K.; Huang, Y. Mobile Open-Source Plant-Canopy Monitoring System. *Mod. Instrum.* **2017**, *6*, 1–13. [CrossRef]
24. Sui, R.; Thomasson, J.A.; Ge, Y. Development of Sensor Systems for Precision Agriculture in Cotton. *Biol. Eng.* **2012**, *5*, 14.
25. Sharma, B.; Ritchie, G.L. High-Throughput Phenotyping of Cotton in Multiple Irrigation Environments. *Crop Sci.* **2015**, *55*, 958–969. [CrossRef]
26. Barker, J.; Zhang, N.; Sharon, J.; Steeves, R.; Wang, X.; Wei, Y.; Poland, J. Development of a Field-Based High-Throughput Mobile Phenotyping Platform. *Comput. Electron. Agric.* **2016**, *122*, 74–85. [CrossRef]
27. Andrade-Sanchez, P.; Gore, M.A.; Heun, J.T.; Thorp, K.R.; Carmo-Silva, A.E.; French, A.N.; Salvucci, M.E.; White, J.W. Development and Evaluation of a Field-Based High-Throughput Phenotyping Platform. *Funct. Plant Biol.* **2013**, *41*, 68–79. [CrossRef]
28. Montazeaud, G.; Langrume, C.; Moinard, S.; Goby, C.; Ducanchez, A.; Tisseyre, B.; Brunel, G. Development of a low cost open-source ultrasonic device for plant height measurements. *Smart Agric. Technol.* **2021**, *1*, 100022. [CrossRef]

29. Mori, U.; Mendiburu, A.; Lozano, J.A. Distance Measures for Time Series in R: The TSdist Package. *R J.* **2016**, *8*, 451. [CrossRef]
30. Liao, W.T. Clustering of time series data—A survey. *Pattern Recogn.* **2005**, *38*, 1857–1874. [CrossRef]
31. Sui, R.; Thomasson, J.A. Ground-based sensing system for cotton nitrogen status determination. *Trans. ASABE* **2006**, *49*, 1983–1991. [CrossRef]
32. Aziz, S.A.; Steward, B.L.; Birrell, S.J.; Shrestha, D.S.; Kaspar, T.C. Ultrasonic sensing for corn plant canopy characterization. In Proceedings of the ASAE Annual Meeting, Ottawa, ON, USA, 15–18 August 2004; p. 041120.
33. Jones, C.L.; Maness, N.O.; Stone, M.L.; Jayasekara, R. Sonar and digital imagery for estimating crop biomass. In Proceedings of the ASAE Annual Meeting, Ottawa, ON, USA, 15–18 August 2004; p. 04306.
34. Fiorentino, C.; Donvito, A.R.; D'Antonio, P.; Lopinto, S. Experimental Methodology for Prescription Maps of Variable Rate Nitrogenous Fertilizers on Cereal Crops. In *Innovative Biosystems Engineering for Sustainable Agriculture, Forestry and Food Production*; Coppola, A., Di Renzo, G., Altieri, G., D'Antonio, P., Eds.; Lecture Notes in Civil Engineering; Springer: Cham, Switzerland, 2020; Volume 67. [CrossRef]
35. Jeon, H.Y.; Zhu, H.; Derksen, R.; Ozkan, E.; Krause, C. Evaluation of ultrasonic sensor for variable-rate spray applications. *Comput. Electron. Agric.* **2011**, *75*, 213–221. [CrossRef]
36. Legg, M.; Bradley, S. Ultrasonic Proximal Sensing of Pasture Biomass. *Remote Sens.* **2019**, *11*, 2459. [CrossRef]
37. Lawson, H.F.; Giri, K.; Thomson, A.L.; Karunaratne, S.B.; Smith, K.F.; Jacobs, J.L.; Morse-McNabb, E.M. Multi-site calibration and validation of a wide-angle ultrasonic sensor and precise GPS to estimate pasture mass at the paddock scale. *Comput. Electron. Agric.* **2022**, *195*, 106786. [CrossRef]
38. Duchemin, B.; Hadria, R.; Erraki, S.; Boulet, G.; Maisongrande, P.; Chehbouni, A.; Escadafal, R.; Ezzahar, J.; Hoedjes, J.C.B.; Kharrou, M.H.; et al. Monitoring wheat phenology and irrigation in Central Morocco: On the use of relationships between evapotranspiration, crops coefficients, leaf area index and remotely-sensed vegetation indices. *Agric. Water Manag.* **2006**, *79*, 1–27. [CrossRef]
39. Zaman, Q.U.; Schumann, A.W.; Hostler, H.K. Quantifying sources of error in ultrasonic measurements of citrus orchards. *Appl. Eng. Agric.* **2007**, *23*, 449–453. [CrossRef]
40. Bitella, G.; Rossi, R.; Bochicchio, R.; Perniola, M.; Amato, M. A Novel Low-Cost Open-Hardware Platform for Monitoring Soil Water Content and Multiple Soil-Air-Vegetation Parameters. *Sensors* **2014**, *14*, 19639–19659. [CrossRef] [PubMed]
41. Lesch, S.M. Sensor-directed response surface sampling designs for characterizing spatial variation in soil properties. *Comput. Electron. Agric.* **2005**, *46*, 153–179. [CrossRef]
42. Zadoks, J.C.; Chang, T.T.; Konzak, C.F. Decimal Code for the Growth Stages of Cereals. *Weed Res.* **1974**, *14*, 415–421. [CrossRef]
43. R Core Team. *R: A Language and Environment for Statistical Computing*; R Foundation for Statistical Computing: Vienna, Austria, 2021. Available online: https://www.R-project.org/ (accessed on 3 September 2023).

plants

Article

Indirect Estimation of Heavy Metal Contamination in Rice Soil Using Spectral Techniques

Liang Zhong [1,2], Shengjie Yang [1,2], Yicheng Rong [1,2], Jiawei Qian [1,2], Lei Zhou [3], Jianlong Li [1,2,*] and Zhengguo Sun [4,*]

1 State Key Laboratory of Pharmaceutical Biotechnology, School of Life Sciences, Nanjing University, Nanjing 210023, China; zhongliang1007@163.com (L.Z.)
2 Department of Ecology, School of Life Sciences, Nanjing University, Nanjing 210023, China
3 Livestock Development and Promotion Center, Linyi 276037, China
4 College of Agro-Grassland Science, Nanjing Agricultural University, Nanjing 210095, China
* Correspondence: jlli2008@nju.edu.cn (J.L.); sunzg@njau.edu.cn (Z.S.)

Abstract: The rapid growth of industrialization and urbanization in China has led to an increase in soil heavy metal pollution, which poses a serious threat to ecosystem safety and human health. The advancement of spectral technology offers a way to rapidly and non-destructively monitor soil heavy metal content. In order to explore the potential of rice leaf spectra to indirectly estimate soil heavy metal content. We collected farmland soil samples and measured rice leaf spectra in Xushe Town, Yixing City, Jiangsu Province, China. In the laboratory, the heavy metals Cd and As were determined. In order to establish an estimation model between the pre-processed spectra and the soil heavy metals Cd and As content, a genetic algorithm (GA) was used to optimise the partial least squares regression (PLSR). The model's accuracy was evaluated and the best estimation model was obtained. The results showed that spectral pre-processing techniques can extract hidden information from the spectra. The first-order derivative of absorbance was more effective in extracting spectral sensitive information from rice leaf spectra. The GA-PLSR model selects only about 10% of the bands and has better accuracy in spectral modeling than the PLSR model. The spectral reflectance of rice leaves has the capacity to estimate Cd content in the soil (relative percent difference [RPD] = 2.09) and a good capacity to estimate As content in the soil (RPD = 2.97). Therefore, the content of the heavy metals Cd and As in the soil can be estimated indirectly from the spectral data of rice leaves. This study provides a reference for future remote sensing monitoring of soil heavy metal pollution in farmland that is quantitative, dynamic, and non-destructive over a large area.

Keywords: rice; soil-crop system; heavy metal contamination; spectral technique; genetic algorithm; indirect estimation

Citation: Zhong, L.; Yang, S.; Rong, Y.; Qian, J.; Zhou, L.; Li, J.; Sun, Z. Indirect Estimation of Heavy Metal Contamination in Rice Soil Using Spectral Techniques. *Plants* **2024**, *13*, 831. https://doi.org/10.3390/plants13060831

Academic Editor: Youzhen Xiang

Received: 1 February 2024
Revised: 29 February 2024
Accepted: 9 March 2024
Published: 14 March 2024

1. Introduction

Soil serves both as the basis for the growth of crops and as a vital natural resource for the sustenance and production of human beings [1]. However, soil environmental pollution has increased significantly in China due to rapid industrialization and urbanization [2]. Among various pollutants, soil heavy metal contamination stands out due to its slow migration, high toxicity, and irreversible nature [3]. Over time, heavy metals accumulate in the food chain and pose severe health risks when ingested and accumulated by humans [4,5]. According to the 2014 Chinese Soil Pollution Status Report, the overall pollution excess rate of soil in China is 16.1% [6]. This alarming statistic has profound implications for China's food security as heavy metal pollution in soil leads to an annual loss of approximately 12 million tons of grain crops [7]. As a result, both the government and scholars have shown widespread concern regarding soil heavy metal pollution [8,9]. Quantitative monitoring of heavy metal content plays an important role in understanding the extent and sources of heavy metal pollution in a region [10]. It also offers a theoretical foundation for the remediation and management of such pollution [11,12]. While traditional chemical analysis methods are highly accurate in detecting soil heavy metal content, they are time-consuming, labor-intensive, and costly. Consequently, they fail to support the demands of real-time and large-scale monitoring for efficient heavy metal content assessment [13]. Fortunately, the advancement of remote sensing technology has paved the way for nondestructive and rapid soil heavy metal monitoring using spectral remote sensing [14,15]. This technique, characterized by its high-resolution and ability to capture details of the object's spectral information, holds great potential in enabling fast and efficient monitoring of soil heavy metal content [16].

Spectral remote sensing applications for monitoring soil heavy metal contamination include two main approaches: direct and indirect. Direct monitoring focuses on the mechanism where soil heavy metal are adsorbed by soil organic matter, iron-manganese oxides and clay minerals [17]. These components affect soil spectral morphology and reflectance, leading to specific soil spectral absorption features. At present, more research has been conducted on direct monitoring [18–22]. The direct monitoring method is not without its limitations, although it results in high model accuracy and stable models. Firstly, it is a time-consuming and cumbersome process that requires field soil sampling and laboratory processing to obtain soil spectral data. In addition, due to soil drying, grinding, and sieving, the spectral features of heavy metals extracted from laboratory soil spectra often differ from those obtained from remote sensing images. This difference complicates the direct application of these models to large-scale soil pollution monitoring using aerospace images [23]. Therefore, a more convenient and practical approach for widespread application is to use spectral data from plant leaves or canopies for indirect estimation of soil heavy metal contamination [24]. The approach is based on the principle that heavy metals move from the soil to the plants, accumulating there [25]. Under the stress of heavy metals, the protein and chlorophyll content of the plants are affected, which leads to a discernible difference in the reflectance spectra [26,27]. Related research has made significant progress. For example, Shi et al. [28] developed a multivariate spectral vegetation index based on rice canopy spectra for the estimation of arsenic (As) in farmland soil. Zhong et al. [29] estimated leaf Cu content using leaf hyperspectra and then inverted the Cu content of other parts of wheat and soil using bioconcentration factors. Wang et al. [30] observed that wheat canopy Cu increased with increasing soil Cu concentration, accompanied by distinct variations in spectral reflectance, providing a foundation for indirect estimation of soil Cu conten. However, the feasibility and accuracy of indirect estimation of soil heavy metals from rice leaves are not yet clear.

In this study, we will explore the potential of rice leaf spectra for estimating soil heavy metals. Firstly, we collected soil samples from farmland and also measured the spectral data of rice leaves in Xushe Town, Yixing City, Jiangsu Province, China. Then, we processed the leaf spectra with various spectral transformations and screened the spectral feature bands using a genetic algorithm (GA). Next, we used partial least squares regression (PLSR) to model soil Cadmium (Cd) and As content for for the spectra after different pre-processing. Finally, we evaluated the accuracy of the models and obtained the best estimation models.

2. Results

2.1. Statistics of Soil Samples

Figure 1 shows the content of the soil Cd and As of the sampling sites in the study area. The average value of soil pH at the sampling site was 5.86. The Cd content was between 0.13 to 0.97 mg kg^{-1} with a average value of 0.29 mg kg^{-1}, and the high values were located in the eastern and central parts of the study area. The As content was between 3.23 to 9.32 mg kg^{-1} with a average value of 5.64 mg kg^{-1}, and the high values were mainly located in the central part of the study area. The correlation coefficient between Cd and As content at the sampling sites was 0.33.

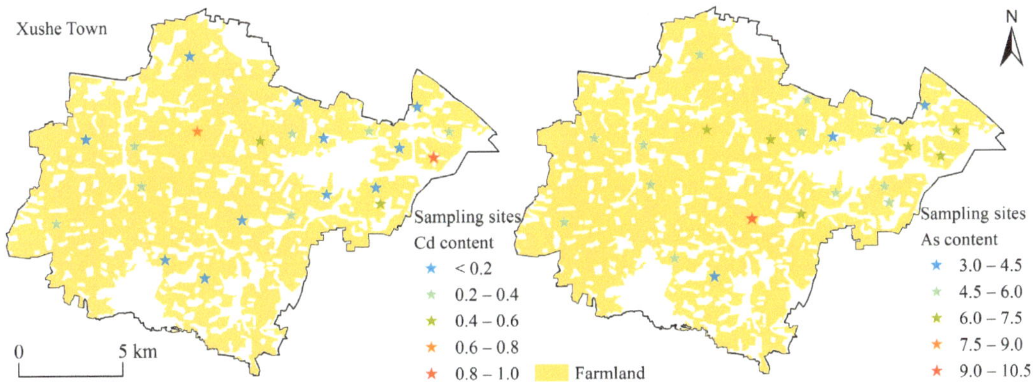

Figure 1. Distribution of Cd and As content at sampling sites in the study area.

2.2. Characterization of Spectral Curves of Rice Leaves

Figure 2 shows the spectra of rice leaves after different pre-processing. From the raw spectra (R) (Figure 2a), it can be seen that the spectral curve of rice leaves is significantly different in the 760–1120 nm band. It has a green light reflection peak at 550 nm, blue-violet light absorption valleys at 450 nm, and red light absorption valleys at 670 nm. This is due to the fact that chlorophyll absorbs weakly in the green band of light and strongly in the blue-violet and red bands of light. There is a high reflectance in the 760 nm to 1120 nm near-infrared band, which may be caused by multiple reflections within the leaf structure.

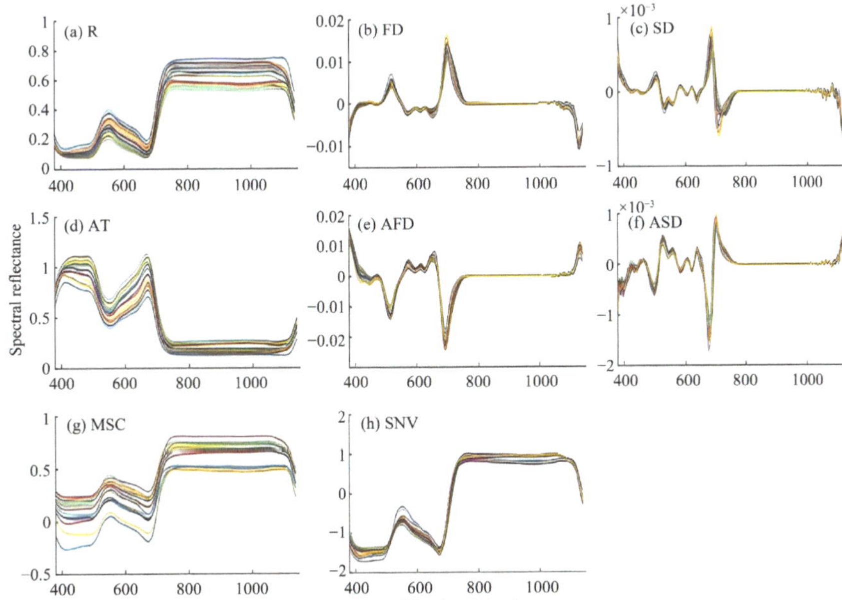

Figure 2. Characteristics of the spectral curves of rice leaves with different pre-processing.

Compared with R, the shape of the first-order derivative (FD) spectrum curve (Figure 2b) changed significantly, with a valley of absorption at 1129 nm and peaks of reflectance at 516 nm and 705 nm. The shape of the second-order derivative (SD) spectrum curve (Figure 2c) also changed significantly, with valleys of absorption at 712 nm and 1119 nm and peaks of reflectance at 503 nm and 686 nm. The absorbance transformation (AT) spectrum (Figure 2d) has high reflectance in the 400–510 nm band, low reflectance in the 730–1100 nm band, a valley of absorption at 553 nm, and a peaks of reflectance at 673 nm. The first-order derivative of absorbance (AFD) spectrum (Figure 2e) has valleys of absorption at 516 nm and 693 nm, and a reflection peak at 572 nm. The second-order derivative of absorbance (ASD) spectrum (Figure 2f) has valleys of absorption at 503 nm and 679 nm, and peaks of reflectance at 446 nm, 526 nm, 560 nm, 709 nm, and 1123 nm. The trend of the multiplicative scatter correction (MSC) spectrum curve (Figure 2g) is the same as that of the original spectrum, but the difference in the spectral curve of the 380–700 nm band is enlarged. The trend of the standard normal variate (SNV) spectrum curve (Figure 2h) is similar to that of the original spectrum, but the spectra are denser, indicating the ability of the SNV transform to reduce background noise.

2.3. Spectral Feature Bands Selected by GA

Rice leaf spectral feature bands were screened using GA as shown in Figure 3. Under different spectral pre-processing, GA selected 17–25 feature bands of soil Cd and 15–30 feature bands of soil As among 230 full bands. The soil Cd feature bands are mostly concentrated in 400–410 nm, 520–580 nm, 630–650 nm, 690–770 nm, 870–925 nm, and 970–1000 nm; the soil Cd feature bands are mostly concentrated in 400–440 nm, 815–860 nm, 890–920 nm, 970–990 nm, 1005–1030 nm, and 1065–1135 nm.

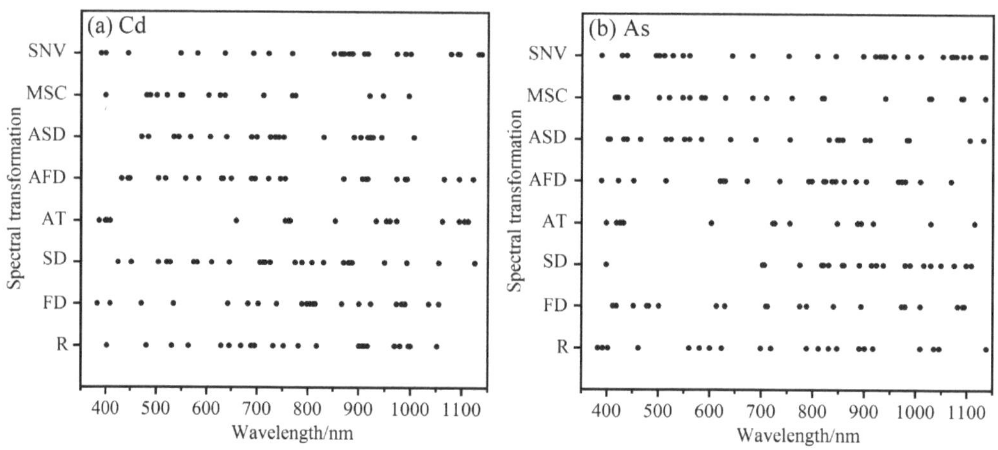

Figure 3. The feature bands of rice leaves spectral screened by GA.

2.4. Comparison of GA-PLSR and PLSR Modeling Results

Using the GA-PLSR and PLSR models, the modeling and analysis of rice leaf spectra under different pre-processing were carried out, and the cross-validation results are shown in Table 1. Compared with the PLSR model established directly using the full band, the PLSR model established by first screening the bands with GA has the same or reduced number of PCs, indicating that GA can select rice leaf spectral bands that are more meaningful to the PLSR model. Compared with the PLSR model, the R^2_{cv} value of estimating soil Cd content under different pre-processing spectra using the GA-PLSR model increased by 6.25~33.96%, and the $RMSE_{cv}$ value decreased by 0.00~50.00%. The R^2_{cv} value of estimating soil As content increased by 14.29~53.19%, and the $RMSE_{cv}$ value decreased by 3.33~69.64%. The results indicate that using GA for spectral wavelength selection before establishing a model for estimating heavy metal content in rice leaf spectra can improve model accuracy and stability.

2.5. Best Estimate Model

A cross-validation and an external validation were carried out on the GA-PLSR model to estimate the heavy metal content of the soil, and the results are presented in Table 2. The results of the soil Cd content estimation model show that, compared with R, the accuracy of each indicator was improved to different degrees in the cross-validation and external validation of the 7 transformed spectra. This indicates that the accuracy and stability of the Cd content estimation model have been improved after different spectral transformations. Among them, the RPD values for R, FD, SD, ASD, and MSC spectral preprocessing are all less than 1.50, suggesting poor accuracy in estimating soil Cd content. The RPD values of AT and SNV spectral preprocessing ranges between 1.50 and 2.00, indicating the possibility to discriminate between soil with high and low Cd content. The AFD spectral preprocessing has the highest model accuracy, with R^2_{cv}, $RMSE_{cv}$, R^2_{ev}, $RMSE_{ev}$, and RPD of 0.71, 0.07 mg kg^{-1}, 0.77, 0.06 mg kg^{-1}, and 2.09, respectively, indicating the ability to approximate soil Cd content estimation.

Table 1. Comparison of the accuracy of GA-PLSR and PLSR models for estimating soil Cd and As content in rice leaves.

Heavy Metal	Pre-Processing	Number of Bands	GA-PLSR			PLSR		
			PC	R^2_{cv}	$RMSE_{cv}$/(mg kg^{-1})	PC	R^2_{cv}	$RMSE_{cv}$/(mg kg^{-1})
Cd	R	22	2	0.34	0.16	2	0.32	0.18
	FD	21	3	0.46	0.15	3	0.39	0.15
	SD	25	1	0.42	0.15	2	0.35	0.17
	AT	17	2	0.53	0.15	2	0.45	0.15
	AFD	25	5	0.71	0.07	8	0.53	0.14
	ASD	22	2	0.47	0.13	4	0.38	0.17
	MSC	16	1	0.47	0.15	3	0.40	0.16
	SNV	26	2	0.52	0.15	2	0.45	0.15
As	R	21	2	0.50	1.18	3	0.41	1.27
	FD	20	1	0.52	1.15	2	0.42	1.22
	SD	21	2	0.55	1.16	3	0.44	1.25
	AT	15	2	0.56	1.16	2	0.49	1.20
	AFD	23	5	0.89	0.34	9	0.61	1.12
	ASD	23	2	0.72	0.82	6	0.47	1.44
	MSC	22	4	0.58	1.15	4	0.42	1.27
	SNV	30	2	0.70	0.93	6	0.51	1.17

Table 2. Accuracy of GA-PLSR models for estimating soil Cd and As content in rice leaves.

Heavy Metal	Pre-Processing	Cross-Validation		External Validation		
		R^2_{cv}	$RMSE_{cv}$/(mg kg^{-1})	R^2_{ev}	$RMSE_{ev}$/(mg kg^{-1})	RPD
Cd	R	0.34	0.16	0.41	0.11	1.30
	FD	0.46	0.15	0.52	0.10	1.44
	SD	0.42	0.15	0.47	0.10	1.37
	AT	0.53	0.15	0.59	0.09	1.56
	AFD	0.71	0.07	0.77	0.06	2.09
	ASD	0.47	0.13	0.53	0.09	1.46
	MSC	0.47	0.15	0.49	0.10	1.40
	SNV	0.52	0.15	0.62	0.08	1.62
As	R	0.50	1.18	0.57	0.65	1.52
	FD	0.52	1.15	0.68	0.56	1.77
	SD	0.55	1.16	0.66	0.58	1.71
	AT	0.56	1.16	0.64	0.60	1.66
	AFD	0.89	0.34	0.89	0.30	2.97
	ASD	0.72	0.82	0.71	0.53	1.86
	MSC	0.58	1.15	0.64	0.60	1.66
	SNV	0.70	0.93	0.76	0.48	2.06

The results of the model for estimating soil As content show that the cross-validation and external validation of the seven transformed spectra improve the accuracy of each index to varying degrees compared to R. This indicates that the different spectral transformations have improved the accuracy and stability of the As estimation models. Among them, the R, FD, SD, AT, ASD, and MSC spectral preprocessing have an RPD values between 1.50 and 2.00, indicating the possibility of distinguishing between soil with high and low As content. The RPD values of SNV spectral preprocessing is 2.06, indicating the ability to approximate soil As content estimation. The AFD spectral preprocessing has the highest level of model precision, with $R^2{}_{cv}$, $RMSE_{cv}$, $R^2{}_{ev}$, $RMSE_{ev}$, and RPD of 0.89, 0.34 mg kg^{-1}, 0.89, 0.30 mg kg^{-1}, and 2.97, respectively, indicating good ability to estimate soil As content.

3. Discussion

3.1. Effect of Spectral Pre-Processing and Feature Selection for Modeling Performance

Modelling results from different spectral preprocessing techniques show that most preprocessed spectra show varying degrees of accuracy improvement over the original spectra. This improvement can be attributed to the fact that external disturbances can introduce noise when collecting spectral data, making it difficult to accurately represent the spectral characteristics of objects [31]. Spectral pre-processing approaches efficiently reduce spectral noise and improve the information about the spectral features [32,33]. For indirect soil heavy metal content estimation using rice leaf spectra, AFD is the optimal spectral transformation method for both Cd and As estimation. This is because of the FD spectral transform, which can effectively extract and enhance the hidden information in the spectrum [34]. Yao et al. [35] found that derivative transformations were able to highlight the spectral features more compared with MSC, SNV and Continuum removal. Meanwhile, the absorbance transformation can further improve the inversion accuracy of As content, which is consistent with our ability to obtain a good estimation of soil As content using AFD spectra.

In this study, the number of feature bands selected by GA in the optimal inversion models for soil Cd and As contents were 25 and 23, respectively, only about 10% of the bands were used. Moreover, higher accuracy was achieved by utilizing GA for selecting spectral characteristic bands compared to PLSR modelling. The reason for this approach is that spectral data have properties of redundancy and collinearity, and direct modelling with PLSR is susceptible to being disturbed by significant amounts of redundancy information [36]. GA improves model quality and stability by successfully filtering feature bands from the full spectrum [37]. The results of Sun and Zhang [38], Sun et al. [39], and Zhong et al. [17], who also showed that GA-PLSR outperforms PLSR in estimating heavy metal concentration using soil spectral data, are consistent with this methodology. It indicates that the method applied to the spectral modelling of heavy metal content in agricultural soils. In addition, Zhang et al. [40] in the estimation of soil heavy metal Cd, the accuracy of R^2 was 0.88 by PLSR modelling using soil spectral features associated with organic matter extracted using GA. Wei et al. [41] in the estimation of soil heavy metal As, the accuracy of R^2 was 0.82 and 0.70 for Honghu City and Daye City in Hubei Province, China, after selecting the feature bands by using the stable competitive adaptive reweighting sampling algorithm coupled the successive projections algorithm followed by PLSR modelling. Bian et al. [42] in the estimation of a variety of soil heavy metals, PLSR and extreme learning machine models obtained the best accuracy for Cd (R^2 of 0.89) and As (R^2 of 0.86), respectively. These studies generally obtained good model accuracies, but they mainly used laboratory soil spectra for direct monitoring of soil heavy metal content. Our study used rice leaf spectra to indirectly estimate soil heavy metal Cd and As contents, which is of significant value in the future development of field soil heavy metal hyperspectral instrumentation and in the exploration of aerospace hyperspectral remote sensing for monitoring soil heavy metal contamination on a large scale.

3.2. Application and Perspectives of Spectral Techniques in Heavy Metal Inversion in Soil-Rice System

Spectral data are characterized by high spectral resolution, which can obtain fine spectral information about the object. Using spectral analysis technology, it can highlight the subtle differences between spectra, which is conducive to extracting spectral features and finally constructing models to invert the information of the object [23]. In the soil-crop system, heavy metals migrate from the soil and accumulate in the crop. When the concentration of heavy metals in the soil increases, the content of heavy metals in different parts of rice usually increases as well, resulting in a stress effect [43]. As the stress of heavy metals in rice increases, some of the cellular structures are damaged, resulting in a decrease in chlorophyll content, which is reflected in differences in leaf spectra [44,45]. Therefore, we successfully inverted the content of heavy metals (Cd and As) in soil by spectral transformation, characteristic analysis, and modeling of rice leaf spectra. This study can help to develop soil heavy metal monitoring instruments in the future directly through rice leaf spectra, and thereby improve the detection efficiency of soil heavy metals. Meanwhile, in the future, attempts can be made to expand this indirect monitoring approach to aerospace spectral remote sensing, further combining space-air-ground spectral remote sensing data. This is conducive to making full use of the characteristics and laws of ground-based spectral indirect monitoring, while giving full play to the characteristics of aerospace spectral remote sensing in terms of dynamics and wide range. This will be helpful to form a multi-scale monitoring and validation system for soil-crop heavy metal pollution [17].

3.3. Limitations and Future Work

In this study, we used rice leaf spectra to indirectly estimate the content of soil heavy metals Cd and As, but there are still some limitations that we would like to explore and improve in future work, such as: (1) We studied and tested our method based on spectral data at only one sampling time, and the stability and applicability of the method need to be further validated. In the future, the indirect estimation method of soil heavy metal content in this study can be further explored for application at different time and spatial scales. (2) We only selected two typical pollutant elements in the study area for our study, and the potential of other heavy metal elements in spectral monitoring needs to be further explored in subsequent studies. (3) The mechanisms of uptake, transport and accumulation of heavy metal elements in the soil-crop system can be further explored in the future, which will provide more basis for the indirect monitoring of soil heavy metal content using crop spectra.

4. Materials and Methods

4.1. Study Area

The study area is in the town of Xushe (31°18′–31°27′ N, 119°31′–119°44′ E), west of Yixing City, Jiangsu Province (Figure 4), and the total area is about 1.8×10^4 hm^2. The region features a subtropical monsoon climate with well-defined seasons, ample rainfall, and an average annual temperature of 16.0 °C, accompanied by precipitation of 1434.0 mm. The topography of the region is characterized by higher elevation in the western areas and lower elevation in the eastern areas, consisting mainly of plains and hills. With a cultivated area of 1.2×10^4 hm^2, Xushe Town is the largest agricultural town in Yixing City, primarily used for cultivating rice and wheat. Paddy, fluvo-aquic and yellow-brown soils are the three main soil types.

4.2. Data Collection and Processing

4.2.1. Soil Sampling and Data Determination

We collected 22 surface (0–20 cm) soil samples in September 2019 in the study area using a five-point mixing method (Figure 3). At the time of sampling, we determined and recorded the location of each sampling point using a hand-held GPS device. The soil samples were then returned to the laboratory in sealed bags. In the laboratory, all soil samples were dried, ground and sieved (0.15 mm pore size) and divided equally into two parts. A part of the soil samples was used for the measurement of the pH by the potentiometric method (NY/T 1377-2007) [46]. The other part was weighed at 0.2 g of soil sample, put into the bottom of the PTFE digestion tank, added 5 mL of nitric acid, 2 mL of hydrogen peroxide, and 2 mL of hydrofluoric acid, and microwaved digested for 15 min. After the digestion solution was clarified, it was fixed to 50 mL and filtered. Finally, the Cd and As contents were determined by inductively coupled plasma mass spectrometry (ICP-MS) [47].

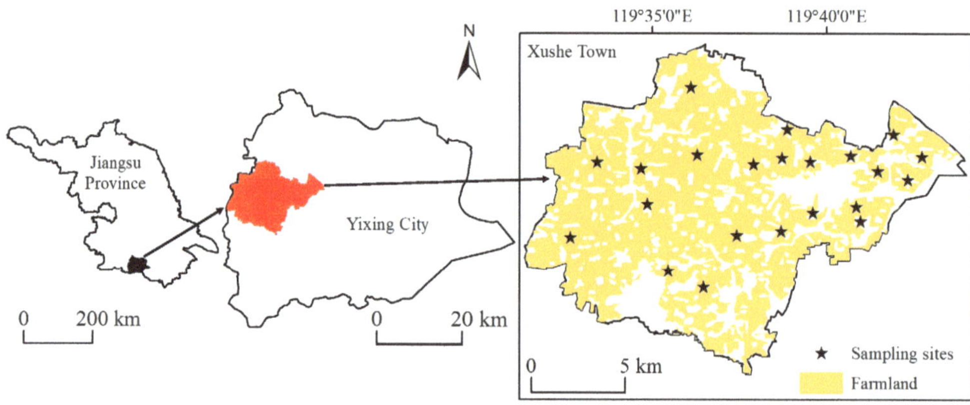

Figure 4. Location of the study area and distribution of sampling sites.

During the soil sampling process, we collected spectral data of rice leaves using a portable field spectrometer (UniSpec, PP systems, Haverhill, MA, USA) between 11:00 a.m. and 2:00 p.m. Beijing time. The spectrometer had a spectral range of 301–1145 nm and a spectral resolution of 3.3 nm. At each sampling site, five rice plants were randomly selected, and three fully expanded leaves from each rice were selected for spectral measurement under sunny and light wind conditions. White calibration was carried out before each spectral measurement and five measurements were repeated. For each sampling site, 75 spectral data points were collected from rice leaves and averaged to obtain the spectral data.

4.2.2. Spectral Pre-Processing

Firstly, the spectral data of the rice leaf are stripped of the noisy edge bands below 380 nm. At each sampling site, the bands between 380 and 1145 nm are selected for spectral data processing and analysis. Next, the leaf spectral data of rice is processed using Savitzky-Golay smoothing. The smoothed spectral data is referred to as raw spectrum R. Finally, based on the R spectra, the spectral pre-processing was carried out by applying mathematical transform methods such as AT, SNV, MSC, FD, AFD, SD, and ASD [48].

4.3. Research Methods

4.3.1. Genetic Algorithm

The GA is an evolutionary algorithm used for solving optimization problems. It simulates the mechanisms of genetics and natural selection, assessing individuals with superior fitness, selecting, crossing, and mutating them using genetic operators to generate individuals in the new generation of the population. In order to find the optimal solution, the iterative process is repeated until the convergence criteria have been met [49]. Simultaneously, GA can avoid the overfitting problem of general iterative methods, which may fall into the local minimum.

Prior to modeling, feature bands are selected using GA in order to reduce redundancies and optimize model performance [50]. In feature selection using GA, each band is treated as a gene, and a specific number of bands are designated as chromosomes. Next, a subset of the samples is taken to form the initial population. Then, crossing and mutation are used to simulate the genetic and evolutionary processes of random populations in nature, while the fitness function is used to assess the model's performance in predicting outcomes. After conducting tests, the GA parameters for population size, crossover probability, mutation probability, and genetic generation were set to 40, 0.5, 0.01, and 100, respectively. We repeated the process 10 times to minimize the influence of randomness. We used the root mean squared error of cross-validation ($RMSE_{cv}$) for PLSR as the fitness criterion. As the individual's fitness increases, the $RMSE_{cv}$ decreases.

4.3.2. Partial Least Squares Regression

The PLSR is one of the most commonly used methods for processing spectral data to estimate soil heavy metal content [51]. In this method, the independent variable and the dependent variable are projected onto a new coordinate system. The principal component, which has the strongest explanatory power, is extracted and used to construct a new linear model. This helps reduce collinearity and noise effects and makes the model more robust [52]. During the process of PLSR modeling, cross-validation is utilized to identify the number of most efficient principal components.

The spectra after different pre-processing were used to select the characteristic bands using GA and estimate the heavy metal content using PLSR. The 22 data samples are divided into two parts, with one sample selected from every 4 samples for validation. In all, 17 samples were used for modeling, and 5 samples were used for validating the accuracy of the model. The GA feature band selection and the PLSR modelling were done in R-Studio 3.5.3 (https://posit.co/products/open-source/rstudio/ (accessed on 6 November 2019)).

4.3.3. Model Assessment

This study used the coefficient of determination (R^2_{cv}) and the $RMSE_{cv}$ for model cross-validation. The R^2_{ev}, $RMSE_{ev}$, and relative percent difference (RPD) were chosen for model external validation. The closer R^2_{cv} and R^2_{ev} are to 1, the lower $RMSE_{cv}$ and $RMSE_{ev}$ are, and a higher RPD indicates a better model fit and accuracy. The five-layer interpretation method proposed by Williams et al. [53] was adopted for the evaluation criteria of RPD. If the RPD exceeds 3.00, the model has excellent ability to estimate. If the RPD ranges from 2.50 to 3.00, the model is considered to have good predictive performance. If the RPD ranges from 2.00 to 2.50, the model can be used for an approximate quantitative estimate. If the RPD ranges from 1.50 and 2.00, the model has the ability to discriminate between high and low values. If the RPD is less than 1.50, the model has a poor ability to estimate.

5. Conclusions

By combining spectral preprocessing, feature selection and modelling methods, this study fully explored the potential of rice leaf spectra for indirect estimation of soil heavy metals Cd and As, and the following conclusions were drawn:

(1) Spectral preprocessing technology enhances the modeling accuracy by revealing hidden information in the spectrum, leading to varying degrees of improvement compared to the original spectrum. When modeling rice leaf spectra, the most effective estimation models for soil Cd and As content are obtained through AFD spectral preprocessing. These results highlight the advantages of mathematical transformations, such as derivative transformation and absorbance, in extracting spectral sensitive information.

(2) The GA-PLSR model demonstrates superior performance compared to the PLSR model in modeling of rice leaf spectra. Specifically, compared to the PLSR model, GA-PLSR used only approximately 10% of the bands and enhanced the R^2_{cv} values for estimating soil Cd and As content by 0.00% to 50.00% and 3.33% to 69.64%, respectively, for the different preprocessed spectra. These findings illustrate that

incorporating a GA for spectral band selection before establishing a model for estimating soil heavy metal content can significantly enhance the accuracy and efficiency of the model.

(3) In the modeling of soil Cd content using rice leaf spectra, the best estimation model is the combination of AFD and GA-PLSR, with $R^2{}_{ev}$, $RMSE_{ev}$, and RPD values of 0.77 0.06 mg kg^{-1}, and 2.09, respectively, which has the ability to approximate estimation The best estimation model for soil As content is also the combination of AFD and GA PLSR, with $R^2{}_{ev}$, $RMSE_{ev}$, and RPD values of 0.89, 0.30 mg kg^{-1}, and 2.97, respectively, which has good estimation ability.

Author Contributions: Conceptualization, L.Z. (Liang Zhong), J.L. and Z.S.; methodology, J.Q. and Y.R.; software, S.Y. and J.Q.; validation, Y.R., S.Y. and L.Z. (Lei Zhou); formal analysis, S.Y. and L.Z. (Lei Zhou); investigation, J.Q. and L.Z. (Lei Zhou); resources, J.L. and Z.S.; data curation, L.Z. (Liang Zhong) and J.L.; writing—original draft preparation, L.Z. (Liang Zhong); writing—review and editing, L.Z. (Liang Zhong), J.L. and Z.S.; visualization, L.Z. (Liang Zhong) and Y.R.; supervision, J.L. and Z.S.; project administration, J.L.; funding acquisition, J.L. All authors have read and agreed to the published version of the manuscript.

Funding: This research was supported by the High-level international cooperation and innovation exchange forum (grant number: 202202), the National Key R&D Plan Project of China (grant number: 2018YFD0800201), and the project of Asia-Pacific network for global change research (grant number: ARCP2015-03CMY-Li).

Data Availability Statement: Data are contained within the article.

Acknowledgments: We are grateful to the anonymous reviewers for their comments and recommendations. We are grateful for the financial support from the fund to guarantee the smooth running of the study. We thank all the members of the research team for their help in this study.

Conflicts of Interest: The authors declare no conflict of interest.

References

1. Sanchez, P.; Ahamed, S.; Carre, F.; Hartemink, A.; Hempel, J.; Huising, J.; Lagacherie, P.; McBratney, A.; McKenzie, N.; Mendonca-Santos, M.; et al. Digital Soil Map of the World. *Science* **2009**, *325*, 680–681. [CrossRef]
2. Jiang, Y.; Huang, M.; Chen, X.; Wang, Z.; Xiao, L.; Xu, K.; Zhang, S.; Wang, M.; Xu, Z.; Shi, Z. Identification and risk prediction of potentially contaminated sites in the Yangtze River Delta. *Sci. Total Environ.* **2022**, *815*, 151982. [CrossRef]
3. Xia, F.; Zhao, Z.; Niu, X.; Wang, Z. Integrated pollution analysis, pollution area identification and source apportionment of heavy metal contamination in agricultural soil. *J. Hazard. Mater.* **2024**, *465*, 133215. [CrossRef]
4. Jiang, Y.; Chen, S.; Hu, B.; Zhou, Y.; Liang, Z.; Jia, X.; Huang, M.; Wei, J.; Shi, Z. A comprehensive framework for assessing the impact of potential agricultural pollution on grain security and human health in economically developed areas. *Environ. Pollut.* **2020**, *263*, 114653. [CrossRef]
5. Yuan, X.; Xue, N.; Han, Z. A meta-analysis of heavy metals pollution in farmland and urban soils in China over the past 20 years. *J. Environ. Sci.* **2021**, *101*, 217–226. [CrossRef]
6. Chen, W.; Yang, Y.; Xie, T.; Wang, M.; Peng, C.; Wang, R. Challenges and countermeasures for heavy metal pollution control in farmlands of china. *Acta Pedol. Sin.* **2018**, *55*, 261–272. [CrossRef]
7. Jiang, S.; Wang, F.; Li, Q.; Sun, H.; Wang, H.; Yao, Z. Environment and food safety: A novel integrative review. *Environ. Sci. Pollut. Res.* **2021**, *28*, 54511–54530. [CrossRef]
8. Yang, Q.; Li, Z.; Lu, X.; Duan, Q.; Huang, L.; Bi, J. A review of soil heavy metal pollution from industrial and agricultural regions in China: Pollution and risk assessment. *Sci. Total Environ.* **2018**, *642*, 690–700. [CrossRef]
9. Zhang, J.; Wang, X.; Li, J.; Luo, J.; Wang, X.; Ai, S.; Cheng, H.; Liu, Z. Bioavailability (BA)-based risk assessment of soil heavy metals in provinces of China through the predictive BA-models. *J. Hazard. Mater.* **2024**, *465*, 133327. [CrossRef]
10. Fei, X.; Lou, Z.; Xiao, R.; Ren, Z.; Lv, X. Source analysis and source-oriented risk assessment of heavy metal pollution in agricultural soils of different cultivated land qualities. *J. Clean. Prod.* **2022**, *341*, 130942. [CrossRef]
11. Ren, S.; Song, C.; Ye, S.; Cheng, C.; Gao, P. The spatiotemporal variation in heavy metals in China's farmland soil over the past 20 years: A meta-analysis. *Sci. Total Environ.* **2022**, *806*, 150322. [CrossRef]
12. Sorrentino, M.; Giordano, S.; Spagnuolo, V. Phytomonitoring and Phytoremediation of Environmental Pollutants. *Plants* **2023**, *13*, 366. [CrossRef]
13. Zhong, L.; Guo, X.; Xu, Z.; Ding, M. Soil properties: Their prediction and feature extraction from the LUCAS spectral library using deep convolutional neural networks. *Geoderma* **2021**, *402*, 115366. [CrossRef]

14. Shi, T.; Chen, Y.; Liu, Y.; Wu, G. Visible and near-infrared reflectance spectroscopy—An alternative for monitoring soil contamination by heavy metals. *J. Hazard. Mater.* **2014**, *265*, 166–176. [CrossRef] [PubMed]
15. Liu, Z.; Lu, Y.; Peng, Y.; Zhao, L.; Wang, G.; Hu, Y. Estimation of Soil Heavy Metal Content Using Hyperspectral Data. *Remote Sens.* **2019**, *11*, 1464. [CrossRef]
16. Shi, T.; Guo, L.; Chen, Y.; Wang, W.; Shi, Z.; Li, Q.; Wu, G. Proximal and remote sensing techniques for mapping of soil contamination with heavy metals. *Appl. Spectrosc. Rev.* **2018**, *53*, 783–805. [CrossRef]
17. Zhong, L.; Chu, X.; Qian, J.; Li, J.L.; Sun, Z. Multi-Scale Stereoscopic Hyperspectral Remote Sensing Estimation of Heavy Metal Contamination in Wheat Soil over a Large Area of Farmland. *Agronomy* **2023**, *13*, 2396. [CrossRef]
18. Liu, P.; Liu, Z.; Hu, Y.; Shi, Z.; Pan, Y.; Wang, L.; Wang, G. Integrating a Hybrid Back Propagation Neural Network and Particle Swarm Optimization for Estimating Soil Heavy Metal Contents Using Hyperspectral Data. *Sustainability* **2018**, *11*, 419. [CrossRef]
19. Pyo, J.; Hong, S.; Kwon, Y.; Kim, M.; Cho, K. Estimation of heavy metals using deep neural network with visible and infrared spectroscopy of soil. *Sci. Total Environ.* **2020**, *741*, 140162. [CrossRef]
20. Guo, F.; Xu, Z.; Ma, H.; Liu, X.; Tang, S.; Yang, Z.; Zhang, L.; Liu, F.; Peng, M.; Li, K. Estimating chromium concentration in arable soil based on the optimal principal components by hyperspectral data. *Ecol. Indic.* **2021**, *133*, 108400. [CrossRef]
21. Zhou, W.; Yang, H.; Xie, L.; Li, H.; Lu, H.; Zhao, Y.; Yue, T. Hyperspectral inversion of soil heavy metals in Three-River Source Region based on random forest model. *Catena* **2021**, *202*, 105222. [CrossRef]
22. Shi, T.; Fu, Z.; Miao, X.; Lin, F.; Ma, J.; Gu, S.; Li, L.; Wu, C.; Luo, Y. Would it be better for partition prediction of heavy metal concentration in soils based on the fusion of XRF and Vis-NIR data? *Sci. Total Environ.* **2024**, *908*, 168381. [CrossRef]
23. Wang, F.; Gao, J.; Zha, Y. Hyperspectral sensing of heavy metals in soil and vegetation: Feasibility and challenges. *ISPRS J. Photogramm. Remote Sens.* **2018**, *136*, 73–84. [CrossRef]
24. Zhou, W.; Zhang, J.; Zou, M.; Liu, X.; Du, X.; Wang, Q.; Liu, Y.; Li, J. Feasibility of using rice leaves hyperspectral data to estimate CaCl2-extractable concentrations of heavy metals in agricultural soil. *Sci. Rep.* **2019**, *9*, 16084. [CrossRef]
25. Zhang, L.; Liu, Z.; Song, Y.; Sui, J.; Hua, X. Advances in the Involvement of Metals and Metalloids in Plant Defense Response to External Stress. *Plants* **2023**, *13*, 313. [CrossRef]
26. Sanaeifar, A.; Yang, C.; De la Guardia, M.; Zhang, W.; Li, X.; He, Y. Proximal hyperspectral sensing of abiotic stresses in plants. *Sci. Total Environ.* **2023**, *861*, 160652. [CrossRef]
27. Falcioni, R.; Gonçalves, J.; De Oliveira, K.; De Oliveira, C.; Reis, A.; Crusiol, L.; Furlanetto, R.; Antunes, W.; Cezar, E.; De Oliveira, R.; et al. Chemometric Analysis for the Prediction of Biochemical Compounds in Leaves Using UV-VIS-NIR-SWIR Hyperspectroscopy. *Plants* **2023**, *12*, 3424. [CrossRef]
28. Shi, T.; Liu, H.; Chen, Y.; Wang, J.; Wu, G. Estimation of arsenic in agricultural soils using hyperspectral vegetation indices of rice. *J. Hazard. Mater.* **2016**, *308*, 243–252. [CrossRef]
29. Zhong, L.; Yang, S.; Chu, X.; Sun, Z.; Li, J. Inversion of heavy metal copper content in soil-wheat systems using hyperspectral techniques and enrichment characteristics. *Sci. Total Environ.* **2024**, *907*, 168104. [CrossRef]
30. Wang, G.; Wang, Q.; Su, Z.; Zhang, J. Predicting copper contamination in wheat canopy during the full growth period using hyperspectral data. *Environ. Sci. Pollut. Res.* **2020**, *27*, 39029–39040. [CrossRef]
31. Sun, T.; Li, Z.; Wang, Z.; Liu, Y.; Zhu, Z.; Zhao, Y.; Xie, W.; Cui, S.; Chen, G.; Yang, W.; et al. Monitoring of Nitrogen Concentration in Soybean Leaves at Multiple Spatial Vertical Scales Based on Spectral Parameters. *Plants* **2024**, *13*, 140. [CrossRef] [PubMed]
32. Tan, K.; Wang, H.; Zhang, Q.; Jia, X. An improved estimation model for soil heavy metal(loid) concentration retrieval in mining areas using reflectance spectroscopy. *J. Soils Sediments* **2018**, *18*, 2008–2022. [CrossRef]
33. Zhong, L.; Guo, X.; Ding, M.; Ye, Y.; Jiang, Y.; Zhu, Q.; Li, J. SHAP values accurately explain the difference in modeling accuracy of convolution neural network between soil full-spectrum and feature-spectrum. *Comput. Electron. Agric.* **2024**, *217*, 108627. [CrossRef]
34. Zhou, W.; Zhang, J.; Zou, M.; Liu, X.; Du, X.; Wang, Q.; Liu, Y.; Liu, Y.; Li, J. Prediction of cadmium concentration in brown rice before harvest by hyperspectral remote sensing. *Environ. Sci. Pollut. Res.* **2019**, *26*, 1848–1856. [CrossRef] [PubMed]
35. Yao, L.; Xu, M.; Liu, Y.; Niu, R.; Wu, X.; Song, Y. Estimating of heavy metal concentration in agricultural soils from hyperspectral satellite sensor imagery: Considering the sources and migration pathways of pollutants. *Ecol. Indic.* **2024**, *158*, 111416. [CrossRef]
36. Arif, M.; Qi, Y.; Dong, Z.; Wei, H. Rapid retrieval of cadmium and lead content from urban greenbelt zones using hyperspectral characteristic bands. *J. Clean. Prod.* **2022**, *374*, 133922. [CrossRef]
37. Hasan, U.; Jia, K.; Wang, L.; Wang, C.; Shen, Z.; Yu, W.; Sun, Y.; Jiang, H.; Zhang, Z.; Guo, J.; et al. Retrieval of Leaf Chlorophyll Contents (LCCs) in Litchi Based on Fractional Order Derivatives and VCPA-GA-ML Algorithms. *Plants* **2023**, *12*, 501. [CrossRef]
38. Sun, W.; Zhang, X. Estimating soil zinc concentrations using reflectance spectroscopy. *Int. J. Appl. Earth Obs. Geoinf.* **2017**, *58*, 126–133. [CrossRef]
39. Sun, W.; Liu, S.; Zhang, X.; Zhu, H. Performance of hyperspectral data in predicting and mapping zinc concentration in soil. *Sci. Total Environ.* **2022**, *824*, 153766. [CrossRef]
40. Zhang, X.; Sun, W.; Cen, Y.; Zhang, L.; Wang, N. Predicting cadmium concentration in soils using laboratory and field reflectance spectroscopy. *Sci. Total Environ.* **2019**, *650*, 321–334. [CrossRef]
41. Wei, L.; Pu, H.; Wang, Z.; Yuan, Z.; Yan, X.; Cao, L. Estimation of Soil Arsenic Content with Hyperspectral Remote Sensing. *Sensors* **2020**, *20*, 4056. [CrossRef]

42. Bian, Z.; Sun, L.; Tian, K.; Liu, B.; Huang, B.; Wu, L. Estimation of multi-media metal(loid)s around abandoned mineral processing plants using hyperspectral technology and extreme learning machine. *Environ. Sci. Pollut. Res.* **2023**, *30*, 19495–19512. [CrossRef] [PubMed]
43. Elazab, D.; Lambardi, M.; Capuana, M. In Vitro Culture Studies for the Mitigation of Heavy Metal Stress in Plants. *Plants* **2023**, *12*, 3387. [CrossRef] [PubMed]
44. Su, Z.; Wang, G.; Xu, L.; Zhang, J.; Liu, X. Effects of Cu stress on physiological, biochemical, and spectral properties of wheat at different growth stages. *Int. J. Agric. Biol. Eng.* **2019**, *12*, 147–153. [CrossRef]
45. Székely, Á.; Szalóki, T.; Jancsó, M.; Pauk, J.; Lantos, C. Temporal Changes of Leaf Spectral Properties and Rapid Chlorophyll—A Fluorescence under Natural Cold Stress in Rice Seedlings. *Plants* **2023**, *12*, 2415. [CrossRef]
46. *NY/T 1377-2007*; De-Termination of pH in Soil. Standards Press of China: Beijing, China, 2007.
47. Le, S.; Duan, Y. Determination of Heavy Metal Elements in Soil by ICP-MS. *Chin. J. Inorg. Anal. Chem.* **2015**, *5*, 16–19. [CrossRef]
48. Shi, Z.; Wang, Q.; Peng, J.; Ji, W.; Liu, H.; Li, X.; Viscarra Rossel, R.A. Development of a national VNIR soil-spectral library for soil classification and prediction of organic matter concentrations. *Sci. China Earth Sci.* **2014**, *57*, 1671–1680. [CrossRef]
49. Mitchell, M. *An Introduction to Genetic Algorithms*; MIT Press: Cambridge, MA, USA, 1998.
50. Du, X.; Li, X.; Liu, Y.; Zhou, W.; Li, J. Genetic algorithm optimized non-destructive prediction on property of mechanically injured peaches during postharvest storage by portable visible/shortwave near-infrared spectroscopy. *Sci. Hortic.* **2019**, *249*, 240–249. [CrossRef]
51. Nawar, S.; Mohamed, E.; Essam-Eldeen Sayed, S.; Mohamed, W.; Rebouh, N.; Hammam, A. Estimation of key potentially toxic elements in arid agricultural soils using vis-NIR spectroscopy with variable selection and PLSR algorithms. *Front. Environ. Sci.* **2023**, *11*, 1222871. [CrossRef]
52. Wold, S.; Sjöström, M.; Eriksson, L. PLS-regression: A basic tool of chemometrics. *Chemom. Intell. Lab. Syst.* **2001**, *58*, 109–130. [CrossRef]
53. Williams, P.; Manley, M.; Antoniszyn, J. *Near Infrared Technology: Getting the Best Out of Light*; African Sun Media: Stellenbosch, South Africa, 2019.

MDPI AG
Grosspeteranlage 5
4052 Basel
Switzerland
Tel.: +41 61 683 77 34

Plants Editorial Office
E-mail: plants@mdpi.com
www.mdpi.com/journal/plants

www.ingramcontent.com/pod-product-compliance
Lightning Source LLC
Chambersburg PA
CBHW041408210926
51850CB00013B/3210
* 9 7 8 3 7 2 5 8 3 6 6 9 7 *